A dictionary of inequalities

Main Editors

H. Brezis, Université de Paris
R.G. Douglas, Texas A&M University
A. Jeffrey, University of Newcastle upon Tyne *(Founding Editor)*

Editorial Board

H. Amann, University of Zürich
R. Aris, University of Minnesota
G.I. Barenblatt, University of Cambridge
H. Begehr, Freie Universität Berlin
P. Bullen, University of British Columbia
R.J. Elliott, University of Alberta
R.P. Gilbert, University of Delaware
R. Glowinski, University of Houston
D. Jerison, Massachusetts Institute of Technology
K. Kirchgassner, Universität Stuttgart
B. Lawson, State University of New York at Stony Brook
B. Moodie, University of Alberta
S. Mori, Kyoto University
L.E. Payne, Cornell University
D.B. Pearson, University of Hull
I. Raeburn, University of Newcastle
G.F. Roach, University of Strathclyde
I. Stakgold, University of Delaware
W.A. Strauss, Brown University
J. van der Hoek, University of Adelaide

π Pitman Monographs and
Surveys in Pure and Applied Mathematics 97

A dictionary of inequalities

P S Bullen
University of British Columbia

 LONGMAN

Addison Wesley Longman Limited
Edinburgh Gate, Harlow
Essex CM20 2JE, England
and Associated companies throughout the world.

*Published in the United States of America
by Addison Wesley Longman Inc.*

© Addison Wesley Longman Limited 1998

The right of Peter S Bullen to be identified as author of this Work has been asserted by him in accordance with the Copyright, Designs and Patents Act 1988.

All rights reserved; no part of this publication may be reproduced, stored in a retrieval system, or transmitted in any form or by any means, electronic, mechanical, photocopying, recording, or otherwise, without the prior written permission of the Publishers, or a licence permitting restricted copying in the United Kingdom issued by the Copyright Licensing Agency Ltd, 90 Tottenham Court Road, London, W1P 9HE.

First published 1998

AMS Subject Classifications 26Dxx

ISSN 0269-3666

ISBN 0 582 32748 2

Visit Addison Wesley Longman on the world wide web at
http://www.awl-he.com

British Library Cataloguing in Publication Data
A catalogue record for this book is
available from the British Library

Printed and bound by Bookcraft (Bath) Ltd

This book is dedicated to the memory of

Professors D S Mitrinović and P M Vasić

CONTENTS

NOTATIONS	1
1. General Notations	1
2. Sequences and n-tuples	2
3. Means	4
4. Continuous Variable Analogues	6
5. Rearrangements and Orders	8
6. Some Functions and Classes of Functions	9
7. Matrices	11
8. Probability and Statistics	11
9. Symbols Indicating Some Inequalities	12
A	13
B	24
C	43
D	69
E	75
F	84
G	93
H	108
I	133
J	139
K	144
L	153
M	165
N	187
O	196
P	201
Q	216
R	220
S	228
T	247
U-V	256
W	259
Y-Z	265
REFERENCES	269
Basic References	269
Books	269
Papers	274

INTRODUCTION

The object of this dictionary is to provide an easy way for researchers to locate an inequality either by name or by subject. Proofs will not be given, although the methods used may be indicated. Instead, references will be given where such details and further information can be found.

The bibliography is not intended to be complete; although an attempt has been made to include all books on inequalities. References mentioned in these books are not repeated; only papers more recent than the books or which seem to have been missed by the other references are included.

Inequalities are not necessarily given in their most general forms; in some cases the intuitive physical version is given rather than the more abstract mathematical formulation that may appear elsewhere under a different guise. Usually the most common version is given, and the more general results are stated later as extensions or variants. In addition, the name of the inequality may be historically that of a variant of the one stated; when this is done it will be made clear.

The history of most inequalities is often obscure, therefore the correct attribution is not always easy to make. Little attempt is made to trace the history of particular inequalities; this topic is covered by many excellent articles by the late D.S.Mitrinović and members of his school. Some of these articles are summarized in the book *Analytic Inequalities, Mitrinović*.

The field of inequalities is vast and some restrictions have had to be made. The whole area of elementary geometric and trigonometric inequalities is omitted because such inequalities are little used except within the field itself. Those interested are referred to the monograph *Geometric Inequalities, Bottema et al*. In addition, very few results from number theory are given; again, those interested are referred to the literature, for instance, *Inequalities in Number Theory, Mitrinović & Popadić*.

In recent years the number of papers on inequalities, as well as the number of journals devoted to this subject, has considerably increased. Most of the inequalities in this book have been subjected to various generalizations and in particular have often been put into an abstract form. Although these abstractions often enable many inequalities to be placed under one roof, so to speak, and are therefore of interest, it is felt that often the beauty and importance of the original is lost. In any case this direction has not been pursued. Those interested in this development are referred to the literature; for example, the books *Classical and New Inequalities in Analysis, Mitrinović et al*, and *Convex Functions, Partial Orderings and Statistical Applications, Pečarić et al*.

Many of the basic inequalities exist at various levels of generality; as statements about real numbers, complex numbers, vectors and so on. In such cases the inequality will be presented in its simplest form with the other forms as extensions, or under a different heading.

Basic concepts will be assumed to be known although less basic ideas will be defined in the appropriate entry.

Most notations taken from real analysis are standard but some that are peculiar to this monograph are in the section, **Notations**; full details can be found in the books in the references.

Transliteration of names from Cyrillic and Chinese causes some difficulties as the actual spelling has varied from time to time and from journal to journal. In the case of Cyrillic the spelling used is given in the old Mathematical Reviews, old Zentralblatt, form. This is very close to that of the British Standards Institute used in the *Encyclopedia of Mathematics*; it has the advantage of being less oriented to the English language. However this is not insisted upon when reference to the literature is made. Other Slavic languages employ various diacritical letters and that spelling has been used. In the case of Chinese, the spelling chosen is that used in the book or paper where the inequality occurs, but other versions used will be given if known; in addition the family name is given first as is usual, except for Chinese who are living abroad and who have adopted the standard western convention. Cross references are given between the various transliterations that have been used.

It is hoped that this dictionary will have the attraction attributed to most successful dictionaries; that on looking up a given item one is led by interest to others, and perhaps never gets to the original object being researched.

An attempt has been made to be complete, accurate and up-to-date. However, new inequalities, variants and generalizations of known inequalities come in every set of new journals that arrive in the library. Readers wishing to keep up with the latest inequalities are urged to go to the latest issues of the reference journals — *Mathematical Reviews, Zentralblatt für Mathematik und ihre Grenzegebiete*, section 26Dxx; *Referantivniĭ Žurnal Matematika (Mehanika etc.)*[1]. Also there are journals devoted to inequalities: *Univerzitet u Beogradu Publikacije Elektrotehničkog Fakulteta. Serija Matematika i Fizika*, (the "*i Fizika*" has been dropped in the more recent issues), all of whose volumes, from the first in the early fifties, are full of papers on inequalities; the recent *Journal for Inequalities and Applications* and *Mathematical Inequalities and Applications*; the *American Mathematical Monthly*, the *Journal of Mathematical Analysis and Applications*, the *Mathematics Gazette*, and the *Mathematics Magazine*, as well as most of the various journals appearing in the countries of the Balkans; and there are the earlier issues of the *Journal, and Proceedings, of the London Mathematical Society*, and the *Quarterly Journal of Mathematics, Oxford*. Some may find a favourite inequality omitted, or mis-quoted. Please let the author know of these errors and omissions and if there is another edition the corrections and additions will be included. Accuracy is difficult as the original papers are often hard to obtain, and most books on inequalities contain errors, a notable exception being *Inequalities*, Hardy, Littlewood & Polyá, the field's classic.

P S Bullen, Department of Mathematics, University of British Columbia,
    Vancouver BC, Canada, V6T 1Z2.   bullen@math.ubc.ca

---

[1] Референтивний Журнал Математика (Механика)

NOTATIONS

1. General Notations

$\mathbb{R}$ is the set of real numbers. $\mathbb{Q}$ the set of rational numbers. $\mathbb{N} = \{0, 1, 2, \ldots\}$, the set of non-negative integers.
$\overline{\mathbb{R}} = \{-\infty\} \cup \mathbb{R} \cup \{\infty\}$. $\mathbb{Z} = \{\ldots -2, -1, 0, 1, 2, \ldots\}$, the set of all integers.
The open, closed(compact), half-open (-closed) intervals in $\mathbb{R}$, or $\overline{\mathbb{R}}$, are written $[a, b]$, $]a, b[$, $[a, b[$, $]a, b]$.
$\mathbb{C}$ is the set of complex numbers; $\overline{\mathbb{C}} = \mathbb{C} \cup \{\infty\}$.
The *open unit disk (ball)* and *closed unit disk (ball)* in $\mathbb{C}$ are respectively $D = \{z; |z| < 1\}$, $\overline{D} = \{z; |z| \leq 1\}$.
In general we will use n, m, i, j, k to be integers, and x, y, to be real, and $z = re^{i\theta}$ to be complex.
However, in a situation where complex numbers are involved $i = \sqrt{-1}$.
If $z \in \mathbb{C}$ to write $z \geq r$ means that z, r are real and the inequality holds.
If $p, q \in \mathbb{R}$ then p *is the conjugate of* q, equivalently p *is the conjugate of* q, if

$$\frac{1}{p} + \frac{1}{q} = 1; \quad \text{or} \quad q = \frac{p}{p-1}.$$

We also say that p, q are *conjugate indices*. Usually $p > 1$, when of course $q > 1$. From time to time $p = 1, q = \infty$ is allowed.
Sometimes the word *dual* is used instead of conjugate.
If (I) denotes the inequality $P < Q$, or $P \leq Q$, then ($\sim$I) will denote the *reverse inequality* $P > Q$, respectively $P \geq Q$. The functions P, Q in the inequality (I) are usually continuous in their variables. If then we restrict these variables to a compact set, the functions $Q - P$, and Q/P attain maxima on that set. This leads to inequalities of the form $Q < A + P$, $Q < BP$, for some constants A, B that depend on the set. Such inequalities are called *converse*, or *inverse inequalities*, to (I).

A *domain*, is an open simply connected set; the closure of such a set is called a *closed domain*.
If A is a set then its *boundary*, or *frontier*, is written ∂A.
If $A \subseteq X$, $f : A \to Y$ then *the image of* A *by* f is $f[A] = \{y; y = f(x), x \in A\}$; if $B \subseteq Y$ then *the pre-image of* B *by* f is $f^{-1}[B] = \{x; f(x) \in B\}$.
Partial derivatives are denoted in various ways: thus $\partial^2 f/\partial x \partial y$ and f''_{12}.

2. Sequences and n-tuples

If $a_i \in \mathbb{R}, i \in \mathbb{N}, i \geq 1$, then we will write

$$\underline{a} = (a_1, a_2, \ldots);$$

for the *real sequence* $\underline{a}$, or if a positive integer n is specified, $\underline{a}$ will denote the *real n-tuple*,

$$\underline{a} = (a_1, a_2, \ldots, a_n).$$

The $a_i, i = 1, \ldots$, are called the *elements*, *entries* or *terms of the sequence or n-tuple* $\underline{a}$. If $\underline{a}$ is a real n-tuple its *length* is,

$$|\underline{a}| = \sqrt{|a_1|^2 + \cdots + |a_n|^2}.$$

If $\underline{b}$ is another real n-tuple,

$$\underline{a}.\underline{b} = \sum_{i=1}^{n} a_i b_i.$$

If $\underline{a}.\underline{b} = 0$ we say that $\underline{a}$ and $\underline{b}$ are *orthogonal*.

These notations extend to sequences providing the relevant series converge.

The set of all real n-tuples is $\mathbb{R}^n = \overbrace{\mathbb{R} \times \cdots \times \mathbb{R}}^{n\ factors}$.
The *unit ball, closed unit ball* in $\mathbb{R}^n$ is $B_n = \{\underline{a}; |\underline{a}| < 1\}$, $\overline{B}_n = \{\underline{a}; |\underline{a}| \leq 1\}$ respectively; if $n = 1$, or the value of n is obvious, the suffix may be omitted.

The partial sums of the sequence or n-tuple $\underline{a}$ will be written

$$A_k = \sum_{i=1}^{k} a_i, \ k = 1, 2, \ldots;$$

and if needed we will put $A_0 = 0$.

$\underline{e}$ will denote the n-tuple, or sequence, with entries all equal to 1. $\underline{0}$ will denote the n-tuple, or sequence, with entries all equal to 0; called *the zero, or null n-tuple, or sequence*.
A sequence, or n-tuple, all of whose terms are equal, will be called *a constant sequence, or n-tuple* or just said to be *constant*; so a real $\underline{a}$ is constant if $\underline{a} = \lambda \underline{e}$ for some $\lambda \in \mathbb{R}$.

The *differences* of $\underline{a}$ are defined inductively as follows:

$$\Delta^0 a_i = a_i; \quad \Delta^1 a_i = \Delta a_i = a_i - a_{i+1} \quad \Delta^2 a_i = a_i + a_{i+2} - 2a_{i+1},$$
$$\Delta^k a_i = \Delta(\Delta^{k-1} a_i), \ k = 2, 3, \ldots,$$

and we write $\Delta \underline{a}, \Delta^k \underline{a}, k > 1$, for the sequences of these differences. We will also write $\tilde{\Delta} a_i = -\Delta^1 a_i$, and then $\tilde{\Delta}^k a_i = (-1)^k \Delta^k a_i, k \geq 2$.

If $1 \leq i \leq n$ then $\underline{a}'_i$ denotes the $(n-1)$-tuple obtained from $\underline{a}$ by omitting the element a_i.

If f, g are suitable functions then:

$$f(\underline{a}) = (f(a_1), f(a_2), \dots);$$
$$g(\underline{a}, \underline{b}) = (g(a_1, b_1), g(a_2, b_2), \dots).$$

For example: $\underline{a}^p = (a_1^p, a_2^p, \dots), \underline{a}\,\underline{b} = (a_1 b_1, a_2 b_2, \dots)$.

If f is a function of k variables, $\underline{a}$ an n-tuple, with $n \geq k$, then $\sum_k ! f(a_{i_1}, \dots, a_{i_k})$ means that the sum is taken over all permutation of k elements from $a_1, \dots, a_n$; in the case $k = n$, we write $\sum ! f(\underline{a})$.

The above notations can easily be modified to allow for complex sequences. In addition many of the concepts apply in much more abstract situations.

If two real n-tuples are such that for some $\lambda, \mu \in \mathbb{R}$, not both zero $\lambda \underline{a} + \mu \underline{b} = \underline{0}$ we will write $\underline{a} \sim \underline{b}$. If in addition $\lambda, \mu \geq 0$, we will write $\underline{a} \sim^+ \underline{b}$. Geometrically $\underline{a} \sim \underline{b}$ means that $\underline{a}, \underline{b}$ are on the same line through the origin, while $\underline{a} \sim^+ \underline{b}$ means that they are on the same ray, or half-line with vertex at the origin.

If for all i, $a_i \leq b_i$ we will write $\underline{a} \leq \underline{b}$; the definitions of other inequalities can be made analogously. In particular if $\underline{a} > \underline{0}, (\geq \underline{0})$ then $\underline{a}$ is said to be *positive, (non-negative)*; similar definitions can be made for *negative, (non-positive)* sequences. If $\Delta \underline{a}, > \underline{0}, (\geq \underline{0})$ then $\underline{a}$ is said to be *strictly decreasing, (decreasing)*; in a similar way we can define *strictly increasing, (increasing)* $\underline{a}$.

When convenient we will write $\underline{a} > M$ to mean that for all $i, a_i > M$, that is $\underline{a} > M\underline{e}$, and similarly for the other inequalities

We will write for $p \in \mathbb{R}, p \neq 0$,

$$||\underline{a}||_p = \left(\sum_{i=1}^n |a_i|^p\right)^{1/p} \quad \text{or} \quad \left(\sum_{i=1}^\infty |a_i|^p\right)^{1/p}.$$

In particular $||\underline{a}||_2 = |\underline{a}|$. As usual when $p = 1$ the suffix is omitted. Of course for negative p we assume that $\underline{a}$ is positive.

The set of all sequences $\underline{a}$ with finite $||\underline{a}||_p$ is denoted by ℓ_p.

$$\max \underline{a} = \max\{a_1, \dots, a_n\} \quad \text{or} \quad \sup\{a_1, \dots\};$$
$$\min \underline{a} = \min\{a_1, \dots, a_n\} \quad \text{or} \quad \inf\{a_1, \dots\}.$$

We write $||\underline{a}||_\infty = \max |\underline{a}|$, and the set of all sequences $\underline{a}$ with finite $||\underline{a}||_\infty$ is denoted by ℓ_∞.

If $\underline{a}, \underline{b}$ are positive sequences then their *convolution*, $\underline{a} \star \underline{b}$, is the sequence $\underline{c}$, with terms

$$c_n = \sum_{i+j=n} a_i b_j, n \in \mathbb{N}.$$

3. Means

These are an important source of inequalities and the more common means are listed first, followed by a few means that are defined for two numbers only. Other means are defined in the entries dealing with their inequalities; see **Arithmetic-geometric Compound Mean Inequalities, Counter Harmonic Mean Inequalities, Difference Means of Gini, Gini-Dresher Mean Inequalities, Hamy Mean Inequalities, Logarithmic Mean Inequalities, Mixed Mean Inequalities, Muirhead Symmetric Function and Mean Inequalities, Nanjundiah's Inverse Mean Inequalities, Quasi-arithmetic Mean Inequalities, Whiteley Mean Inequalities**.

The *arithmetic, geometric, harmonic mean, of order n*, of the positive sequence $\underline{a}$ with positive *weight* $\underline{w}$ is respectively:

$$\mathfrak{A}_n(\underline{a};\underline{w}) = \frac{1}{W_n}\sum_{i=1}^n w_i a_i\,, \quad \mathfrak{G}_n(\underline{a};\underline{w}) = \left(\prod_{i=1}^n a_i^{w_i}\right)^{\frac{1}{W_n}}, \quad \mathfrak{H}_n(\underline{a};\underline{w}) = \left(A_n(\underline{a}^{-1};\underline{w})\right)^{-1}.$$

It is worth noting that
$$\log \mathfrak{G}(\underline{a};\underline{w}) = \mathfrak{A}(\log \underline{a};\underline{w}).$$

If $p \in \overline{\mathbb{R}}$ the *p power mean, of order n*, of the positive sequence $\underline{a}$ with positive weight $\underline{w}$ is

$$\mathfrak{M}_n^{[p]}(\underline{a};\underline{w}) = \begin{cases} \left(\dfrac{1}{W_n}\displaystyle\sum_{i=1}^n w_i a_i^p\right)^{1/p}, & \text{if } p \in \mathbb{R}\setminus\{0\}, \\ \mathfrak{G}_n(\underline{a};\underline{w}), & \text{if } p = 0, \\ \max \underline{a}, & \text{if } p = \infty, \\ \min \underline{a}, & \text{if } p = -\infty. \end{cases}$$

The definitions for $p=0, \pm\infty$ are justified as limits ; see [*MI, p.133*]. In particular,

$$\mathfrak{M}_n^{[-1]}(\underline{a};\underline{w}) = \mathfrak{H}_n(\underline{a};\underline{w}), \qquad \mathfrak{M}_n^{[1]}(\underline{a};\underline{w}) = \mathfrak{A}_n(\underline{a};\underline{w}).$$

In the case of equal weights, reference to the weights will be omitted from the notation, thus $\mathfrak{A}_n(\underline{a})$ etc.

The n-tuples in the above notations may be written explicitly, $\mathfrak{A}_n(a_1,\ldots,a_n;w_1,\ldots,w_n)$, or $\mathfrak{A}_n(a_i;w_i; 1 \le i \le n)$, for instance.

Most of these mean definitions are well-defined for more general sequences and will be used in that way when needed. If the weights are allowed k zero elements, $1 \le k < n$, this is equivalent to reducing n by k.

Given two sequences $\underline{a}, \underline{w}$, the sequence $\{\mathfrak{A}_1(\underline{a};\underline{w}), \mathfrak{A}_2(\underline{a};\underline{w}),\ldots\}$ has an obvious meaning and will be written $\mathfrak{A}(\underline{a};\underline{w})$, although the reference to $\underline{a}, \underline{w}$ will be omitted when convenient. Similar notations will be used for other means.

The various means introduced above can be regarded, for given sequences $\underline{a}, \underline{w}$, as functions of $n \in \mathbb{N}$. They can then, by extension, be regarded as functions on non-empty finite subsets of $\mathbb{N}$, the *index sets*. For instance if $\mathcal{I}$ is such a subset of $\mathbb{N}$

$$\mathfrak{A}_\mathcal{I}(\underline{a};\underline{w}) = \frac{1}{W_\mathcal{I}}\sum_{i\in\mathcal{I}} w_i a_i,$$

where of course $W_\mathcal{I} = \sum_{i \in \mathcal{I}} w_i$.
This notation will be used in analogous situations without explanation.
A function ϕ on the index sets is respectively, *positive, increasing, superadditive*, if:

$$\phi(\mathcal{I}) > 0, \quad \mathcal{I} \subseteq \mathcal{J} \longrightarrow \phi(\mathcal{I}) \leq \phi(\mathcal{J}), \quad \mathcal{I} \cap \mathcal{J} = \emptyset \Longrightarrow \phi(\mathcal{I} \cup \mathcal{J}) \geq \phi(\mathcal{I}) + \phi(\mathcal{J}),$$

where $\mathcal{I}, \mathcal{J}$ are index sets; a definition of convexity of such functions can be found in **Hlwaka-Type Inequalities**, COMMENTS (iii).

If $\underline{a}$ is a positive n-tuple, $r \in \mathbb{Z}$ then:
(a) the rth *elementary symmetric functions* of $\underline{a}$ are

$$e_n^{[r]}(\underline{a}) = \frac{1}{r!} \sum_r ! \prod_{j=1}^r a_{i_j}, \quad p_n^{[r]}(\underline{a}) = \frac{e_n^{[r]}(\underline{a})}{\binom{n}{r}}, \quad 1 \leq r \leq n;$$

$$e_n^{[0]}(u\underline{a}) = p_n^{[0]}(\underline{a}) = 1; \quad e_n^{[r]}(\underline{a}) = p_n^{[r]}(\underline{a}) = 0, \quad r > n, \ r < 0;$$

(b) the rth *symmetric mean* of $\underline{a}$ is

$$\mathfrak{P}_n^{[r]}(\underline{a}) = \left(p_n^{[r]}(\underline{a})\right)^{1/r}, 1 \leq r \leq n; \quad \mathfrak{P}_n^{[0]}(\underline{a}) = 1; \quad \mathfrak{P}_n^{[r]}(\underline{a}) = 0, \ r > n, \ r < 0.$$

It is easily seen that

$$\mathfrak{P}_n^{[1]}(\underline{a}) = \mathfrak{A}_n(\underline{a}), \quad \mathfrak{P}_n^{[n]}(\underline{a}) = \mathfrak{G}_n(\underline{a}), \quad \mathfrak{P}_n^{[n-1]}(\underline{a}) = \left(\frac{\mathfrak{G}_n^n(\underline{a})}{\mathfrak{H}_n(\underline{a})}\right)^{1/(n-1)}.$$

For other relations see **Hamy Mean Inequalities, Mixed Mean Inequalities, Muirhead Symmetric Function and Mean Inequalities**.
The symmetric functions can be generated as follows:

$$\prod_{i=1}^n (1 + a_i x) = \sum_{i=1}^n e_n^{[i]}(\underline{a}) x^i = \sum_{i=1}^n \binom{n}{i} p_n^{[i]}(\underline{a}) x^i; \tag{1}$$

or

$$e_n^{[r]}(\underline{a}) = \sum \left(\prod_{i=1}^n a_i^{i_j}\right),$$

where the sum is over the $\binom{n}{r}$ n-tuples $(i_1, \ldots, i_n)$ with $i_j = 0$ or 1, $1 \leq j \leq n$, and $\sum_{j=1}^n = r$.
In many references $p_n^{[r]}$ is called the symmetric mean.
The *complete symmetric functions* are defined for $r \neq 0$ by

$$c_n^{[r]}(\underline{a}) = \sum \left(\prod_{i=1}^n a_i^{i_j}\right) \quad \text{and} \quad q_n^{[r]}(\underline{a}) = \frac{c_n^{[r]}(\underline{a})}{\binom{n+r-1}{r}},$$

where the sum is over the $\binom{n+r-1}{r}$ n-tuples $(i_1, \ldots, i_n)$ with $\sum_{j=1}^{n} = r$. In addition $c_n^{[0]}(\underline{a})$ is defined to be 1. Equivalently

$$\prod_{i=1}^{n}(1-a_i x)^{-1} = \sum_{i=1}^{n} c_n^{[i]}(\underline{a}) x^i = \sum_{i=1}^{n} \binom{n}{i} q_n^{[i]}(\underline{a}) x^i.$$

The rth *complete symmetric mean* of $\underline{a}$ is

$$\mathfrak{Q}_n^{[k]}(\underline{a}) = \left(q_n^{[r]}(\underline{a})\right)^{1/r}.$$

A common extension of these symmetric functions and means can be seen under the entry **Whitely Symmetric Inequalities**.

If $-\infty < p < \infty$ and $a, b > 0$, $a \neq b$ the *generalized logarithmic mean* of a, b, is

$$\mathfrak{L}^{[p]}(a,b) = \begin{cases} \left(\dfrac{(b^{p+1} - a^{p+1})}{(p+1)(b-a)}\right)^{1/p}, & \text{if } p \neq, -1, 0, \\ \dfrac{(b-a)}{(\log b - \log a)}, & \text{if } p = -1, \\ \dfrac{1}{e}\left(\dfrac{b^b}{a^a}\right)^{1/b-a}, & \text{if } p = 0. \end{cases}$$

The case $p = -1$ is called the *logarithmic mean* of a, b, and will be written $\mathfrak{L}(a,b)$, while the case $p = 0$ is the *identric mean* of a, b, written $\mathfrak{I}(a,b)$.
The definition is completed by putting $\mathfrak{L}^{[p]}(a,a) = a$.
There are several easy identities involving these means:

$$\mathfrak{L}^{[-2]}(a,b) = \mathfrak{G}_2(a,b), \quad \mathfrak{L}^{[-1/2]}(a,b) = \mathfrak{M}_2^{[1/2]}(a,b) = \frac{\mathfrak{A}_2(a,b) + \mathfrak{G}_2(a,b)}{2},$$

$$\mathfrak{L}^{[1]}(a,b) = \mathfrak{A}_2(a,b), \quad \mathfrak{L}^{[-3]}(a,b) = \sqrt[3]{\mathfrak{H}_2(a,b)\mathfrak{G}_2^2(a,b)}.$$

4. Continuous Variable Analogues

Many of the above concepts can be extended to functions
If (X, μ), or $(X, \mathcal{M}, \mu)$ is a measure space we will also write μ for the outer measure defined on all the subsets of X that extends the measure on $\mathcal{M}$. If then $f : X \to \mathbb{R}$, $p > 0$,

$$\|f\|_{p,\mu,X} = \left(\int_X |f|^p \, d\mu\right)^{1/p},$$

and we write $\mathcal{L}_\mu^p(X)$ for the class of real functions on X with $\|f\|_{p,\mu,X} < \infty$; [Hewitt & Stromberg, pp. 188–189, 347; Rudin, 1964, p. 250]. As usual if $p = 1$ it is omitted from the notations.

In addition if $p = \infty$:

$$\|f\|_{\infty,\mu,X} = \mu\text{-essential upper bound of } |f| = \inf_{N \in \mathcal{N}} \sup_{x \in X \setminus N} |f(x)|,$$

where $\mathcal{N} = \{N; \mu(N) = 0\}$. Then the set of functions for which this quantity is finite is written $\mathcal{L}^\infty_\mu(X)$.

In all cases if the measure or the space is clear it may be omitted from the notation; thus $\|f\|_{p,X}$, or just $\|f\|_p$.

We will write λ_n for Lebesgue measure in $\mathbb{R}^n$ and if $E \subseteq \mathbb{R}^n$ write $|E|_n$ for $\lambda_n(E)$; further we will write $\int_E f(\underline{x}) \, \mathrm{d}\lambda_n$ as $\int_E f(\underline{x}) \, \mathrm{d}\underline{x}$. In the case $n = 1$, or if its value is clear, the suffix is omitted. When using Lebesgue measure it is omitted from the notation, thus $\mathcal{L}^p(X) = \mathcal{L}^p_{\lambda_n}(X)$.

The *meaures* of the volume of the B_n and of the surface area of B_n are, respectively,

$$v_n = \frac{\pi^{n/2}}{(n/2)!}, \qquad a_n = n v_n. \tag{1}$$

If $f \in \mathcal{L}^r(\mathbb{R}^p), g \in \mathcal{L}^s(\mathbb{R}^p)$, where $1 \le r, s \le \infty$ and then the *convolution function*

$$f \star g(x) = \int_{\mathbb{R}^p} f(x-t) g(t) \, \mathrm{d}t$$

is defined almost everywhere.

Various of the means defined above have integral analogues. Assuming that $\mu([a,b]) \ne 0$, the various integrals exist, and in case of negative p are not zero;

$$\mathfrak{M}^{[p]}_{[a,b]}(f; \mu) = \begin{cases} \left(\dfrac{\int_a^b |f|^p \, \mathrm{d}\mu}{\mu([a,b])} \right)^{1/p}, & \text{if } p \in \mathbb{R}, p \ne 0; \\[2ex] \exp\left(\dfrac{\int_a^b \log \circ |f| \, \mathrm{d}\mu}{\mu([a,b])} \right), & \text{if } p = 0; \\[2ex] \mu\text{-essential upper bound of } f, & \text{if } p = \infty; \\ \mu\text{-essential lower bound of } f, & \text{if } p = -\infty. \end{cases}$$

As usual we write $\mathfrak{H}_{[a,b]}(f; \mu), \mathfrak{G}_{[a,b]}(f; \mu), \mathfrak{A}_{[a,b]}(f; \mu)$ if $p = -1, 0, 1$, respectively. In the case that $\mu = \lambda$ reference to the measure in the notation will be omitted. Further if μ is absolutely continuous, when $f \mathrm{d}\mu$ can be written $f(x) w(x) \mathrm{d}x$ say, the notation will be $\mathfrak{M}^{[p]}_{[a,b]}(f; w)$. The interval $[a,b]$ can be replaced by a more general μ-measurable set, of positive μ-measure.

It is useful to note that if $i(x) = x, a \le x \le b$ then

$$\mathfrak{L}^{[p]}(a,b) = \mathfrak{M}^{[p]}_{[a,b]}(i), \ p \in \mathbb{R}.$$

5. Rearrangements and Orders

Other important sources of inequalities, and of proofs of inequalities, are the concepts of rearrangements and orders.

In general X or $(X, \preceq)$ is an *ordered*, or *partially ordered*, *set* if $\preceq$ is an *order* on X; that is for all $x, y, z \in X$: (i) $x \preceq x$, (ii) $x \preceq y$ and $y \preceq z$ implies that $x \preceq z$, (iii) $x \preceq y$ and $y \preceq x$ implies that $x = y$; if (iii) does not hold then $\preceq$ is called a *pre-order* on X: if in addition $x \vee y = \sup\{x, y\}$ and $x \wedge y = \inf\{x, y\}$ always exist then X is called a *lattice*; finally if $(x \vee y) \wedge z = (x \wedge z) \vee (x \wedge z)$ then X is a *distributive lattice*. The previous notation is extended to $A \vee B = \{z; z = x \vee y, x \in A, y \in B\}$, $A \wedge B = \{z; z = x \wedge y, x \in A, y \in B\}$, where of course $A, B \subseteq X$; see various entries in [EM].

The two real n-tuples $\underline{a}, \underline{b}$ are said to be *similarly ordered* when

$$(a_i - a_j)(b_i - b_j) \geq 0 \quad \text{for all} \quad i, j, 1 \leq i, j \leq n. \tag{1}$$

If ($\sim$1) holds then we say the n-tuples are *oppositely ordered*.

Equivalently $\underline{a}, \underline{b}$ are similarly ordered when under a simultaneous permutation both become increasing.

Given a real n-tuple $\underline{a}$ then $\underline{a}^*$, $(\underline{a}_*)$, will denote the n-tuple derived from $\underline{a}$ by rearranging the elements in decreasing, (increasing), order. We then will write

$$\underline{a}^* = (a_{[1]}, \ldots, a_{[n]}), \qquad \underline{a}_* = (a_{(1)}, \ldots, a_{(n)}).$$

We define an pre-order on the set of n-tuples by saying that, $\underline{b}$ *precedes* $\underline{a}$, $\underline{b} \prec \underline{a}$, if

$$\sum_{l=1}^{k} b_{[i]} \leq \sum_{l=1}^{k} a_{[i]}, \ 1 \leq k < n, \quad \text{and} \quad \sum_{l=1}^{n} b_i = \sum_{l=1}^{n} a_i.$$

This pre-order is an order on the set of decreasing, or on the set of increasing, n-tuples.

Suppose that $\underline{a} = (a_{-n}, \ldots, a_n)$ then two rearrangements of $\underline{a}$, $\underline{a}^+ = (a_{-n}^+, \ldots, a_n^+)$, $^+\underline{a} = (^+a_{-n}, \ldots, ^+a_n)$ are defined by

$$a_0^+ \geq a_1^+ \geq a_{-1}^+ \geq a_2^+ \geq a_{-2}^+ \geq \cdots; \quad {^+a_0} \geq {^+a_{-1}} \geq {^+a_1} \geq {^+a_{-2}} \geq {^+a_2} \geq \cdots.$$

If every value of the elements of $\underline{a}$, except the largest, occurs an even number of times, and the largest value occurs an odd number of times then $\underline{a}^+ = {^+\underline{a}}$. We say that $\underline{a}$ is *symmetrical*, and write $\underline{a}^{(*)}$ for either of the rearrangements just defined; that is $\underline{a}^{(*)} = (a_{-n}^{(*)}, \ldots, a_n^{(*)})$ where

$$a_0^{(*)} \geq a_1^{(*)} = a_{-1}^{(*)} \geq a_2^{(*)} = a_{-2}^{(*)} \geq \cdots$$

So $\underline{a}^{(*)}$ is a *symmetrically decreasing rearrangement* of $\underline{a}$; while the other rearrangements are as symmetrical as possible.

Two functions $f, g : I \to \mathbb{R}$ are *similarly ordered* if

$$\bigl(f(x) - f(y)\bigr)\bigl(g(x) - g(y)\bigr) \geq 0, \quad \text{for all} \quad x, y \in I. \tag{1}$$

If ($\sim$1) holds we say the functions are *oppositely ordered*.
This can easily be extended to functions of several variables; see [Pólya & Szegö, 1951, p. 151].
If f, g are two decreasing functions defined on $[a, b]$ and if

$$\int_a^x f \leq \int_a^x g, \quad a \leq x < b, \quad \text{and} \quad \int_a^b f = \int_a^b g$$

we will say that f precedes g, $f \prec g$ on $[a, b]$.

Given a non-negative measurable function f on $[0, a]$ that is finite almost everywhere then $m_f(y) = |E_y| = |\{x;\ f(x) > y\}|$, defines the *distribution function* m_f of f. Two functions with the same distribution function are said to be *equidistributed* or *equimeasurable*. Every function f has a decreasing right continuous function f^*, and an increasing right continuous function f_*, that are equidistributed with f, called the *increasing*, *decreasing*, *rearrangement of f*.

If $f : \mathbb{R}^n \to \mathbb{R}$ then the function $f^{(*)}$ defined by

$$f^{(*)}(\underline{x}) = \sup \left\{ y;\ |E_y| > v_n |\underline{x}|^n \right\},$$

is called the *spherical*, or *symmetrical*, *decreasing rearrangement* of f; v_n is defined in **4** (1).

A discussion of various rearrangements can be found in [HLP], [Kawohl], [Lieb & Loss].

6. Some Functions and Classes of Functions

Various types of *convex functions* are defined as part of the entries for their inequalities; **Convex Function Inequalities, Convex Matrix Function Inequalities, n-Convex Function Inequalities, Log-convex Function Inequalities, Q-class Function Inequalities, Quasi-convex Function Inequalities, Schur Convex Function Inequalities, Strongly Convex Function Inequalities, Subadditive Function Inequalities, Subharmonic Function Inequalities**.

Functions of *bounded variation* are defined in [Rudin, 1964, p. 117]; we write $V(f; a, b)$ for the *variation* of f on $[a, b]$.

If f is continuous on E we will write $f \in C(E)$; if, in addition, E is unbounded set in $\mathbb{R}$ and $\lim_{x \to \infty, x \in E} f(x) = 0$, we write $f \in C_0(E)$; these functions are said to be "zero at infinity", and such functions are obviously bounded.

If f is continuous and X is compact, or if f has compact support, or if $f \in C_0(E)$, then $\|f\|_{\infty, X} = \|f\|_\infty = \max_{x \in X} |f(x)|$.

If a real valued function f has continuous derivatives of order up to and including p, $n = 1, 2, \ldots$ on a set E, we will write $f \in C^n(E)$.

Further classes of functions are defined under the entry **Sobolev's Inequalities**.

If Ω is an open set in $\mathbb{C}$ then a differentiable function $f : \Omega \to \mathbb{C}$ is said to be *analytic in* Ω; the terms *holomorphic, regular* are also used in the literature.

An injective analytic function is said to be *univalent*, otherwise it is *multivalent*; specifically *m-valent* if it assumes each value in its range at most m times.

If a univalent function has as its domain $\mathbb{C}$ it is said to be an *entire function*.

If f is analytic in D and if $0 < r < 1, 0 \leq p \leq \infty$ then

$$M_p(f;r) = \begin{cases} \exp\left(\dfrac{1}{2\pi}\displaystyle\int_{-\pi}^{\pi} \log^+ |f(re^{i\theta})|\, d\theta\right), & \text{if } p = 0, \\ \left(\dfrac{1}{2\pi}\displaystyle\int_{-\pi}^{\pi} |f(re^{i\theta})|^p\, d\theta\right)^{1/p}, & \text{if } 0 < p < \infty, \\ \sup_{-\pi \leq \theta \leq \pi} |f(re^{i\theta})|, & \text{if } p = \infty. \end{cases}$$

We write $\mathcal{H}^p(D)$ for the class of functions with $\sup_{0 \leq r < 1} M_p(h;r) < \infty$; see **Analytic Function Inequalities** COMMENTS (i).

This notation can easily be extended to other classes of functions, and to functions with domain any disk centred at the origin, or to an annulus such as $\{z;\, a < |z| < b\}$. Related functions are defined in **Conjugate Harmonic Function Inequalities, Harmonic Function Inequalities, Quasi-conformal Function Inequalities, Subharmonic Function Inequalities**.

If the sequence of functions $\phi_n : [a, b] \to \mathbb{C}, n \in \mathbb{N}$ has the properties

$$\int_a^b \phi_n \overline{\phi}_m = \begin{cases} 0 & \text{if } n \neq m, \\ 1 & \text{if } n = m, \end{cases}$$

for $n, m \in \mathbb{N}$, then $\phi_n, n \in \mathbb{N}$ is called *an orthonormal sequence of complex valued functions defined on an interval* $[a, b]$. Given such a sequence of functions the sequence $\underline{c}$ defined by

$$c_n = \int_a^b f\overline{\phi}_n, \quad n \in \mathbb{N},$$

is the *sequence of Fourier coefficients of* f *with respect to* $\phi_n, n \in \mathbb{N}$. If, in addition, $\sup_{a \leq x \leq b} |\phi_n(x)| \leq M, n \in \mathbb{N}$, the orthonormal sequence is said to be *uniformly bounded*. Other index sets, such as $\mathbb{Z}$ are possible, and real orthonormal sequences can be considered; see [Zygmund, vol.I, pp. 5–7].

$[\cdot]$ denotes the *greatest integer function*,

$$[x] = \max\{n;\, n \leq x < n+1, n \in \mathbb{Z}\}.$$

If $x > 0$,

$$\log^+ x = \max\{0, \log x\} = \begin{cases} \log x, & \text{if } x \geq 1, \\ 0, & \text{if } 0 < x \leq 1. \end{cases}$$

The *indicator function of a set* A is

$$1_A(x) = \begin{cases} 1, & \text{if } x \in A, \\ 0, & \text{if } x \notin A. \end{cases}$$

The *factorial or gamma function* is $x!$, or $\Gamma(x) = (x-1)!$; $0! = 1! = 1$, if $n = 2, 3, \ldots$
$n! = n(n-1)\ldots\ldots 2 \cdot 1 = \mathfrak{S}_n^n(\underline{n})$, $\underline{n} = \{1, 2, \ldots, n\}$; and $(1/2)! = \sqrt{\pi}/2$. The *domain* of this function is $\mathbb{C} \setminus \{-1, -2, \ldots\}$, and if both z and $z-1$ are in this domain $z! = z(z-1)!$.
Related functions are defined in **Digamma Function Inequalities, Vietoris's Inequality**.

Other functions are defined in the following entries: **Bernšteĭn Polynomial Inequalities, Beta Function Inequalities, Bieberbach's Conjecture, Čebišev Polynomial Inequalities, Copula Inequalities, Internal Function Inequalities, Laguerre Function Inequalities, Lipschitz Function Inequalities, N-function Inequalities, Segre's Inequalities, Semi-continuous Function Inequalities, Starshaped Function Inequalities, Totally Positive Function Inequalities, Trigonometric Polynomial Inequalities**.

7. Matrices

An $m \times n$ *matrix* is $A = (a_{ij})_{\substack{1 \le i \le m \\ 1 \le j \le n}}$; the a_{ij} are the *entries* and if they are real, complex then A is said to be *real, complex*. The *transpose* of A is $A^T = (a_{ji})_{\substack{1 \le i \le m \\ 1 \le j \le n}}$; $A^* = (\overline{a}_{ji})_{\substack{1 \le i \le m \\ 1 \le j \le n}}$ is the *conjugate transpose* of A.
If $A = A^T$ then A is *symmetric*; if $A = A^*$ then A is *Hermitian*.
$\det A = \det(A)$, $\operatorname{tr} A = \operatorname{tr}(A)$, $\operatorname{rank} A = \operatorname{rank}(A)$, $\operatorname{per} A = \operatorname{per}(A)$ denote the *determinant, trace, rank* and *permanent* of A respectively. I_n denotes $n \times n$ *unit matrix*; just written I if the value of n is obvious.
If A is an $n \times n$ matrix, $n \ge 2$, and if $1 \le i_1 < i_2 < \cdots < i_m \le n$, $1 \le m \le n$ then $A_{i_1, \ldots, i_m}$ denotes the *principal submatrix* of A made up of the intersection of the rows and columns $i_1, \ldots, i_m$ of A.
If A is a matrix and if A_i' denotes the sub-matrix obtained by deleting the i-th row and column.
The *eigenvalues* of a square matrix A will be written $\lambda_s(A)$, and when real, by the notation in 5, $\lambda_{[s]}(A), \lambda_{(s)}(A)$ will denote them in descending, ascending, order; if they are complex, by a slight extension of that notation. $|\lambda_{[s]}(A)|, |\lambda_{(s)}(A)|$ will denote their absolute values in descending, ascending, order,
Various concepts of order, and norm will be defined in the relevant entries.

Full details can be found in [*Marcus & Minc*], and [*Horn & Johnson*].

8. Probability and Statistics

If Ω is a non-empty set, $\mathcal{A}$ a σ-field of subsets of Ω, called *events*, and P a *probability* on $\mathcal{A}$ then $(\Omega, \mathcal{A}, P)$ is a *probability space* and an $\mathcal{A}$-measurable function $X : \Omega \to \mathbb{R}$ is called a *random variable*.
The *(mathematical) expectation* of X is $EX = EX(\Omega) = \int_\Omega X \, \mathrm{d}P$.
The *variance* of X is $\sigma^2 X = \sigma^2 X(\Omega) = E(X - EX)^2$; σX is the *standard deviation*, the *coefficient of variation* is $CV(X) = \sigma X/EX$, and mX is the *median* of X.
The probability of the event $\{\omega;\ X(\omega) > r\}$ will be written $P(X > r)$.
The quantities EX^r, $E|X|^r$ are the *k-th moment, absolute moment*, of X respectively.
If the σ-field $\mathcal{A}$ is finite then Ω is called a *finite probability space*.
If $\mathcal{B} \subseteq \mathcal{A}$ is another σ-field then $E(X|\mathcal{B})$ denotes the *conditional expectation* of X with respect to B.

If $T = \mathbb{N}$, or $[0, \infty[$ then *stochastic process* $\mathcal{X} = (X_t, \mathcal{F}_t, t \in T)$ defined on the probability space $(\Omega, \mathcal{F}, P)$ is a *martingale* if $\mathcal{F}_t$ is a σ-field, $\mathcal{F}_s \subseteq \mathcal{F}_t \subseteq \mathcal{F}, s \leq t, s,t \in T$, $EX_t < \infty, t \in T$, X_t is $\mathcal{F}_t$-measurable, $t \in T$, and if almost surely, that is with probability one,

$$E(X_t|\mathcal{F}_s) = X_s, s \leq t,\ s,t \in T. \qquad (1)$$

If in (1) $=$ is replaced by $\geq, (\leq)$, then we have a *sub-martingale*, (*super-martingale*).
There are various functions involved with the *Gaussian*, or *normal distribution*.

$$\mathfrak{z}(x) = \frac{1}{\sqrt{2\pi}} e^{-x^2/2}; \quad \mathfrak{p}(x) = \int_{-\infty}^{x} \mathfrak{z}; \quad \mathfrak{q}(x) = \int_{x}^{\infty} \mathfrak{z}; \quad \mathfrak{a}(x) = \int_{-x}^{x} \mathfrak{z}; \quad \mathfrak{r} = \frac{\mathfrak{q}}{\mathfrak{z}}.$$

The last is usually called the *Mills' Ratio*.
There are some simple relations between these functions:

$$\mathfrak{p} + \mathfrak{q} = 1; \qquad \mathfrak{p}(-x) = \mathfrak{q}(x); \qquad \mathfrak{p} = \frac{1+\mathfrak{a}}{2}.$$

For more details see [*Feller*], and [*Loève*].

9. Symbols Indicating Some Inequalities
Certain inequalities will be referred to by a symbol to make referencing easier. They are listed here.

Bernoulli's Inequality .... (B)
Cauchy's Inequality .... (C)
Čebyšev's[1] Inequality .... (Č)
Geometric-Arithmetic Mean Inequality .... (GA)
Harmonic-Arithmetic Mean Inequality.. (HA)
Harmonic-Geometric Mean Inequality .... (HG)
Hölder's Inequality .... (H)
Jensen's Inequality .... (J)
Minkowski's Inequality .... (M)
Power Mean Inequality .... (r;s)
Symmetric Mean Inequality .... S(r;s)
Triangle Inequality .... (T)

---

[1] П Л Чебышев. Also transliterated as Chebyshev, Tchebyshev, Tchebichev.

A

Abel's Inequalities (a) If $\underline{w}, \underline{a}$ are real n-tuples and $\underline{a}$ is monotonic then

$$\left|\sum_{i=1}^{n} w_i a_i\right| \leq \max_{1 \leq i \leq n} \{|W_i|\}(|a_1| + 2|a_n|).$$

(b) Let $\underline{a}$ be a real n-tuple with $A_i > 0, 1 \leq i \leq n$, then

$$\frac{A_n}{A_1} \leq \exp\left(\sum_{i=2}^{n} \frac{a_i}{A_{i-1}}\right).$$

(c) If $p < 0$, q the conjugate index, and $\underline{a}$ is a positive sequence, then

$$\sum_{i=2}^{\infty} \frac{a_i}{A_i^{1/q} A_i^{1/q}} < -p a_1^{1/p}. \tag{1}$$

COMMENTS (i) (a), known as *Abel's lemma*, is a simple consequence of *Abel's summation formula*, an "integration by parts formula" for sequences:

$$\sum_{i=1}^{n} w_i a_i = W_n a_n + \sum_{i=1}^{n-1} W_i \Delta a_i. \tag{2}$$

(ii) If $\underline{a}$ is non-negative and decreasing then we have the simpler result:

$$a_1 \min_{1 \leq i \leq n} |W_i| \leq \sum_{i=1}^{n} w_i a_i \leq a_1 \max_{1 \leq i \leq n} |W_i|. \tag{3}$$

EXTENSIONS (a) [BROMWICH] If $\underline{w}, \underline{a}$ are real n-tuples with $\underline{a}$ decreasing, and if

$$\overline{W}_k = \max\{W_i, 1 \leq i \leq k-1\}, \quad \overline{W}'_k = \max\{W_i, k \leq i \leq n\},$$
$$\underline{W}_k = \min\{W_i, 1 \leq i \leq k-1\}, \quad \underline{W}'_k = \min\{W_i, k \leq i \leq n\},$$

for $1 \leq k \leq n$ then

$$\underline{W}_k(a_1 - a_k) + \underline{W}'_k a_k \leq \sum_{i=1}^{n} w_i a_i \leq \overline{W}_k(a_1 - a_k) + \overline{W}'_k a_k. \tag{4}$$

(b) [REDHEFFER] (i) Under the hypotheses of (b) above

$$\frac{A_n}{A_1} \leq \left(1 + \frac{1}{n-1}\sum_{i=2}^{n}\frac{a_i}{A_{i-1}}\right)^{n-1},$$

with equality if and only if for some $\lambda > 0$ $A_i = \lambda^{i-1}A_1$, $1 \leq i \leq n$.
(ii) Under the hypotheses of (c) above and $\underline{w}$ a positive sequence with $w_1 = 1$

$$\sum_{i=2}^{\infty} w_i A_i^{1/p} < q \sum_{i=1}^{\infty} w_i^{1/q} a_i^{1/p}.$$

COMMENTS   (iii) The last inequality reduces to (1) on putting $w_i = a_i/A_i$ for all i.
(iv) See also **Integral Inequalities** DISCRETE ANALOGUES.

RELATED RESULTS   (a) [KALAJDŽIĆ] If $\underline{a}$, $\underline{w}$ are real n-tuples with $W_n = 0$ and if sgn $W_k = (-1)^{k-1}$sgn w_1, $1 \leq k \leq n$, then

$$\left|\sum_{i=1}^{n} w_i a_i\right| \leq \frac{1}{2} \max_{1 \leq k \leq n-1} |\Delta a_{k-1}| \sum_{i=1}^{n-1} |w_i|.$$

(b) [BENNETT] If $\underline{a}, \underline{b}, \underline{c}$ are non-negative sequences, with $\underline{c}$ decreasing then

$$A_n \leq B_n, \; n = 1, 2, \ldots \implies \sum_{i=1}^{n} a_i c_i \leq \sum_{i=1}^{n} b_i c_i, \; n = 1, 2, \ldots.$$

COMMENTS   (v) This last result follows by two applications of (2), and is related to **Steffensen's Inequalities** (b).
(vi) Integral analogues of (3) and (4) are given by Bromwich; see **Integral Mean Value Theorems** (b), COMMENTS (ii).

REFERENCES   [AI, pp. 32–33], [EM, vol.1, pp. 5, 8], [MPF, pp. 333–337]; [Bennett, 1996, p. 9], [Bromwich, pp. 57–58, 473–475], [Zygmund, vol.I, pp. 3–4]; [Kalajdžić], [Redheffer, pp. 690, 696].

Absolutely and Completely Monotonic Function Inequalities

A continuous function $f : [a, b] \to \mathbb{R}$ with derivatives of all orders is said to be *absolutely, completely, monotonic* if, respectively,

$$f^{(k)} \geq 0, \qquad (-1)^k f^{(k)} \geq 0, \quad a < x < b, \; k \in \mathbb{N}.$$

(a) If f is absolutely monotonic on $]-\infty, 0]$ then

$$\sqrt{f^{(k)} f^{(k+2)}} \geq f^{(k+1)}, \quad k \in \mathbb{N}.$$

(b) If f is absolutely monotonic on $]-\infty, 0[$ then

$$\det\left(\left(f^{(i+j-2)}\right)_{\substack{1\leq i\leq n \\ 1\leq j\leq n}}\right) \geq 0 \quad \text{and} \quad \det\left(\left(f^{(i+j-1)}\right)_{\substack{1\leq i\leq n \\ 1\leq j\leq n}}\right) \geq 0, \qquad n\geq 1.$$

(c) [FINK] If $\underline{a}, \underline{b}$ are non-negative n-tuples of integers with $\underline{a} \prec \underline{b}$ then:
(i) if f is absolutely monotonic on $[0, \infty[$,

$$\prod_{i=1}^n f^{(a_i)} \leq \prod_{i=1}^n f^{(b_i)},$$

with equality if for some $a > 0$, $f(x) = e^{ax}$;
(ii) if f is completely monotonic on $[0, \infty[$,

$$\prod_{i=1}^n (-1)^{a_i} f^{(a_i)} \leq \prod_{i=1}^n (-1)^{b_i} f^{(b_i)},$$

with equality if for some $a > 0$, $f(x) = e^{-ax}$.

COMMENTS (i) (a) is a simple consequence of the integral analogue of (Č), or (GA), and a well known integral representation for the class of absolutely monotonic functions.
(ii) The first inequality in (b) generalizes that in (a).
REFERENCES [MPF, pp. 365–377]; [Widder, pp. 167–168].

Absolute Value Inequalities
(a) If $a \in \mathbb{R}$ then

$$-|a| \leq a \leq |a|,$$

with equality on the left-hand side if and only if $a \leq 0$, and on the right-hand side if and only if $a \geq 0$.
(b) If $b \geq 0$ then: $\quad |a| \leq b \iff -b \leq a \leq b$.
(c) If $a, b \in \mathbb{R}$ then:
(i)

$$ab \leq |a||b|,$$

with equality if and only if $ab \geq 0$;
(ii)

$$|a+b| = \begin{cases} |a|+|b|, & \text{if } ab \geq 0, \\ ||a|-|b||, & \text{if } ab \leq 0; \end{cases}$$

(iii)

$$|a+b| \leq |a|+|b|, \qquad (1)$$
$$|a-b| \geq ||a|-|b||, \qquad (2)$$

with equality if and only if $ab \geq 0$.
(d) If $a, b, c \in \mathbb{R}$ then
$$|a - c| \leq |a - b| + |b - c|, \qquad (3)$$
with equality if and only if b is between a and c.

COMMENTS (i) The proofs are by examining the cases in the definition of absolute value:
$$|a| = \begin{cases} a, & \text{if } a \geq 0, \\ -a, & \text{if } a \leq 0. \end{cases}$$

Alternatively (c)(iii) follows from (c)(i) by considering $(a \pm b)^2$, using the definition
$$|a| = \sqrt{a^2};$$
and (d), which is a special case of (T), follows from (1).
(ii) The condition $ab \geq 0$ can be written as $a \sim^+ b$; see **Triangle Inequality** (a), (b).

EXTENSIONS AND OTHER RESULTS (a) If $a_j \in \mathbb{R}, 1 \leq j \leq n$, then
$$|a_1 + \cdots + a_n| \leq |a_1| + \cdots + |a_n|,$$
with equality, if and only if all, the non-zero a_j have the same sign.
(b) If $a_j \in \mathbb{R}$ with $|a_j| \leq 1, 1 \leq j \leq 4$, then
$$|a_1(a_3 + a_4) + a_2(a_3 - a_4)| \leq 2. \qquad (4)$$

COMMENTS (iii) The right-hand side of (4) is attained when $a_j = 1, 1 \leq j \leq 4$.
(iv) If the numbers in (b) are taken to be complex, the upper bound in (4) is $2\sqrt{2}$, an upper bound that is attained when $a_1 = i, a_2 = 1, a_3 = (1+i)/\sqrt{2}, a_4 = (1-i)/\sqrt{2}$.
(v) See also **Complex Number Inequalities** (a), (b) EXTENSIONS (a), **Triangle Inequality**.

REFERENCES [EM, vol.1, p. 22]; [Apostol, vol.I, pp. 31–33], [Halmos, p. 15], [Hewitt & Stromberg, pp. 36–37].

Aczel's Inequality

If $\underline{a}, \underline{b}$ are real n-tuples with $a_1^2 - \sum_{i=2}^{n} a_i^2 > 0$ then
$$\left(a_1^2 - \sum_{i=2}^{n} a_i^2 \right) \left(b_1^2 - \sum_{i=2}^{n} b_i^2 \right) \leq \left(a_1 b_1 - \sum_{i=2}^{n} a_i b_i \right)^2,$$
with equality if and only if $\underline{a} \sim \underline{b}$.

COMMENTS (i) This result can be considered as an analogue of (C) in a non-Euclidean geometry. It is an example of a reverse inequality; see **Reverse Inequalities**.

EXTENSIONS   (a) [POPOVICIU] If $p, q > 1$ are conjugate indices and $\underline{a}, \underline{b}$ are real n-tuples with $a_1^p - \sum_{i=2}^n a_i^p > 0$, $b_1^q - \sum_{i=2}^n b_i^q > 0$ then

$$\left(a_1^p - \sum_{i=2}^n a_i^p\right)^{1/p} \left(b_1^p - \sum_{i=2}^n b_i^p\right)^{1/q} \le a_1 b_1 - \sum_{i=2}^n a_i b_i. \qquad (1)$$

If $0 < p < 1$ then (~ 1) holds.

(b) [BELLMAN] If $p \ge 1$, or $p < 0$ and $\underline{a}, \underline{b}$ are real n-tuples with $a_1^p - \sum_{i=2}^n a_i^p > 0$, $b_1^p - \sum_{i=2}^n b_i^p > 0$ then

$$\left(a_1^p - \sum_{i=2}^n a_i^p\right)^{1/p} + \left(b_1^p - \sum_{i=2}^n b_i^p\right)^{1/p} \le \left((a_1+b_1)^p - \sum_{i=2}^n (a_i+b_i)^p\right)^{1/p}. \qquad (2)$$

If $0 < p < 1$ then (~ 2) holds.

COMMENTS   (ii) The results of Aczel and Popoviciu can be deduced from $(\sim J)$, while that of Bellman is a deduction from (M).

REFERENCES   [AI, pp. 57–59], [BB, pp. 38–39], [MPF, pp. 117–121], [PPT, pp. 124–126].

Adamović's Inequality   The following inequality holds if all the factors are positive:

$$\prod_{i=1}^n a_i \ge \prod_{i=1}^n (A_n - (n-1)a_i),$$

with equality if and only if $\underline{a}$ is constant.

COMMENTS   (i) This inequality has been generalized by Klamkin.
(ii) If a, b, c are the sides of a triangle then the case $n = 3$ gives

$$abc \ge (a+b-c)(b+c-a)(c+a-b);$$

this is known as *Padoa's inequality*.

REFERENCES   [AI, pp. 208–209]; [Bottema et al, p. 12]; [Klamkin, 1976, 1996].

Agarwal's Inequality   If $R = [0, a] \times [0, b]$ and $f \in C^2(R)$, with $f(x, 0) = f(0, y) = 0, 0 \le x \le a, 0 \le y \le b$, then

$$\int_R |f||f''_{12}| \le \frac{ab}{2\sqrt{2}} \int_R |f''_{12}|^2.$$

COMMENTS   The best value of the constant on the right-hand side is not known; but it is greater than $(3 + \sqrt{13})ab/24$.

REFERENCES   [General Inequalities, vol. 3, pp. 501–503].

Ahlswede-Daykin Inequality Let $f_i : X \to [0, \infty[,\ 1 \leq i \leq 4$, where X is a distributive lattice. If for all $a, b \in X$

$$f_1(a)f_2(b) \leq f_3(a \vee b)f_4(a \wedge b), \tag{1}$$

then for all $A, B \subseteq X$,

$$\sum_{a \in A} f(a) \sum_{b \in B} f(b) \leq \sum_{x \in A \vee B} f(x) \sum_{y \in A \wedge B} f(y).$$

COMMENTS This is also known as *the Four Functions inequality*. it is an example of a *correlation inequality*; see also **Holley's Inequality**, COMMENTS.

REFERENCES [EM, Supp., p. 201].

Aleksandrov-Fenchel Inequality
See **Mixed-volume Inequalities** EXTENSIONS.

Almost Symmetric Function Inequalities
See **Segre's Inequalities**.

Alternating Sum Inequalities [BORWEIN & BORWEIN] If f is a non-negative function on n-tuples of non-negative integers that is increasing in each variable, and if $\underline{a}, \underline{b}$ are two such n-tuples with $A_n \leq B_n$ then

$$f(\underline{a}) \geq (-1)^{A_n} \sum (-1)^{C_n} f(\underline{c}),$$

where the sum is over all n-tuples of non-negative integers, $\underline{c}$, with $A_n \leq C_n \leq B_n$.

COMMENTS See also **Opial's Inequalities** (b), **Szegö's Inequality**.
REFERENCES [Borwein, D & Borwein, J M].

Alzer's Inequalities (a) If f is convex on $[a, b]$ with a bounded second derivative then

$$\frac{f(a) + f(b)}{2} - f\left(\frac{a+b}{2}\right) \leq \frac{(b-a)^2}{8} \sup_{a \leq x \leq b} f''(x).$$

(b) If $\underline{n} = \{1, 2, \ldots, n\}, n > 1$, then

$$\frac{1}{e^2} < \mathfrak{G}_n^2(\underline{n}) - \mathfrak{G}_{n-1}(\underline{n-1})\mathfrak{G}_{n+1}(\underline{n+1}).$$

The constant on the left-hand side is best possible.
(c) If $\underline{a}, \underline{b}$ are real sequences satisfying $b_i \leq \min\{a_i, a_{i+1}\}, i = 1, 2, \ldots$, then with $p \geq 1$,

$$\sum_{i=0}^{n} a_i^p - \sum_{i=0}^{n-1} b_i^p \leq \left(\sum_{i=0}^{n} a_i - \sum_{i=0}^{n-1} b_i\right)^p.$$

COMMENTS (i) (a) is an easy consequence of

$$[a, \frac{a+b}{2}, b; f] = \frac{f''(c)}{2}$$

for some $c, a < c < b$; this notation is defined in **n-Convex Function Inequalities** (1).
(ii) (c) implies **Székely, Clark & Entringer Inequality** (a).

EXTENSIONS [KIVINUKK] If $f \in \mathcal{L}^p([a,b])$, $1 \le p \le \infty$, is convex, and if q is the conjugate index

$$\overline{1-\lambda}f(a)+\lambda f(b)-f\bigl(\overline{1-\lambda}\,a+\lambda b\bigr) \le 2^{1/p}(q+1)^{-1/q}\lambda(1-\lambda)(b-a)^{1+1/q}\|f''\|_{p,[a,b]}.$$

COMMENTS   (iii) The case $p=\infty$, $\lambda=1/2$ is (a) above.

REFERENCES   [Alzer, 1994[(3)]], [Kivinukk].

Analytic Function Inequalities

(a) If f is analytic in D and if $0 \le r_1 \le r_2 < 1$ then,

$$M_p(f;r_1) \le M_p(f;r_2), \quad 0 \le p \le \infty. \qquad (1)$$

(b) If f,g are analytic in D and if $1 \le p \le \infty$, then

$$M_p(f+g;r) \le M_p(f;r) + M_p(g;r), \ 0 \le r < 1.$$

(c) If $f(z) = \sum_{n=0}^{\infty} a_n z^n$ is bounded and analytic in D with $|f(z)| < M$, $|z| < 1$ and if

$$s_n(z) = \sum_{k=0}^{n} a_k z^k, \ n \in \mathbb{N}; \quad \sigma_n(z) = \frac{1}{n}\sum_{k=0}^{n-1} s_k(z), n \ge 1,$$

then:

$$\begin{aligned} |\sigma_n(z)| &\le M, & \text{if } |z| < 1; \\ |s_n(z)| &\le KM \log n, & \text{if } |z| < 1; \\ |s_n(z)| &\le M, & \text{if } |z| < 1/2; \end{aligned}$$

where K is an absolute constant.

COMMENTS   (i) From (1) if $f \in \mathcal{H}^p(D)$ then $\|f\|_p = \lim_{r \to 1} M_p(f;r)$ exists and is finite. If $1 \le p \le \infty$, $\|f\|_p$ is a norm on the space $\mathcal{H}^p(D)$, as (b) readily proves; (b) is an easy consequence of (M). Further properties of $M_p(f;r)$ are given in **Hardy's Analytic Function Inequality**.
(ii) The results in (a) and (b) readily extend to harmonic functions.
(iii) The converse of the first inequality in (c) holds, in the sense that if $\sigma_n(z) \le M$, $|z| < 1$ then $|f(z)| < M$, $|z| < 1$.
(iv) See also **Area Theorems, Bieberbach's Conjecture, Bloch's Constant, Borel-Carathéodory Inequality, Cauchy-Hadamard Inequality, Cauchy Transform Inequality, Distortion Theorems, Entire Function Inequalities, Fejér-Riesz Theorem, Gabriel's Problem, Hadamard's Three Circle Theorem, Landau's Constant, Lebedev-Milin Inequalities, Littlewood-Paley Inequalities, Maximum-Modulus Principle, Phragmén-Lindelöf Inequality, Picard-Schottky Theorem, Rotation Theorems, Subordination Inequalities.**

REFERENCES   [Rudin, 1966, pp. 330–331], [Titchmarsh, 1939, pp. 235–238].

Arc Length Inequality If f, g are continuous and of bounded variation on $[a, b]$ and L is the arc length of the curve $\{(x, y); x = f(t), y = g(t), a \leq t \leq b\}$ then

$$\int_a^b \sqrt{f'^2 + g'^2} \leq L \leq V(f; [a, b]) + V(g; [a, b]).$$

There is equality on the left if and only if f, g are absolutely continuous; there is equality on the right if and only if $f'g' = 0$ almost everywhere, and for some sets A, B with $A \cup B = \{x; f'(x) = \infty, \text{ and } g'(x) = \infty\}$ we have $|f[A]| = |g[B]| = 0$.

REFERENCES [Saks, pp. 121–125]; [Cater].

Area Theorems [GRONWALL; BIEBERBACH] If f is univalent in $\{z; |z| > 1\}$ with $f(z) = z + \sum_{n=0}^{\infty} b_n z^{-n}$ then

$$\sum_{n=1}^{\infty} n|b_n|^2 \leq 1.$$

COMMENTS (i) The name of this result follows from the proof which computes the area outside $f[\{z; |z| > 1\}]$.

EXTENSIONS [GRUNSKY, GOLUSIN] If f is analytic and m-valent for $|z| > 1$ with $f(z) = \sum_{n=-m}^{\infty} b_n z^{-n}, b_{-m} \neq 0$ then

$$\sum_{n=1}^{\infty} n|b_n|^2 \leq \sum_{n=1}^{m} n|b_{-n}|^2;$$

$$\sum_{n=1}^{m} n|b_n||b_{-n}|^2 \leq \sum_{n=1}^{m} n|b_{-n}|^2.$$

COMMENTS (ii) See also **Bieberbach's Conjecture**.

REFERENCES [EM, vol.1, pp. 245–246]; [Ahlfors, 1973, pp. 82–87], [Conway, vol.II, pp. 56–58], [Dictionary, vol.III, p. 1666].

Arithmetic's Basic Inequalities (a) If $a, b \in \mathbb{R}$ then:

$$a > b \text{ if and only if } a - b \text{ is positive};$$
$$a < b \text{ if and only if } b - a \text{ is positive}.$$

(b) If $a, b, c \in \mathbb{R}$ and if $a > b$ then

$$a + c > b + c, \qquad ac \begin{cases} > bc & \text{if } c > 0, \\ < bc & \text{if } c < 0, \end{cases} \text{ and if } a, b > 0 \quad a^{r/s} > b^{r/s},$$

where r, s are positive integers.
(c) If $a, b, c, d \in \mathbb{R}$ and if $a > b, c > d$ then

$$a + c > b + d,$$

and if all are positive then
$$ac > bd, \quad \frac{a}{d} > \frac{b}{c}.$$

(d) If $a, b \in \mathbb{R}$ then
$$a > b > 0 \quad \text{or} \quad 0 > a > b \implies \frac{1}{a} < \frac{1}{b};$$
$$a > 0 > b \implies \frac{1}{a} > \frac{1}{b}.$$

COMMENTS (i) (a) is just the definition of inequality, and all of the other results follow from it.
(ii) These inequalities are called *strict inequalities*, and we also have the following.

EXTENSIONS [WEAK INEQUALITIES] (a) If $a, b \in \mathbb{R}$ then:
$a \geq b$ if and only if $a - b$ is positive or zero;
$a \leq b$ if and only if $b - a$ is positive or zero.

(b) If $a, b, c \in \mathbb{R}$ and if $a \geq b$ then
$$a + c \geq b + c \quad \text{and if } a, b > 0 \quad a^{r/s} \geq b^{r/s},$$
where r, s are positive integers, with equality if and only if $a = b$;
$$ac \begin{cases} \geq bc, & \text{if } c \geq 0, \\ \leq bc, & \text{if } c \leq 0, \end{cases}$$
with equality if either $c = 0$ or $a = b$.
(c) If $a \geq b$ and if $b > c$, or if $a > b$ and $b \geq c$ then $a > c$.

COMMENTS (iii) Of course the last inequality in (b) will hold if r/s is replaced by any positive real number, see **Exponential Function Inequalities** (a).
Most of these inequalities can be extended by induction.

EXTENSIONS CONTINUED (d) If $a_i \geq b_i, 1 \leq i \leq n$ then
$$a_1 + \cdots + a_n \geq b_1 + \cdots + b_n,$$
and this inequality is strict if at least one of the inequalities $a_i \geq b_i, 1 \leq i \leq n$ is strict.
(e) If $a_i \geq b_i \geq 0, 1 \leq i \leq n$ then
$$a_1 a_2 \ldots a_n \geq b_1 b_2 \ldots b_n.$$

COMMENTS (iv) See also **Mediant Inequalities** (a).
REFERENCES [Apostol, vol.I, pp. 15–17].

Arithmetic-geometric Compound Mean Inequalities

If $0 < a = a_0 \leq b = b_0$ define the sequences $\underline{a}, \underline{b}$ by
$$a_n = \mathfrak{G}_2(a_{n-1}, b_{n-1}), \quad b_n = \mathfrak{A}_2(a_{n-1}, b_{n-1}), \quad n \geq 1.$$
The arithmetic-geometric compound mean of a and b is
$$\mathfrak{A} \otimes \mathfrak{G}(a, b) = \lim_{n \to \infty} a_n = \lim_{n \to \infty} b_n. \tag{1}$$
This is also sometimes called the *arithmetico-geometric mean*.

If $0 < a \leq b$

$$a < \mathfrak{G}_2(a,b) < \mathfrak{L}(a,b) < \mathfrak{A} \otimes \mathfrak{G}(a,b) < \mathfrak{A}_2(a,b) < b.$$

COMMENTS   In an analogous way we can define the *geometric-harmonic compound mean*, $\mathfrak{G} \otimes \mathfrak{H}(a,b)$, and the *arithmetic-harmonic compound mean*, $\mathfrak{A} \otimes \mathfrak{H}(a,b)$.
In general any two means can be *compounded* in this way provided the associated limits in (1) exist.
Not all the means obtained in this way are new; in particular $\mathfrak{A} \otimes \mathfrak{H}(a,b) = \mathfrak{G}_2(a,b)$.

REFERENCES   [MI, pp. 40–41, 359–367], [MPF, pp. 47–48]; [Borwein & Borwein, 1987, pp. 1–5]; [Borwein & Borwein, $1987^{(2)}$].

Arithmetic-Geometric Mean Inequality
See **Geometric-Arithmetic Mean Inequality**.

Arithmetic Mean Inequalities
(a) If $\underline{a}, \underline{w}$ are positive n-tuples then

$$\min \underline{a} \leq \mathfrak{A}_n(\underline{a}; \underline{w}) \leq \max \underline{a}, \qquad (1)$$

with equality if and only if $\underline{a}$ is constant.

(b) [DIANANDA] If $\underline{a}, \underline{w}$ are positive n-tuples such that for some positive integer k, $\underline{a} \geq k$ then

$$\mathfrak{A}_n(\underline{a};\underline{w}) \leq \left(\prod_{i=1}^n \frac{a_i}{[a_i]}\right) \mathfrak{A}_n([\underline{a}];\underline{w}) < \left(1 + \frac{1}{k}\right)^{n-1} \mathfrak{A}_n(\underline{a};\underline{w}). \qquad (2)$$

In particular if $\underline{a} \geq n - 1$ then

$$\left(\prod_{i=1}^n \frac{a_i}{[a_i]}\right) \mathfrak{A}_n([\underline{a}];\underline{w}) < e\, \mathfrak{A}_n(\underline{a};\underline{w}).$$

Further there is equality on the left of (2) if and only if all the entries in $\underline{a}$ are integers.

(c) [PREŠIĆ] If $\underline{a}$ is a real n-tuple then

$$\min_{1 \leq i < j \leq n}(a_i - a_j)^2 \leq \frac{12}{n^2 - 1}\left(\mathfrak{A}_n(\underline{a}^2) - (\mathfrak{A}_n(\underline{a}))^2\right).$$

(d) [REDHEFFER] If $\underline{a}$ is an increasing positive sequence and if $p < 0$, q the conjugate index, then

$$\sum_{i=1}^\infty \frac{a_i}{\mathfrak{A}_i^{1/q}(\underline{a})} < (1-p) \sum_{i=1}^\infty \mathfrak{A}_i^{1/p}(\underline{a}).$$

(e) If $\mathfrak{A}_n(\underline{a};\underline{w}) = 1$ then

$$\mathfrak{A}_n(\underline{a} \log \underline{a}; \underline{w}) \geq 0,$$

with equality if and only if $\underline{a}$ is constant.

(f) [KLAMKIN] If $\underline{a}$ is a positive n-tuple,
$$r \sum_{r}! \frac{a_1 \cdots a_r}{a_1 + \cdots + a_r} \leq \binom{n}{r} (\mathfrak{A}_n(\underline{a}))^{r-1}.$$

COMMENTS (i) Variants of (c) have been given by Lupaş.
(ii) Inequality (d) is a recurrent inequality; see **Recurrent Inequalities**.
(iii) Inequality (f) is proved using Schur convexity; see **Schur Convex Function Inequalities**.

EXTENSIONS [CAUCHY] If $\underline{a}, \underline{b}, \underline{w}$ are positive n-tuples
$$\min\left(\underline{a}\,\underline{b}^{-1}\right) \leq \frac{\mathfrak{A}_n(\underline{a};\underline{w})}{\mathfrak{A}_n(\underline{b};\underline{w})} \leq \max\left(\underline{a}\,\underline{b}^{-1}\right).$$

COMMENTS (iv) Inequality (1) and its extension above, are properties common to most means; see **Basic Mean Inequalities** (1), EXTENSIONS.

INTEGRAL ANALOGUES [FAVARD] If $f \geq 0$ on $[a,b]$ is concave then
$$\mathfrak{A}_{[a,b]}(f) \geq \frac{1}{2} \max_{a \leq x \leq b} f.$$

COMMENTS (v) The last inequality is a limiting case of **Favard's Inequalities** (b).
(vi) For an integral analogue of the basic inequality (1) see **Integral Mean Value Theorems** (a).
(vii) See also **Binomial Function Inequalities** (j), **Čebyšev's Inequality**, **Convex Sequence Inequalities** (d), **Geometric-Arithmetic Mean Inequality** (5), **Grusses' Inequalities** (a), DISCRETE ANALOGUES, **Harmonic Mean Inequalities** (c) and COMMENTS (iv), **Increasing function Inequalities** (2) and EXTENSIONS, **Kantorović's Inequality**, **Karamata's Inequality**, **Levinson's Inequality**, **Logarithmic Mean Inequalities** COROLLARIES (b), (c), EXTENSIONS (a), **Mitrinović & Djoković's Inequality**, **Mixed Mean Inequalities** SPECIAL CASES (b), **Muirhead Symmetric Function and Mean Inequalities** COMMENTS (ii), **Nanson's Inequality**, **n-convex Sequence Inequalities** (b), EXTENSIONS, **Statistical Inequalities** (c).

REFERENCES [AI, pp. 204, 340–341], [BB, p. 44], [MI, pp. 35,37–38,128, 167], [MO, p. 91]; [Lupaş], [Redheffer, p. 689].

Askey-Karlin Inequalities

If $f(x) = \sum_{n \in \mathbb{N}} a_n x^n$, $1 \leq x < 1$, $A_{n+1} = \sum_{k=0}^{n} a_k$, $\Phi : [0,\infty[\to [0,\infty[$ increasing and convex, and $\beta > -2$ then
$$\int_0^1 \Phi \circ |f|(x)(1-x)^\beta \, dx \leq (\beta + 1)! \sum_{n \in \mathbb{N}} \Phi(|A_{n+1}|) \frac{n!}{(n+\beta+2)!};$$
$$\int_0^1 \frac{\Phi \circ |f|(x)}{(1-x)^2} \, dx \leq \sum_{n \in \mathbb{N}} \Phi\left(\frac{|A_{n+1}|}{n+1}\right).$$

COMMENTS These results have been generalized by Pachpatte.

REFERENCES [Pachpatte, 1988$^{(2)}$].

- end of A -

B

Backward Hölder Inequality
See **Young's Convolution Inequality** COMMENTS (iii).

Banach Algebra Inequalities
If X is a Banach algebra then for all $x, y \in B$
$$||xy|| \leq ||x||\,||y||.$$

COMMENTS (i) This is the defining inequality for a topological algebra that is a Banach space; for a definition see **Norm Inequalities** COMMENTS (i).
(ii) For particular cases see **Bounded Variation Function Inequalities** (b), **Matrix Norm Inequalities**.

REFERENCES [EM, vol. 1, pp. 329–332]; [Hewitt & Stromberg, pp. 83–84].

Banach Space Inequalities
See **Kallman-Rota Inequality, von Neumann & Jordan Inequality**.

Barnes's Inequalities
(a) If $\underline{a}, \underline{b}$ are non-negative concave n-tuples, $\underline{a}$ increasing and $\underline{b}$ decreasing, then

$$\sum_{i=1}^{n} a_i b_i \geq \frac{n-2}{2n-1} \left(\sum_{i=1}^{n} a_i^2 \right)^{1/2} \left(\sum_{i=1}^{n} b_i^2 \right)^{1/2}. \tag{1}$$

The inequality is sharp, and is strict unless $a_i = n - i$, $b_i = i - 1$, $1 \leq i \leq n$.

(b) Let $\underline{a}, \underline{b}$ be non-negative non-null n-tuples with p, q conjugate indices and $||\underline{a}||_p \neq 0$, $||\underline{b}||_q \neq 0$, define

$$H_{p,q}(\underline{a}, \underline{b}) = \frac{\underline{a}.\underline{b}}{||\underline{a}||_p ||\underline{b}||_q}.$$

If $\underline{c}$ is an increasing n-tuple with $\underline{c} \prec \underline{a}$, and if $\underline{d}$ is a decreasing n-tuple with $\underline{b} \prec \underline{d}$, and if $p, q \geq 1$

$$H_{p,q}(\underline{a}, \underline{b}) \geq H_{p,q}(\underline{c}, \underline{d}). \tag{2}$$

If $\underline{c}$ and $\underline{d}$ are increasing with $\underline{c} \prec \underline{a}$ and $\underline{d} \prec \underline{b}$, and if $p, q \leq 1$ inequality ($\sim$1) holds.

COMMENTS (i) Inequality (1) is a converse of (C) and (2) is a converse of (H).
(ii) Extensions of (2) to more than two n-tuples, and other situations have been given; see the references.
(iii) For another inequality of Barnes see **Grüss-Barnes's Inequality**.

REFERENCES [AI, p. 386], [MPF, pp. 148–156].

Basic Mean Inequalities   If $\mathfrak{M}(\underline{a})$ is a mean of $\underline{a}$ then

$$\min \underline{a} \leq \mathfrak{M}(\underline{a}) \leq \max \underline{a}, \qquad (1)$$

with equality only if $\underline{a}$ is constant.

COMMENTS   In **Notations 3** there are several examples of means, and references to others.. Most means have the properties:

$$\mathfrak{M}(\lambda \underline{a}) = \lambda \mathfrak{M}(\underline{a}); \qquad \underline{a} \leq \underline{b} \Longrightarrow \mathfrak{M}(\underline{a}) \leq \mathfrak{M}(\underline{b}), \text{ with equality } \iff \underline{a} = \underline{b}.$$

EXTENSIONS   If a mean has the properties mentioned above then,

$$\min\left(\underline{a}\,\underline{b}^{-1}\right) \leq \frac{\mathfrak{M}(\underline{a})}{\mathfrak{M}(\underline{b})} \leq \max\left(\underline{a}\,\underline{b}^{-1}\right).$$

REFERENCES   [MI, pp. 35, 37, 372–375].

Beckenbach's Inequalities   (a) If $p, q > 1$ are conjugate indices, $\underline{a}, \underline{b}$ positive n-tuples, and if $1 \leq m < n$ define,

$$\tilde{a}_i = \begin{cases} a_i, & \text{if } 1 \leq i \leq m, \\ \left(\dfrac{b_i \sum_{j=1}^{m} a_j^p}{\sum_{j=1}^{m} a_j b_j}\right)^{q/p}, & \text{if } m+1 \leq i \leq n. \end{cases}$$

Then given $a_j, 1 \leq j \leq m$, and $\underline{b}$

$$\frac{\left(\sum_{i=1}^{n} a_i^p\right)^{1/p}}{\sum_{i=1}^{n} a_i b_i} \geq \frac{\left(\sum_{i=1}^{n} \tilde{a}_i^p\right)^{1/p}}{\sum_{i=1}^{n} \tilde{a}_i b_i},$$

for all choices of $a_i, m < i \leq n$, with equality if and only if $a_i = \tilde{a}_i, m < i \leq n$.
(b) If $p, q > 1$ are conjugate indices, α, β, γ positive real numbers, $f \in \mathcal{L}^p([a,b])$, $g \in \mathcal{L}^q([a,b])$, then

$$\frac{\left(\alpha + \gamma \int_a^b f^p\right)^{1/p}}{\beta + \gamma \int_a^b fg} \geq \frac{\left(\alpha + \gamma \int_a^b h^p\right)^{1/p}}{\beta + \gamma \int_a^b hg},$$

where $h = (ag/b)^{q/p}$. There is equality if and only if $f = h$ almost everywhere.

COMMENTS   (i) The case $m = 1$ of (a) is just (H).
(ii) Integral analogues of (a) have been given.
(iii) A discrete analogue of (b) is easily stated; and if the $+$ signs are replaced by $-$ signs in all four places the inequality is reversed. This is called the *Beckenbach-Lorentz inequality*.
(iv) For another inequality of Beckenbach see **Counter-Harmonic Mean Inequalities** (d).

REFERENCES   [AI, p. 52], [MI, pp. 156–157], [MPF, pp. 156–163], [PPT, pp. 122–124].

Beckenbach-Lorentz Inequality
See **Beckenbach's Inequalities** COMMENTS (iii).

Bellman-Bihari Inequalities
See **Gronwall's Inequality** COMMENTS (iii).

Bendixson's Inequalities
See **Hirsch's Inequalities**.

Bennett's Inequalities   (a) If $n > 0$ then for all real $r > 0$,

$$1 \le \left(\frac{(n+1) \sum_{i=1}^{n} i^{-1}}{n \sum_{i=1}^{n+1} i^{-1}} \right)^{1/r} \le \frac{((n+1)!)^{1/(n+1)}}{(n!)^{1/n}}; \tag{1}$$

and

$$\frac{1}{n} \sum_{i=1}^{n} \left(\frac{n+1-i}{i} \right)^r < \frac{1}{n+1} \sum_{i=1}^{n+1} \left(\frac{n+2-i}{i} \right)^r. \tag{2}$$

(b) If $0 < r < 1$ and $\underline{a}$ is a non-negative n-tuple then, (i):

$$\sum_{i=1}^{n} \left(\frac{1}{i} \sum_{j=i}^{n} a_j \right)^r \le a_n(r) \sum_{i=1}^{n} \max_{1 \le i \le n} a_i^r, \tag{3}$$

where $a_n(r)$ is the left-hand side of (2), and (ii):

$$\sum_{i=1}^{\infty} \left(\frac{1}{i} \sum_{j=i}^{\infty} a_j \right)^r \le \frac{\pi r}{\sin \pi r} \sum_{i=1}^{\infty} \sup_{k \ge i} a_k^r;$$

the constant on the right-hand side is best possible, and the inequality strict unless $\underline{a} = \underline{0}$.

(c) If $\underline{a}$ is a non-negative sequence then

$$\sum_{n=1}^{\infty} \left(\frac{1}{n} \sum_{i=1}^{n} a_i^p \right)^p \begin{cases} \le \zeta(p) \sum_{n=1}^{\infty} \left(\sum_{i=1}^{n} \frac{a_i}{i} \right)^p, & \text{if } p \ge 2, \\ \ge (p-1)^{-1/p} \sum_{n=1}^{\infty} \left(\sum_{i=1}^{n} \frac{a_i}{i} \right)^p, & \text{if } 1 < p \le 2. \end{cases}$$

The first inequality is strict unless $a_2 = a_3 = \cdots = 0$, the second inequality is strict unless $\underline{a} = \underline{0}$. The constants on the right-hand sides are best possible. [ζ denotes

the zeta function].
(d) If $\underline{a}$ is a non-negative sequence and $1 < p < \infty$ then,

$$\sum_{n=1}^{\infty}\left(\frac{1}{n}\sum_{i=1}^{n}a_i^p\right)^p \leq \sum_{i=1}^{\infty}\left(\sum_{j=1}^{\infty}\frac{|a_j|}{i+j-1}\right)^p \leq \left(\frac{\pi}{q\sin\pi/p}\right)^p \sum_{n=1}^{\infty}\left(\frac{1}{n}\sum_{i=1}^{n}a_i^p\right)^p,$$

where q is the conjugate index. The constants are best possible. The inequality on the left is strict unless $\underline{a}$ has at most one non-zero entry; the inequality on the right is strict unless $\underline{a} = \underline{0}$.
(e) If $\underline{a}$ is a non-negative sequence and $1 < p < \infty$ then,

$$\sum_{n=1}^{\infty}\left(\sum_{i=1}^{n}\frac{a_i}{i}\right)^p \leq (p-1)\sum_{i=1}^{\infty}\left(\sum_{j=1}^{\infty}\frac{|a_j|}{i+j-1}\right)^p.$$

The constant is best possible, and the inequality is strict unless $\underline{a} = \underline{0}$.

COMMENTS (i) (1) is a refinement of Bennett's result due to Alzer, and the constants are best possible.
(ii) The simple proof of (2) given by Alzer depends on the strict convexity of the function $f(x) = x^{-r}(1-x)^r + x^r(1-x)^{-r}$. Alzer has also given an inequality converse to (1).
(iii) The proof of (3) is quite complex.
(iv) There is a left-hand side for the inequality in (d) but the constant is not known; see [Bennett, 1996].
(v) For other inequalities due to Bennett see **Abel's Inequalities** RELATED RESULTS (b), **Copson's Inequality** EXTENSIONS, **Hardy's Inequalities** EXTENSIONS (c), **Hilbert's Inequalities** EXTENSIONS (b), **Littlewood's Conjecture**, **Zeta Function Inequalities**.
REFERENCES [Bennett, 1996, pp. 47–56]; [Alzer, 1993[(2)],1994, 1994[(2)]], [Bennett, 1988].

Benson's Inequalities (a) Given three functions $u(x), P(u,x) > 0, G(u,x)$ write $p(x) = P(u(x), x)$, $g(x) = G(u(x), x)$. If $u, p, g \in C^1([a,b])$, and $n \geq 1$ then

$$\int_a^b \left\{ p(x)u'^{2n}(x) + (2n-1)p^{\frac{-1}{2n-1}}(x)(G_1')^{\frac{2n}{2n-1}}(u(x), x) + 2nG_2'(u(x), x) \right\} dx$$

$$\geq 2n[g(b) - g(a)].$$

There is equality if and only if $u' = (G_1'/P)^{1/(2n-1)}$.
(b) Given functions $u(x), P(u', u, x) > 0, G(u', u, x)$ write $p(x) = P(u'(x), u(x), x)$, $g(x) = G(u'(x), u(x), x), g_i'(x) = G_i'(u'(x), u(x), x), i = 1, 2, 3$. If $u \in C^2([a,b])$ and $p, g \in C^1([a,b])$ then

$$\int_a^b \left\{ p(x)u''(x)^2 + \frac{g_1'^2(x)}{p(x)} + 2u'(x)g_2'(x) + 2g_3'(x) \right\} dx \geq 2[g(b) - g(a)].$$

There is equality if and only if $u'' = (G'_1/P)$.

COMMENTS   These results give many important special cases, including the **Heisenberg-Weyl Inequality**, and **Wirtinger's Inequality**.

REFERENCES   [AI, pp. 126–129].

Bergh's Inequality   If $f : [0, \infty[\to [0, \infty[$ satisfies
$$0 \leq f(x) \leq \max\{1, xy^{-1}\} f(y), \qquad \text{for all } 0 < x, y < \infty, \tag{1}$$
and if $0 < p < q \leq \infty$, $0 < r < 1$ then
$$\left(\int_0^\infty \frac{1}{x} \left(\frac{f(x)}{x^r} \right)^q dx \right)^{1/q} \leq \left(\frac{1-r}{qr} \right)^{1/q} \left(\frac{pr}{1-r} \right)^{1/p} \left(\int_0^\infty \frac{2}{x} \left(\frac{f(x)}{x^r} \right)^p dx \right)^{1/p}.$$
The constant is best possible, and there is equality only if $f(x) = \min\{1, x\}$.

COMMENTS   (i) Functions satisfying (1) have been called *quasi-concave functions*.
(ii) This inequality has been extended to other classes of functions; see the reference.

REFERENCES   [Pečarić & Persson].

Bernoulli's Inequality   If $x \geq -1$, $x \neq 0$, and if $\alpha > 1$ or if $\alpha < 0$ then
$$(1 + x)^\alpha > 1 + \alpha x; \tag{B}$$
if $0 < \alpha < 1$ (~B) holds.

COMMENTS   (i) The inequality (B) in the case $\alpha = 2, 3, \ldots$, is Bernoulli's inequality. It is a standard elementary example used when teaching mathematical induction. Further (B) also holds if $-2 \leq x < -1$.
(ii) When $\alpha < 0$, when of course $x \neq -1$, the inequality follows from the case $\alpha > 0$.
(iii) The case $\alpha = 1/(n+1), n = 1, 2, \ldots$ of (~B) follows from properties a certain polynomial; see **Polynomial Inequalities**, COMMENTS (i).
(iv) In general the result follows by a simple application of Taylor's Theorem.

VARIANTS   (a) If $a, b > 0, a \neq b$, and $\alpha > 1$ or $\alpha < 0$,
$$\frac{a^\alpha}{b^{\alpha-1}} > \alpha a - (\alpha - 1) b; \tag{1}$$
and (~1) holds when $0 < \alpha < 1$.
(b) If a, b and α are as in (a) and $\alpha + \beta = 1$ then
$$a^\alpha b^\beta > \alpha a + \beta b; \tag{2}$$
and if $\alpha > 0, \beta > 0$ and $\alpha + \beta = 1$ then (~2) holds.
(c) If $x \geq 0$, $x \neq 1$ and if $\alpha > 1$ or if $\alpha < 0$ then
$$x^\alpha - 1 > \alpha(x - 1), \tag{3}$$
while if $0 < \alpha < 1$ then (~3) holds.

COMMENTS   (v) Putting $x = \frac{a}{b} - 1$ in (B) gives (1), a symmetric form of (B).
(vi) (2) is an immediate consequence of (1).
(vii) Replacing $1 + x$ by x in (B), gives (3).
(viii) (B) is equivalent to (GA). In particular (~2) is the case $n = 2$ of (GA).

EXTENSIONS (a) If $a_i > -1, 1 \leq i \leq n$, and are all positive or all negative then

$$\prod_{i=1}^{n}(1+a_i) > 1 + A_n.$$

(b) If $x > 1$, $0 < \alpha < 1$ then

$$\frac{1}{2}\alpha(1-\alpha)\frac{x}{(1+x)^2} < 1 + \alpha x - (1+x)^\alpha < \frac{1}{2}\alpha(1-\alpha)x^2(1+x).$$

CONVERSE INEQUALITIES If $\underline{a}$ is a positive n-tuple and $n > 1$ then

$$\prod_{i=1}^{n}(1+a_i) < \sum_{i=0}^{n}\frac{A_n^i}{i!}.$$

COMMENTS (ix) A discussion of the history of Bernoulli's inequality can be found in [Mitrinović & Pečarić, 1993].
(x) Other extensions are given in **Binomial Function Inequalities (c)**, **Gerber's Inequality** COMMENTS (ii), **Kaczmarz & Steinhaus's Inequalities (a)**, **Leindler's Inequality**, **Weierstrass's Inequalities**.

REFERENCES [*AI*, pp. 34–36], [*HLP*, pp. 39–43, 60, 103, 107], [*MI*, pp. 4–6, 170], [*MPF*, pp. 65–81]; [Mitrinović & Pečarić, 1993].

Bernstein's Inequalities
See **Bernšteĭn's Inequality**; **Bernšteĭn Polynomial Inequalities**; **Bernšteĭn's Probability Inequality**.

Bernšteĭn's[1] Inequality
(a) [TRIGONOMETRIC POLYNOMIAL CASE] If T_n is a trigonometric polynomial of degree at most n then

$$\|T_n^{(r)}\|_{\infty,[-\pi,\pi]} \leq n^r \|T_n\|_{\infty,[-\pi,\pi]}, \quad r = 1, 2, \ldots.$$

(b) [POLYNOMIAL CASE] If p_n is a polynomial of degree at most n, then

$$\|p_n'\|_{\infty,[a,b]} \leq \frac{n\|p\|_{\infty,[a,b]}}{\sqrt{(x-a)(b-x)}}, \quad a < x < b.$$

COMMENTS (i) For a definition of *trigonometric polynomial of degree at most n* see **Trigonometric Polynomial Inequalities**.
(ii) The results in (a) cannot be improved as the case $T_n(x) = \cos nx$ shows.
(iii) (a) is a particular case of the **Entire Function Inequalities (b)**.
(iv) (b) is a consequence of a proof of **Markov's Inequality**.

---

[1] С Н Бернштейн. Also transliterated as Bernshteĭn, Bernstein.

(v) Extension to other norms have been made; see [AI].

REFERENCES   [AI, pp. 228, 260], [EM, vol.1, pp. 365–366]; [Zygmund, vol.II, p. 11].

Bernšteĭn[1] Polynomial Inequalities

If $f : [0,1] \to \mathbb{R}, n \in \mathbb{N}$ then

$$B_n(f,x) = \sum_{k=0}^{n} \binom{n}{k} f\left(\frac{k}{n}\right) x^k (1-x)^{n-k},$$

is called the n-th Bernšteĭn polynomial of f.

(a) For all $x \in [0,1]$
$$B_n^2(f,x) \le B_n(f^2, x).$$

(b) A continuous function $f : [0,1] \to \mathbb{R}$ is convex if and only if either of the following holds:

(i) [TEMPLE] $B_{n+1}(f,x) \le B_n(f,x), \qquad n \in \mathbb{N}, \quad 0 \le x \le 1;$

(ii) [PÓLYA & SCHOENBERG]
$$f(x) \le B_n(f,x), \qquad n \in \mathbb{N}, \quad 0 \le x \le 1;$$

The following inequality was used to obtain basic results for these polynomials.

If $x \in [0,1]$ and $n \ge 1$ then

$$0 \le \sum_{k=0}^{n} \binom{n}{k} \left(\frac{k}{n} - x\right)^2 x^k (1-x)^{n-k} \le \frac{1}{4n}. \tag{1}$$

REFERENCES   [PPT, pp. 292–294]; [Kazarinoff, pp. 55–56].

Bernšteĭn's[1] Probability Inequality

If $X_i, 1 \le i \le n$, are independent random variables with
$$EX_i = 0, \quad \sigma^2 X_i = b_i, \quad 1 \le i \le n,$$

and if for $j > 2$,
$$E|X_i|^j \le \frac{b_i}{2} H^{j-2} j!,$$

then
$$P(|X_1 + \cdots + X_n| > r) \le 2\exp\left(-\frac{r^2}{2(B_n + Hr)}\right). \tag{1}$$

---

[1] С Н Бернштейн. Also transliterated as Bernshteĭn, Bernstein.

COMMENTS (i) This is a sharpening of the **Čebišev Probability Inequality**.
(ii) In particular if the random variables are identically distributed and bounded, for instance let $EX_i = 0$, $EX_i^2 = \sigma^2$, $|X_i| \leq M$, $1 \leq i \leq n$, then (1) reduces to

$$P(|X_1 + \cdots + X_n| > r\sigma\sqrt{n}) \leq 2\exp\left(-\frac{r^2}{2(1+\alpha)}\right),$$

where $\alpha = Mr/3\sigma\sqrt{n}$.
(iii) A lower estimate for the left-hand side of (1) has been given by Kolmogorov.

REFERENCES [EM, vol.1, p. 365, vol. 2, p. 120].

Berry-Esseen Inequality
Let $X_i, 1 \leq i \leq n$, be equally distributed random variables with

$$EX_i = 0, \ EX_i^2 = \sigma^2, \ E|X_i|^3 < \infty, \ 1 \leq i \leq n,$$

then

$$\sup_x \left| P\left\{ \frac{1}{\sigma\sqrt{n}} \sum_{j=1}^n X_j \leq x \right\} - \mathfrak{p}(x) \right| \leq \frac{33}{4} \frac{E|X_j|^3}{\sigma^3 \sqrt{n}}.$$

REFERENCES [EM, vol.1, pp. 369–370]; [Feller, vol. II, pp. 542–546], [Loève, pp. 282–288].

Berwald's Inequality
See **Favard's Inequalities** EXTENSIONS, **Geometric-Arithmetic Mean Inequality** INTEGRAL ANALOGUES (b).

Bessel Function Inequalities
(a) [GIORDANO & LAFORGIA] If J_ν is the Bessel function of first kind, $\nu > -1$, and if $\underline{a}$ is a positive n-tuple with $\underline{a} \leq j_{\nu_1}$, the smallest positive zero of J_ν, then

$$\mathfrak{G}_n(J_\nu(\underline{a})) \leq J_\nu(\mathfrak{M}_n^{[2]}(\underline{a})).$$

(b) [EBERT & LAFORGIA; MAKAI] If $\nu > 0$ and $j(\nu) = j_{\nu_k}$ is the k-th positive zero of J_ν then

$$(\nu + j(\nu))j''(\nu) > \frac{\nu j'(\nu)^2}{j(\nu)} - j'(\nu), \text{ and } \left(\frac{j(\nu)}{\nu}\right)' \leq 0,$$

(c) [Ross D K] If $x \in \mathbb{R}$ then

$$J_{\nu+1}^2(x) \geq J_\nu(x) J_{\nu+2}(x).$$

COMMENTS (i) Inequality (c) is related to the similar inequalities **Legendre Polynomial Inequalities** (b), **Ultraspherical Polynomial Inequalities** (b).

COROLLARIES (a) If $0 < a_i < \pi, 1 \leq i \leq n$ then

$$\mathfrak{G}_n(\sin \underline{a}) \leq \sin\left(\mathfrak{M}_n^{[2]}(\underline{a})\right).$$

(b) *With the above notation the function $f(\nu) = \nu/j(\nu)$ is concave.*

COMMENTS (ii) Corollary (a) follows by taking $\nu = 1/2$; it should be compared with **Trigonometric Function Inequalities** (r). The result implies geometric inequalities for triangles and convex polygons.
(iii) See also **Enveloping Series Inequalities** COMMENTS (iii), **Mahajan's Inequality**.

REFERENCES [General Inequalities, vol.1, pp. 35–38; vol.5, pp. 139–150].

Bessel's Inequality *If $f \in \mathcal{L}^2([a,b])$ and if $\phi_n, n \in \mathbb{N}$ is an orthonormal sequence of complex valued functions defined on $[a,b]$ with $\underline{c}$ the sequence of Fourier coefficients of f with respect to $\phi_n, n \in \mathbb{N}$, then*

$$\|\underline{c}\|_2 \leq \|f\|_2. \tag{1}$$

COMMENTS (i) This result remains valid, with the same proof if $\mathcal{L}^2([a,b])$ is replaced by any Hilbert space H, the orthonormal sequence by any orthonormal n-tuple, or sequence in H, $\underline{x}_k, k \in I$ say, and the sequence $\underline{c}$ by $c_k = \underline{x}.\underline{x}_k, k \in I$. Then (1) becomes

$$\sum_{k \in I} |\underline{x}.\underline{x}_k|^2 \leq \underline{x}.\underline{x} = \|\underline{x}\|^2 \tag{2}$$

(ii) In the notation of (i) the proof of (2) depends on the identity

$$\left\|\underline{x} - \sum_{k \in I} \lambda_k \underline{x}_k\right\|^2 = \|\underline{x}\|^2 - \sum_{k \in I} |(\underline{x}.\underline{x}_k)|^2 + \sum_{k \in I} |\lambda_k - (\underline{x}.\underline{x}_k)|^2,$$

where $\lambda_k, k \in I$, is any set of complex numbers.
(iii) If the orthonormal sequence is complete then (1), and (2), becomes an equality known as *Parseval's equality*.

EXTENSIONS [BOMBIERI] *If $\underline{a}, \underline{b}_k, 1 \leq k \leq m$, are complex n-tuples, then*

$$\sum_{k=1}^{m} |\underline{a}.\underline{b}_k|^2 \leq |\underline{a}|^2 \max_{1 \leq k \leq m} \sum_{j=1}^{m} |\underline{b}_k.\underline{b}_j|, \tag{3}$$

with equality if and only if $\underline{a}, \underline{b}_k, 1 \leq k \leq m$, are linearly dependent.

COMMENTS (iv) The case $m = 1$ of (3) is just (C).
(iv) For a further extension see **Ostrowski's Inequalities** COMMENTS (i), **Hausdorff-Young Inequalities, Paley's Inequalities**.

REFERENCES [EM, vol. 1, pp. 373–374; vol. 7, pp. 93–94], [MPF, pp. 391–405]; [Courant & Hilbert, pp. 51–52], [Loeb & Loss, pp. 66–67], [Zygmund, vol.I, p. 13].

Beta Function Inequalities

If $\Re m > -1, \Re n > -1$ then the *Beta function* is defined by

$$B(m+1, n+1) = \frac{m!n!}{(m+n+1)!}.$$

If $x, y, z > 0$ then

$$(x+y)B(z, x+y) \geq xB(z,x)yB(z,y).$$

COMMENTS (i) This is a deduction from **Stolarsky's Inequality**.
(ii) See also **Vietoris's Inequality**.

REFERENCES [Stolarsky].

Beth & van der Corput Inequality   If $z, w \in \mathbb{C}$ and $p \geq 2$ then

$$|w+z|^p + |w-z|^p \geq 2\left(|w|^p + |z|^p\right). \tag{1}$$

COMMENTS (i) This is a converse of **Clarkson's Inequalities** (1).

EXTENSIONS [KLAMKIN] If $\underline{a}_j, 1 \leq j \leq m$, are real n-tuples and $p > 2$ or $p < 0$ then

$$\sum |\pm \underline{a}_1 \pm \cdots \pm \underline{a}_m|^p \geq 2^m \left(\sum_{j=1}^m |\underline{a}_j|^2\right)^{p/2} \geq 2^m \sum_{j=1}^m |\underline{a}_j|^p, \tag{2}$$

where the sum on the left-hand side is over all 2^n permutations of the $\pm$ signs. If $0 < p < 2$ then($\sim$ 2) holds; and for $p = 0, 2$ (2) is an identity.

COMMENTS (ii) If $n = 2$ and $p \geq 2$ an application of (r;s) shows that (2) generalizes (1).
(iii) In the case $p = 1$ (2) has the geometric interpretation:

  amongst all paralleletopes of given edge lengths, the rectangular one has the greatest sum of lengths of diagonals.

(iv) Again in the case $p = 1$, and with $m = 3$, (2) can be regarded as a converse of **Hlawka's Inequality**; put $2\underline{a}_1 = \underline{a} + \underline{b}, 2\underline{a}_2 = \underline{b} + \underline{c}, 2\underline{a}_3 = \underline{c} + \underline{a}$;

$$|\underline{a}| + |\underline{b}| + |\underline{c}| + |\underline{a}+\underline{b}+\underline{c}| \leq 2\left(|\underline{a}+\underline{b}|^2 + |\underline{b}+\underline{c}|^2 + |\underline{c}+\underline{a}|^2\right)^{1/2}.$$

(v) An extension of the (1) to inner product spaces has been made by Dragomir & Sándor.

REFERENCES [AI, p. 322], [MPF, pp. 523, 544–551], [PPT, pp. 134–135].

Bieberbach's Conjecture If $f(z) = z + \sum_{n \geq 2} a_n z^n$ is univalent in D then
$$|a_n| \leq n.$$
There is equality if and only if f is a rotation of the Koebe function.

COMMENTS (i) This famous conjecture was proved by de Branges in 1984.
(ii) The hypothesis of injectivity cannot be omitted as the function $z + 3z^2$ shows.
(iii) The Koebe function is defined in **Distortion Theorems** COMMENTS (i).

REFERENCES [Ahlfors, 1973, pp. 82–91], [Conway, vol. II, pp. 63–64, 148], [General Inequalities, vol. 5, pp. 3–16], [Gelbaum & Olmsted, pp. 184–185]; [Pommerenke].

Bienaymé-Chebyshev Inequality
See Čebišev Probability Inequality COMMENTS (i).

Bihari-Bellman Inequalities
See Gronwall's Inequality COMMENTS (iii).

Bilinear Form Inequalities [M RIESZ] Let $p, q \geq 1$, p', q' the conjugate indices of p, q respectively. If $\underline{a} = \{a_i, i \in \mathbb{Z}\}$, $\underline{b} = \{a_j, j \in \mathbb{Z}\}$, $\underline{c} = \{c_{ij}, i, j \in \mathbb{Z}\}$ are nonnegative, and $\underline{\alpha} = \{\sum_{i \in \mathbb{Z}} c_{ij} a_i, j \in \mathbb{Z}\}$, $\underline{\beta} = \{\sum_{j \in \mathbb{Z}} c_{ij} b_j, i \in \mathbb{Z}\}$ then the following are equivalent:
$$\sum_{i,j \in \mathbb{Z}} c_{ij} a_i b_j \leq C \|\underline{a}\|_p \|\underline{b}\|_q, \text{ for all } \underline{a}, \underline{b};$$
$$\|\underline{\alpha}\|_{q'} \leq C \|\underline{a}\|_p, \text{ for all } \underline{a};$$
$$\|\underline{\beta}\|_{q'} \leq C \|\underline{b}\|_p, \text{ for all } \underline{b}.$$

COMMENTS (i) These equivalencies can be stated in terms of strict inequalities, and they also hold for finite sequences; see [HLP].
(ii) See also **Aczel's Inequality, Grothendieck's Inequality, Hardy-Littlewood-Pólya-Schur Inequalities, Hilbert's Inequalities, Multilinear Form Inequalities, Quadratic Form Inequalities.**

REFERENCES [HLP, pp. 204–225].

Binomial Coefficient Inequalities (a) If $n > 2$ then
$$\frac{2^{2n}}{n+1} < \binom{2n}{n} < \frac{(2n+2)^n}{(n+1)!}.$$

(b) If $n > 1$
$$\binom{2n}{n} > \frac{4^n}{2\sqrt{n}}.$$

(c) [ÅSLUND] If $x > n$ and $y = (1 + 1/n)^n$ then
$$\binom{x}{n} \leq \frac{x^x}{y n^n (x-n)^{x-n}}.$$

(d) If $x^n + y^n = z^n$ then
$$\binom{x}{n} + \binom{y}{n} < \binom{z}{n} < \binom{x}{n} + \binom{y}{n} + \binom{z-1}{n-1}.$$

COMMENTS   Many other inequalities can be found in the first reference.
REFERENCES   [AI, pp. 194–196], [MI, p. 338].

Binomial Function Inequalities   (a) If $-1 < u < v$, then,

$$(1+v)^{1/v} < (1+u)^{1/u}, \tag{1}$$

$$(1+v)^{1+\frac{1}{v}} > (1+u)^{1+\frac{1}{u}}. \tag{2}$$

(b) If $x > 0$, $0 \le k < n$ then,

$$(1+x)^n > \sum_{i=0}^{k} \binom{n}{i} x^i. \tag{3}$$

(c) If $n > 1$ and $-1 < x < 1/(n-1)$, $x \ne 0$, then,

$$(1+x)^n < 1 + \frac{nx}{1-(n-1)x}.$$

(d) [KARANICOLOFF] If $0 < q < p$, $0 < m < n$ and $0 < x < 1$ then,

$$(1-x^p)^m > (1-x^q)^n. \tag{4}.$$

If $0 < p < q$ and $0 < x < x_0$, x_0 being the unique root of $(1-x^p)^m = (1-x^q)^n$ in $]0,1[$, then (~4) also holds, while if $x_0 < x < 1$ (4) holds.
(e) If $x+y = 1$, $0 < x < 1$ and if $r, s > 1$ then

$$(1-x^r)^s + (1-y^s)^r > 1.$$

(f) [BENNETT] If $x \in \mathbb{R}$, $m, n \in \mathbb{N}$ with $m, n > x$ then

$$\left(1 + \frac{x}{m}\right)^m \left(1 - \frac{x}{n}\right)^n < 1.$$

(g) [MALEŠEVIĆ] If $1 \le k \le n$ then

$$\left(1 + \frac{1}{n}\right)^k \le 1 + \frac{k}{n} + \frac{(k-1)^2}{n^2},$$

while if $0 \leq k \leq n, n \geq 2$,

$$\left(1+\frac{1}{n}\right)^k \leq \frac{n+1}{n+1-k}.$$

(h) If $x, y > 0, x \neq y$ and either $r > 1$ or $r < 0$ then

$$rx^{r-1}(x-y) > x^r - y^r > ry^{r-1}(x-y); \tag{5}$$

if $0 \stackrel{<}{\sim} r < 1$ then ($\sim$ 5) holds.

(j) If $x, q > 0, q \in \mathbb{Q}$, and if $m = \min\{q, qx^{q-1}\}, M = \max\{q, qx^{q-1}\}$ then

$$m < \frac{x^q - 1}{x-1} < \mathfrak{H}_2(m, M) \text{ if } \frac{1}{2} < q < 1;$$

$$\mathfrak{H}_2(m, M) < \frac{x^q - 1}{x-1} < \mathfrak{G}_2(m, M) \text{ if } 0 < q < \frac{1}{2};$$

$$\mathfrak{G}_2(m, M) < \frac{x^q - 1}{x-1} < \mathfrak{A}_2(m, M) \text{ if } q > 2;$$

$$\mathfrak{A}_2(m, M) < \frac{x^q - 1}{x-1} < M \qquad \text{if } 1 < q < 2.$$

(k) [MADEVSKI] If $x > 1, p \geq 1$ then

$$(x-1)^p \leq x^p - 1.$$

COMMENTS (i) (a) is an easy consequence of the strict concavity or convexity of the functions on the left-hand sides.

SPECIAL CASES (a) If $n = 1, 2, \ldots$,

$$\left(1+\frac{1}{n+1}\right)^{n+1} > \left(1+\frac{1}{n}\right)^n, \tag{6}$$

$$\left(1+\frac{1}{n+1}\right)^{n+2} < \left(1+\frac{1}{n}\right)^{n+1}, \tag{7}$$

$$\left(1-\frac{1}{n}\right)^n < \left(1-\frac{1}{n+1}\right)^{n+1}. \tag{8}$$

(b) If $x > 0, 0 < p < q$, then

$$\left(1+\frac{x}{p}\right)^p < \left(1+\frac{x}{q}\right)^q.$$

COMMENTS (ii) Inequalities (6) and (7), particular cases of (1), (2) respectively can be obtained using (GA) and (HG); and (8) is an easy consequence of (6).

(iii) Inequalities (6), (7) are important in the theory of the exponential function; see **Exponential Function Inequalities** (1), (2).

(iv) Inequalities (6), (7) can be extended by inserting a variable x. Thus: $(1 + x/n)^n$ increases, as a function of n, and $(1 + x/n)^{n+1}$ decreases as a function of n. An approach to the first can be made by using **Chong's Inequalities** (2) and the fact that

$$\left(1 + \frac{x}{n}, 1 + \frac{x}{n}, \cdots, 1 + \frac{x}{n}\right) \prec \left(1 + \frac{x}{n-1}, 1 + \frac{x}{n-1}, \cdots, 1 + \frac{x}{n-1}, 1\right).$$

(v) (b) is obtained from (1) by putting $v = x/p, u = x/q$, with $x > 0, 0 < p < q$. This inequality holds under the above conditions for x satisfying $0 > x > -p$. In addition the inequality is valid if either $0 > q > p$ and $x < -q$, or if $q < 0 < p, -p < x < -q$.

(vi) See also **Brown's Inequalities**, **Gerber's Inequality**, **Kacmarz & Steinhaus's Inequalities** (a), **Leindler's Inequality**, **Polynomial Inequalities**, **Series Inequalities** (d).

REFERENCES   [AI, pp. 34–35, 278, 280, 356, 365, 384], [HLP, pp. 39–42, 102–103], [MI, pp. 8, 21,64], [MPF, pp. 68, 95]; [Apostol et al, p. 444], [Melzak, 1996; pp. 64–65]; [Malešević], [Savov].

Biplanar Mean Inequalities
See **Gini-Dresher Mean Inequalities** COMMENTS (i).

Blaschke-Santaló Inequality
If $K \subset \mathbb{R}^n$ is convex, compact with $\mathring{K} \neq \emptyset$, and centroid the origin, then

$$|K||K^*| \leq v_n^2,$$

with equality if and only if K is the unit ball.

COMMENTS   (i) K^* is the polar of K, that is

$$K^* = \{\underline{a}; \underline{a}.\underline{b} \leq 1, \underline{b} \in K\}.$$

(ii) The cases $n = 2, 3$ are due to Blaschke; the other cases were given, much later, by Santaló.

(iii) For a converse inequality see **Mahler's Inequalities**.

REFERENCES   [EM, Supp., pp. 129–130].

Bloch's Constant
If f is analytic in the D with $|f'(0)| = 1$ and B_f is the radius of the largest open disk contained in a sheet of the Riemann surface of f and if $B = \inf B_f$, where the inf is taken over all such f, then

$$\frac{\sqrt{3}}{4} \leq B \leq 0.472, \quad \text{and} \quad B \leq L,$$

where L is Landau's constant.

COMMENTS   (i) B is called *Bloch's constant*.

(ii) See also **Landau's Constant**.

REFERENCES   [EM, vol.1, p. 406]; [Ahlfors, 1973, pp. 14–15], [Conway, vol.I, pp. 297–298].

Block Type Inequalities [REDHEFFER] If $f \in \mathcal{C}^1(\mathbb{R})$, not identically zero, with $f, f' \in \mathcal{L}^2(\mathbb{R})$ then

$$u(x) < \left(\int_{\mathbb{R}} f'^2\right)^{1/4} \left(\int_{\mathbb{R}} f^2\right)^{1/4}, \quad x \in \mathbb{R}.$$

COMMENTS   This is a typical example of a class of inequalities introduced by Block. It gives a bound on the function in terms of its integral and the integral of its derivative.

REFERENCES   [Inequalities II, pp. 262, 282–283].

Boas's Inequality   If $f : [-\pi, \pi] \to \mathbb{R}$; $\tilde{f}(x) = f(x)\text{sign}(x), -\pi \leq x \leq \pi$; and if $f, \tilde{f}$ have absolutely convergent Fourier series with coefficients $a_n, n \in \mathbb{N}, b_n, n \geq 1$ and $\tilde{a}_n, n \in \mathbb{N}, \tilde{b}_n, n \geq 1$, respectively; then,

$$\int_{-\pi}^{\pi} \frac{|f(x)|}{x} \, dx \leq \frac{1}{2}a_0 + \sum_{n=1}^{\infty} (|a_n| + |b_n|) + \frac{1}{2}\tilde{a}_0 + \sum_{n=1}^{\infty} (|\tilde{a}_n| + |\tilde{b}_n|).$$

COMMENTS   (i) This has been extended to higher dimensions by Zadereĭ[2].
(ii) For another inequality by Boas see **Function Inequalities** (a).
(ii) See **Wirtinger's Inequality** EXTENSIONS (b) for another inequality involving Fourier coefficients.

REFERENCES   [Boas, 1956], [Zadereĭ].

Bohr-Favard Inequality   If f is a function of period 2π with absolutely continuous $(r-1)$-st derivative, then for each n there is a trigonometric polynomial T_n of order n such that

$$\|f - T_n\|_{\infty,[0,2\pi]} \leq C_{n,r} \|f^{(r)}\|_{\infty,[0,2\pi]},$$

where

$$C_{n,r} = \frac{4}{\pi(n+1)^r} \sum_{k \in \mathbb{N}} \frac{(-1)^{k(r-1)}}{(2k+1)^{r+1}}.$$

COMMENTS   This result is best possible; the case $r = 1$ is due to Bohr. It is part of a large theory on trigonometric approximation.

REFERENCES   [EM, vol.1, p. 415, vol.3, p. 480]; [Zygmund, vol.I, p. 377].

Bohr's Inequality   If $z_1, z_2 \in \mathbb{C}$ and if $c > 0$ then

$$|z_1 + z_2|^2 \leq (1+c)|z_1|^2 + (1 + \frac{1}{c})|z_2|^2,$$

with equality if and only if $z_2 = cz_1$.

COMMENTS   This is a deduction from (C).

---

[2] П В Задерей.

EXTENSIONS   (a) [ARCHBOLD] If $z_i \in \mathbb{C}, 1 \leq i \leq n$, and if $\underline{w}$ is a positive n-tuple such that $W_n = 1$ then

$$\left|\sum_{i=1}^n z_i\right|^2 \leq \sum_{i=1}^n \frac{|z_i|^2}{w_i}.$$

(b) [VASIĆ & KEČKIĆ] If $z_i \in \mathbb{C}, 1 \leq i \leq n$, $\underline{w}$ a positive n-tuple and $r > 1$ then

$$\left|\sum_{i=1}^n z_i\right|^r \leq \left(\sum_{i=1}^n w_i^{1/(r-1)}\right)^{r-1} \sum_{i=1}^n \frac{|z_i|^r}{w_i},$$

with equality if and only if $|z_1|^{r-1}/w_1 = \cdots = |z_n|^{r-1}/w_n$, and $z_k \bar{z}_j \geq 0$, $k, j = 1, \ldots, n$

REFERENCES   [AI, pp.312–315, 338–339], [HLP, p.61], [MPF, pp.499–505], [PPT, p. 131].

Bonferroni's Inequalities   Let $A_i, 1 \leq i \leq n$, be events in the probability space $(\Omega, \mathcal{A}, P)$; denote by S_j the probability that $j, j \geq 1$, events occur simultaneously, and put $S_0 = 0$; and if $0 \leq m \leq n$, let P_m be the probability that at least m events occur, and P_m^e the probability that exactly m events occur. Then for p, q satisfying $0 \leq 2p + 1 \leq n - m, 0 \leq 2q \leq n - m$

$$\sum_{j=0}^{2p+1}(-1)^j\binom{m+j-1}{j}S_{m+j} \leq P_m \leq \sum_{j=0}^{2q}(-1)^j\binom{m+j-1}{j}S_{m+j},$$

$$\sum_{j=0}^{2p+1}(-1)^j\binom{m+j}{j}S_{m+j} \leq P_m^e \leq \sum_{j=0}^{2q}(-1)^j\binom{m+j}{j}S_{m+j}.$$

COMMENTS   Clearly $S_j = \sum_{1 \leq i_1 < i_2 \cdots < i_n \leq n} P(A_{i_1} \cap \ldots \cap A_{i_j})$.

REFERENCES   [EM, Supp. pp. 142–143]; [Feller, vol. I, pp. 88–101], [Galambos & Simonelli].

Bonnensen's Inequality   If A is the area of a convex domain in $\mathbb{R}^2$, with L the length of its boundary then,

$$L^2 - 4\pi A \geq \pi^2(R-r)^2,$$

where r is the inner radius of the domain, and R is the outer radius of the domain. The left-hand side is always positive except when the domain is a disk, and then $R = r$.

COMMENTS   (i) Definitions of inner and outer radius can be found in **Isodiametric Inequality**
COMMENTS (ii).
(ii) See also **Gale's Inequality**.

REFERENCES   [EM, vol.1, p. 420].

Borel-Carathéodory Inequality If f is analytic on $\{z; |z| \leq R\}$ and if $A(r) = \max_{|z|=r} \Re f(z)$, then for $r < R$

$$M_\infty(r, f) \leq \frac{2r}{R-r} A(R) + \frac{R+r}{R-r} |f(0)|. \tag{1}$$

In particular if $A(R) \geq 0$ then

$$M_\infty(r, f) \leq \frac{R+r}{R-r} (A(R) + |f(0)|).$$

EXTENSIONS Under the above conditions and with $A(R) \geq 0$, and $n \geq 1$,

$$M_\infty(r, f^{(n)}) \leq \frac{2^{n+2} n! R}{(R-r)^{n+1}} \Big(A(R) + |f(0)| \Big).$$

REFERENCES [Conway, p. 129], [Titchmarsh, 1939, pp. 174–176].

Bounded Variation Function Inequalities (a) If f, g are of bounded variation on $[a, b]$ then so is $f + g$ and

$$V(f + g; a, b) \leq V(f; a, b) + V(g; a, b).$$

(b) If in addition $f(a) = g(a) = 0$ then

$$V(fg; a, b) \leq V(f; a, b) V(g; a, b).$$

(c) If on $[a, b]$, f is continuous, g of bounded variation, or f, g are both of bounded variation and g is continuous, then

$$\left| \int_a^b f \, dg \right| \leq \int_a^b |f(x)| \, dV(g; a, x) \leq \sup_{a \leq x \leq b} |f(x)| \, V(g; a, b).$$

(d) If on $[a, b]$, f is continuous, g of bounded variation and if $F(x) = \int_a^x f \, dg$, $a \leq x \leq b$ then

$$V(F; a, x) \leq \int_a^x |f| \, |dg|, \quad a \leq x \leq b.$$

(e) If f is of bounded variation on $[0, 1]$ then

$$\left| \int_0^1 f - \frac{1}{n} \sum_{i=1}^n f\left(\frac{i}{n}\right) \right| \leq \frac{V(f; 0, 1)}{n}, \quad n = 1, 2, \ldots.$$

COMMENTS (i) Results (a) and (b) have been extended by A.M. Russell to functions of bounded higher order variation. The integrals above are Riemann-Stieltjes integrals.
(ii) Of course under the hypotheses of (a) fg is also of bounded variation but the inequality in (b) need not hold; we can only get

$$V(fg; a, b) \leq \|f\|_{\infty, [a,b]} V(g; a, b) + \|g\|_{\infty, [a,b]} V(f; a, b).$$

(iii) (c) follows from **Integral Inequalities** EXTENSIONS (a).

EXTENSIONS [KARAMATA]   If f is of bounded variation on $[a,b]$, g bounded on $[a,b]$, with $f,g \in \mathcal{L}_\mu([a,b])$, μ bounded, then

$$\left|\int_a^b fg\,d\mu\right| \leq \Big(|f(b)| + V(f;a,b)\Big) \sup_{a \leq x \leq b}\left|\int_a^x g\,d\mu\right|;$$

$$\left|\int_a^b fg\,d\mu\right| \leq \Big(|f(a)| + V(f;a,b)\Big) \sup_{a \leq x \leq b}\left|\int_x^b g\,d\mu\right|.$$

COMMENTS   (iv) See also **Arc Length Inequality, Variation Inequalities**.

REFERENCES   [MPF, p. 337]; [Pólya & Szegö,1972, p. 49], [Rudin,1964, pp. 118–119, 122], [Widder, pp. 8–10]; [Bullen, 1983], [Russell, A M].

Brown's Inequalities   If $0 \leq x \leq 1$ and $s,t \geq 1$ then

$$(1+x+x^2) \geq (1+x^s)^{1/s}(1+x^t)^{1/t} \quad \text{if} \quad \frac{1}{s} + \frac{1}{t} = \frac{\log 3}{\log 2}, \quad \text{and} \quad s+t \leq \frac{8}{3}.$$

COMMENTS   This inequality has applications in measure theory.

EXTENSIONS   [BROWN] (a) If $0 \leq x \leq 1$ and $s,t \geq 1$ then

$$(1+x+x^2) \geq (1+x^s)^{1/s}(1+x^t+x^{2t})^{1/t}, \quad \text{if} \quad \frac{1}{s}\frac{\log 3}{\log 2} + \frac{1}{t} = 1.$$

(b) If $0 \leq x \leq 1$ and $s,t \geq 1$ then

$$(1+x+x^2+x^3) \geq \begin{cases} (1+x^s)^{1/s}(1+x^t+x^{2t}+x^{3t})^{1/t}, & \text{if } \dfrac{1}{2s}+\dfrac{1}{t}=1, \\ (1+x^s+x^{2s})^{1/s}(1+x^t+x^{2t})^{1/t}, & \text{if } \dfrac{1}{s}+\dfrac{1}{t}=\dfrac{1}{\log_4 3}, \\ (1+x^s+x^{2s})^{1/s}(1+x^t+x^{2t}+x^{3t})^{1/t}, & \text{if } \dfrac{\log_4 3}{s}+\dfrac{1}{t}=1. \end{cases}$$

(c) [ALZER] If $0 \leq x \leq 1$ and $s,t \geq 1$ then

$$(1+x+x^2) \geq \left(1+\frac{tx^s+sx^t}{s+t}\right)^{\frac{1}{s}+\frac{1}{t}}, \quad \text{if} \quad \frac{1}{s}+\frac{1}{t} = \frac{\log 3}{\log 2} \quad \text{and} \quad s+t \leq \frac{4}{3}+\frac{\log 4}{\log 3}.$$

REFERENCES   [Alzer, unpubl.], [Brown], [Kemp].

Brunn-Minkowski Inequalities   (a) If K_0, K_1 are convex sets in $\mathbb{R}^n$ define

$$K_t = \{\underline{x};\ \underline{x} = (1-t)\underline{x}_0 + t\underline{x}_1, \underline{x}_0 \in K_0, \underline{x}_1 \in K_1\},\ 0 \leq t \leq 1.$$

If $V(t) = |(K_t)|^{1/n}$ then

$$V((1-s)t_1 + st_2) \geq (1-s)V(t_1) + sV(t_2), \ 0 \leq s, t_1, t_2 \leq 1;$$

that is, V is a concave function. There is equality if and only if K_0, K_1 are homothetic

(b) [BRUNN-MINKOWSKI-LUSTERNIK[3]] If A, B are non-empty measurable sets in $\mathbb{R}^n$ define

$$A + B = \{\underline{x}; \ \underline{x} = \underline{a} + \underline{b}, \ \underline{a} \in A, \underline{b} \in B\}.$$

then

$$\lambda_{n,*}^{1/n}(A + B) \geq |A|^{1/n} + |B|^{1/n}.$$

There is equality if either of A, B is a singleton, or if they are convex and homothetic.

COMMENTS (i) The set K_t is a *linear combination* of K_0, K_1. It is easy to extend this linear combinations of several sets, and the above result then generalizes. See also **Mixed-volume Inequalities**.
(ii) $A + B$ is called the *sum set* of A and B. It need not be measurable even if both A and B are; so $\lambda_{n,*}$ denotes the inner n-dimensional Lebesgue measure.
(iii) Extensive generalizations can be found in **Prékopa-Leindler Inequalities** COMMENTS (iii).
REFERENCES [EM, vol.1, p. 484; Supp., p. 81], [MPF, pp. 174–178].

Brunn-Minkowski-Lusternik Inequality
See **Brunn-Minkowski Inequality** (b).

Burkholder-Davis-Gundy Inequality
See **Martingale Inequalities** COMMENTS (iii).

Burkill's Inequality
See **Hlawka-Type Inequalities** EXTENSIONS.

Bushell & Okrasinski's Inequality
If $0 < b \leq 1, p \geq 1$ and if $f \in \mathcal{C}([0,b])$ is non-negative and increasing then

$$\int_0^x (x-t)^{p-1} f^p(t) \, dt \leq \left(\int_0^x f\right)^p, \quad 0 \leq x \leq b.$$

REFERENCES [General Inequalities, vol. 6, pp. 495–496].

- end of B -

---

[3] Л Лустерник.

C

Čakalov's[1] Inequality *If $\underline{a}$ is an increasing positive non-constant n-tuple, $\underline{w}$ a positive n-tuple, $n > 2$, then*

$$\lambda_n\{\mathfrak{A}_n(\underline{a};\underline{w}) - \mathfrak{G}_n(\underline{a};\underline{w})\} \geq \lambda_{n-1}\{\mathfrak{A}_{n-1}(\underline{a};\underline{w}) - \mathfrak{G}_{n-1}(\underline{a};\underline{w})\},$$

where

$$\lambda_n = \frac{W_n^2}{(W_{n-1} - w_1)}.$$

COMMENTS (i) The proof, due to Čakalov, is the same as that of the more general result in the reference.

(ii) Since it is easily seen that $\tilde{\lambda}_n = \lambda_n/\lambda_{n-1} > W_{n-1}/W_n$, this inequality generalizes **Rado's Geometric-Arithmetic Mean Inequality Extension** for this class of sequences; there is no analogous extension of **Popoviciu's Geometric-Arithmetic Mean Inequality Extension**.

(iii) In general $\tilde{\lambda}_n$ is an unattained lower bound, unlike the lower bound W_n/W_{n-1} in Rado's geometric-arithmetic mean inequality extension. However $\tilde{\lambda}_n$ is best possible in that for any $\tilde{\lambda}'_n > \tilde{\lambda}_n, n = 1, 2, \ldots$ there are sequences $\underline{a}$ for which the inequality would fail.

(iv) The geometric mean can be replaced by a large class of quasi-arithmetic $\mathfrak{M}$-means; those for which M^{-1} is increasing, convex and 3- convex; for the definition of these terms see **Quasi-arithmetic Mean Inequalities, n-Convex Sequence Inequalities**.

REFERENCES [MI, pp. 243–245].

Capacity Inequalities (a) [POINCARÉ] *If V is the volume of a domain in $\mathbb{R}^3$ and C its capacity then*

$$C^3 \geq \frac{3V}{4\pi},$$

with equality only when the domain is spherical.
If A is the area of a domain in $\mathbb{R}^2$ and C its capacity then

$$C^2 \geq \frac{4A}{\pi^3},$$

with equality only when the domain is spherical.
(b) *If Ω_1, Ω_2 are two domains in $\mathbb{R}^3$ and C is a capacity, then*

$$C(\Omega_1 \cup \Omega_2) + C(\Omega_1 \cap \Omega_2) \leq C(\Omega_1) + C(\Omega_2); \tag{1}$$

---

[1] В Чакалов. Also transliterated as Tchakalof, Chakalov.

and if $\Omega_1 \subseteq \Omega_2$ then $C(\Omega_1) \leq C(\Omega_2)$.

COMMENTS (i) In $\mathbb{R}^3$ capacity can be defined by

$$C = C(\partial\Omega) = \frac{1}{4\pi} \inf \int_{\mathbb{R}^3 \setminus \Omega} |\nabla f|^2,$$

where the inf is over all f such that $f(\underline{x}) = 1, \underline{x} \in \partial\Omega$, and $f(\underline{x}) \to 0$ as $|\underline{x}| \to \infty$.
An analogous definition can be given in $\mathbb{R}^n, n > 3$. The case $n = 2$ is a little different. In addition the whole theory of capacity can be given in a very abstract setting.
(ii) The set function property in (1) is called *strong sub-additivity*; it is a property of abstract capacities.
(iii) The inequalities in (a) are examples of **Symmetrization Inequalities**.
(iv) See also **Logarithmic Capacity Inequalities**.

REFERENCES [EM, vol. 2, pp. 14–17]; [Conway, vol. 2, pp. 331–336], [Pólya & Szegö, 1951, pp. 1, 8–13, 42–44], [Protter & Weinberger, pp. 121–128].

Cardinal Number Inequalities (a) If α is any infinite cardinal then

$$\alpha \geq \aleph_0.$$

(b) If α is any cardinal number then

$$2^\alpha > \alpha.$$

(c) If α, β are two ordinal numbers then

$$\alpha < \beta \implies \aleph_\alpha < \aleph_\beta.$$

COMMENTS The *generalized continuum hypothesis* states that for all ordinals α, $2^{\aleph_\alpha} = \aleph_{\alpha+1}$; in particular the case $\alpha = 0$, that is $2^{\aleph_0} = \aleph_1$, is the *continuum hypothesis*.

REFERENCES [EM, vol. 1, p. 69; vol. 3, pp. 23–24, 390–391]; [Hewitt & Stromberg, pp. 22–24].

Carleman's Inequality If $\underline{a}$ is a positive sequence,

$$\sum_{i=1}^{\infty} \mathfrak{G}_i(\underline{a}) < e \sum_{i=1}^{\infty} a_i. \qquad (1)$$

The constant is best possible.

EXTENSIONS (a) If $0 < p < 1$, and $\underline{a}$ is a non-negative non-null sequence

$$\sum_{i=1}^{\infty} \mathfrak{M}_i^{[p]}(\underline{a}) < \left(\frac{1}{1-p}\right)^{1/p} \sum_{i=1}^{\infty} a_i. \qquad (2)$$

(b) [REDHEFFER] *If $\underline{a}, \underline{b}$ are non-negative n-tuples then*

$$\sum_{i=1}^{n} i(b_i - 1)\mathfrak{G}_i(\underline{a}) + n\mathfrak{G}_n(\underline{a}) \le \sum_{i=1}^{n} a_i b_i^i.$$

There is equality if and only if $a_i b_i^i = \mathfrak{G}_{i-1}(\underline{a}), 2 \le i \le n$.
(c) [ALZER] *If $\underline{a}, \underline{w}$ are positive sequences then*

$$\sum_{i=1}^{\infty} w_i \mathfrak{G}_i(\underline{a}; \underline{w}) + \frac{1}{2} \sum_{i=1}^{\infty} \frac{w_i^2}{W_i} \mathfrak{G}_i(\underline{a}; \underline{w}) < e \sum_{i=1}^{\infty} w_i a_i.$$

COMMENTS (i) Inequality (2) is just **Hardy's Inequality** (1) with a change of notation. Letting $p \to 0$ in this extension gives the weaker form of (1), with $<$ replaced by $\le$.
(ii) The finite case of inequality (1) follows from Redheffer's extension by taking $b_i = 1 + 1/i, i = 1, 2, \ldots$ and using **Exponential Function Inequalities** (1). This extension is an one of Redheffer's **Recurrent Inequalities**.
(iii) Alzer's extension implies a weighted version of (1), due in a weaker form to Pólya.
(iv) There are other extensions of the finite sum case; see **Redheffer's Inequalities** (1), (2), **Kaluza-Szegö Inequality**, **Nanjundiah's Mixed Mean Inequalities** COMMENTS (ii). The finite sum cases do not of course always imply the strict inequality in the series form.

INTEGRAL ANALOGUES [KNOPP] *Unless $f \ge 0$ is zero almost everywhere*

$$\int_0^\infty \mathcal{G}_{[0,x]}(f) \, \mathrm{d}x = \int_0^\infty \exp\left(\frac{1}{x} \int_0^x \log \circ f(t) \, \mathrm{d}t\right) \mathrm{d}x < e \int_0^\infty f.$$

COMMENTS (v) See also **Heinig's Inequality** COMMENTS (iii).
REFERENCES [AI, p. 131], [EM, vol. 2, p. 25], [HLP, pp. 249–250], [MI, pp. 115–116, 273], [PPT, pp. 231, 234]; [Bennett, 1996, pp. 39–40], [General Inequalities, vol. 3, pp. 123–140]; [Alzer, 1993], [Bullen, 1997].

Carleman's Integral Inequality *If $B = \{\underline{x}; 0 < |\underline{x}| < 1\} \subset \mathbb{R}^2$, $f \in C^2(B)$, $p \in \mathbb{R}$ then*

$$\int_B |\underline{x}|^p |f(\underline{x})|^2 \, \mathrm{d}\underline{x} \le C \int_B |\underline{x}|^p |\nabla^2 f(\underline{x})|^2 \, \mathrm{d}\underline{x}.$$

COMMENTS This has been extended to higher dimensions by Meshkov[2].
REFERENCES [Meshkov].

Carlson's[3] Inequalities
See **Mixed Mean Inequalities**, **Muirhead Symmetric Function and Mean Inequalities** (d).

---

[2] В З Мешков.
[3] This is B.C. Carlson.

Carlson's[4] Inequality If $\underline{a}$ is a non-negative sequence that is not identically zero then

$$\left(\sum_{i=1}^{\infty} a_i\right)^4 < \pi^2 \left(\sum_{i=1}^{\infty} a_i^2\right)\left(\sum_{i=1}^{\infty} i^2 a^2\right).$$

The constant π^2 is best possible.

INTEGRAL ANALOGUES If $f > 0$ and $xf(x) \in \mathcal{L}^2([0,\infty[)$ then

$$\left(\int_0^{\infty} f\right)^4 \leq \pi^2 \left(\int_0^{\infty} f^2\right)\left(\int_0^{\infty} x^2 f^2(x)\,\mathrm{d}x\right).$$

The constant π^2 is best possible.

EXTENSIONS If $f : [0,\infty[\to]0,\infty[$ and if p, q, λ, μ are positive real numbers with $\lambda < p+1, \mu < q+1$ then

$$\left(\int_0^{\infty} f\right)^{\mu(p+1)+\lambda(q+1)}$$
$$\leq C_{p,q,\lambda,\mu} \left(\int_0^{\infty} x^{p-\lambda} f^{p+1}(x)\,\mathrm{d}x\right)^{\mu} \left(\int_0^{\infty} x^{q+\mu} f^{q+1}(x)\,\mathrm{d}x\right)^{\lambda}.$$

COMMENTS (i) The best value of the constant has been found by V I Levin[5]
(ii) Many other extensions are given in the first two references.

REFERENCES [AI, pp. 370–372], [BB, pp. 175–177], [EM, vol. 2, p. 27].

Cauchy-Hadamard Inequality (a) If f is analytic in $\{z; |z-z_0| \leq r\}$ and if $M(r)$ is the maximum of $|f(z)|$ on the circle $|z - z_0| = r$ then

$$|f^{(k)}(z_0)| \leq k!\frac{M(r)}{r^k}, \; k \in \mathbb{N}. \qquad (1)$$

(b) If f is analytic in the domain Ω then

$$\limsup_{k\to\infty} \left(\frac{f^{(k)}(z)}{k!}\right)^{1/k} \leq \frac{1}{\rho(z,\partial\Omega)},$$

where $\rho(z,\partial\Omega)$ denotes the distance of z from the boundary of Ω.

COMMENTS (i) The first inequality is known as *Cauchy's Inequality* or *Estimate*.
(ii) Another form of (1) is

$$|c_k| \leq \frac{M(r)}{r^k}, \; k \in \mathbb{N},$$

where $f(z) = \sum_{k=0}^{\infty} c_k(z - z_0)^k$.

REFERENCES [EM, vol.2, p. 62]; [Ahlfors, 1966, p. 122], [Conway, vol. I, p. 73], [Rudin, 1966, pp. 213–214], [Titchmarsh, 1939, pp. 84–85].

---

[4] This is F. Carlson.
[5] В И Левин.

Cauchy's Inequality If $\underline{a}, \underline{b}$ are positive n-tuples then

$$\sum_{i=1}^{n} a_i b_i \leq \left(\sum_{i=1}^{n} a_i^2 \right)^{1/2} \left(\sum_{i=1}^{n} b_i^2 \right)^{1/2}, \qquad (C)$$

with equality if and only if $\underline{a} \sim \underline{b}$.

COMMENTS   (i) This is a special case of (H), to which it is equivalent. However it can be proved independently as any book on linear algebra will show. In particular (C) is implied by either the identity

$$\sum_{i=1}^{n} (a_i x + b_i)^2 = x^2 \left(\sum_{i=1}^{n} a_i^2 \right) + 2x \sum_{i=1}^{n} a_i b_i + \sum i = 1^n b_i^2,$$

or by the *Lagrange Identity*,

$$\left(\sum_{i=1}^{n} a_i^2 \right) \left(\sum_{i=1}^{n} b_i^2 \right) - \left(\sum_{i=1}^{n} a_i b_i \right)^2 = \sum_{i,j=1}^{n} (a_i b_j - a_j b_i)^2.$$

(ii) (C) can be written
$$\underline{a} . \underline{b} \leq |\underline{a}||\underline{b}|.$$

(iii) In proving (C) there would be no loss of generality in assuming that $|\underline{a}| = |\underline{b}| = 1$.
(iv) (C) is the case $m = 1$ of **Bessel's Inequality** (3).

EXTENSIONS   (a) If $\underline{a}, \underline{b}$ are complex n-tuples then

$$\left| \sum_{i=1}^{n} a_i \bar{b}_i \right| \leq \sum_{i=1}^{n} |a_i||b_i| \leq \left(\sum_{i=1}^{n} |a_i|^2 \right)^{1/2} \left(\sum_{i=1}^{n} |b_i|^2 \right)^{1/2}, \qquad (1)$$

with equality if and only if $\underline{a} \sim \bar{\underline{b}}$.
(b) If $\underline{a}, \underline{b}$ are real n-tuples, $\underline{w}$ a positive n-tuple then

$$\sum_{i=1}^{n} w_i |a_i||b_i| \leq \left(\sum_{i=1}^{n} w_i a_i^2 \right)^{1/2} \left(\sum_{i=1}^{n} w_i b_i^2 \right)^{1/2}. \qquad (2)$$

(c) [WAGNER] If $\underline{a}, \underline{b}$ are positive n-tuples and $0 \leq x \leq 1$, then

$$\left(\sum_{i=1}^{n} a_i b_i + \left(x \sum_{i \neq j; i,j=1}^{n} a_i b_j \right) \right)^2$$

$$\leq \left(\sum_{i=1}^{n} a_i^2 + \left(2x \sum_{i<j; i,j=1}^{n} a_i a_j \right) \right) \left(\sum_{i=1}^{n} b_i^2 + \left(2x \sum_{i<j; i,j=1}^{n} b_i b_j \right) \right).$$

(d) [KLAMKIN] If $\underline{a}, \underline{b}$ are positive n-tuples,

$$\left(\sum_{i=1}^n a_i b_i\right)^2 \le \left(\sum_{i=1}^n a_i^{2-\frac{n-1}{n}} b_i^{\frac{n-1}{n}}\right)\left(\sum_{i=1}^n a_i^{\frac{n-1}{n}} b_i^{2-\frac{n-1}{n}}\right) \le \cdots$$

$$\cdots \le \left(\sum_{i=1}^n a_i^{2-\frac{k}{n}} b_i^{\frac{k}{n}}\right)\left(\sum_{i=1}^n a_i^{\frac{k}{n}} b_i^{2-\frac{k}{n}}\right) \le \cdots$$

$$\cdots \le \left(\sum_{i=1}^n a_i^{2-\frac{1}{n}} b_i^{\frac{1}{n}}\right)\left(\sum_{i=1}^n a_i^{\frac{1}{n}} b_i^{2-\frac{1}{n}}\right) \le \left(\sum_{i=1}^n a_i^2\right)\left(\sum_{i=1}^n b_i^2\right).$$

(e) [CALLEBAUT] If either $1 \le z \le y \le 2$ or $0 \le y \le z \le 1$, and if $\underline{a}, \underline{b}$ are positive n-tuples then

$$\left(\sum_{i=1}^n a_i^z b_i^{2-z}\right)\left(\sum_{i=1}^n a_i^{2-z} b_i^z\right)$$
$$\le \left(\sum_{i=1}^n a_i^y b_i^{2-y}\right)\left(\sum_{i=1}^n a_i^{2-y} b_i^y\right). \tag{3}$$

(f) [MCLAUGHLIN] If $\underline{a}, \underline{b}$ are real 2n-tuples then

$$\left(\sum_{i=1}^{2n} a_i b_i\right)^2 \le \left(\sum_{i=1}^{2n} a_i^2\right)\left(\sum_{i=1}^{2n} b_i^2\right) - \left(\sum_{i=1}^n a_{2i} b_{2i-1} - a_{2i-1} b_{2i}\right)^2.$$

(g) [DRAGOMIR] If $\underline{a}, \underline{b}, \underline{c}$ are real n-tuples with $|\underline{c}| = 1$ then

$$|\underline{a}.\underline{b}| \le |\underline{a}.\underline{b} - (\underline{a}.\underline{c})(\underline{c}.\underline{b})| + |(\underline{a}.\underline{c})(\underline{c}.\underline{b})| \le |\underline{a}||\underline{b}|.$$

(h) [ALZER] If $\underline{a}, \underline{b}$ are decreasing positive n-tuples then

$$\sum_{i=1}^n a_i b_i \le \min\left\{\sum_{i=1}^n a_i, \sum_{i=1}^n b_i, \left(\sum_{i=1}^n a_i^2\right)^{1/2}\left(\sum_{i=1}^n b_i^2\right)^{1/2}\right\}.$$

(j) [RYSER] If $\underline{a}, \underline{b}$ are real n-tuples and if $\underline{m} = \min\{\underline{a}, \underline{b}\}, \underline{M} = \max\{\underline{a}, \underline{b}\}$ then

$$|\underline{a}.\underline{b}| \le |\underline{M}||\underline{m}| \le |\underline{a}||\underline{b}|.$$

COMMENTS   (v) It is easy to see that (2) and (C) are the same.
(vi) The inequality of Callebaut, (3) above, interpolates (C) in the sense that if $z = 1, y = 2$, or $y = 0, z = 1$ it reduces to (C). It says that if $f(z)$ is the left-hand side of (2) then f is decreasing on $[0, 1]$, and increasing on $[1, 2]$, and takes as value the left-hand side of (C) when $z = 1$, and the right-hand side of (C) if $z = 0$ or 2. For a similar interpolation of (GA) see **Chong's Inequalities** (c).

INTEGRAL ANALOGUES If $f, g \in \mathcal{L}^2([a,b])$ then $fg \in \mathcal{L}([a,b])$ and

$$\|fg\| \leq \|f\|_2 \|g\|_2.$$

There is equality if and only if $Af = Bg$ almost everywhere, where not both of the constants A, B are zero.

CONVERSE INTEGRAL ANALOGUES [BELLMAN] If f, g are non-negative and concave on $[0,1]$ then

$$\int_0^1 fg \geq \frac{1}{2} \left(\int_0^1 f^2 \right)^{1/2} \left(\int_0^1 g^2 \right)^{1/2}.$$

COMMENTS (vii) There are many converse inequalities for (C) that are discussed elswhere; see for instance **Barnes's Inequalities** (1), **Pólya & Szegö's Inequality, Zagier's Inequality**.
(viii) The result of Bellman has been generalized by Alzer.
(ix) For another inequality also known as Cauchy's inequality see **Cauchy-Hadamard Inequality** (1).
(x) See also **Aczel's Inequality** COMMENTS (i), **Complex Number Inequalities**, EXTENSIONS, **Determinant Inequalities** (d), **Gram Determinant Inequalities** (3), **Inner Product Inequalities** (2), **Ostrowski's Inequalities** (a), EXTENSIONS (a), **Quaternion Inequalities, Trace Inequalities** (b).

REFERENCES [AI, pp. 30–32, 41–44], [EM, vol. 1, p. 485, vol. 2, p. 61], [HLP, pp. 16,132–134], [MI, pp. 140–143, 155, 159],[MPF, pp. 488–491], [PPT, p. 118]; [Ahlfors, 1966, pp. 10–11], [Hewitt & Stromberg, p. 190], [Pólya & Szegö, 1972, p. 68]; [Alzer, 1991$^{(2)}$], [Dragomir].

Cauchy-Schwarz-Bunyakovskiĭ[6] Inequality
See **Cauchy's Inequality**.

Cauchy-Schwarz Inequality
See **Cauchy's Inequality**.

Cauchy Transform Inequality [AHLFORS & BEURLING] If $K \subset \mathbb{C}$ is compact with $|K| > 0$ then

$$\left| \int_K \frac{1}{z-w} \, dw \right| \leq \sqrt{\pi |K|}.$$

COMMENTS The integral is with respect to Lebesgue measure in the plane. It is a special case of the *Cauchy transform of a measure*; in general the Lebesgue measure on K is replaced by any measure on $\mathbb{C}$ having compact support.

REFERENCES [Conway, vol. II, pp. 192–196].

---

[6] В Я Буняковский. Also transliterated as Buniakovsky.

Čebyšev's[7] Inequality

If $\underline{a}, \underline{b}$ are similarly ordered real n-tuples then

$$\mathfrak{A}_n(\underline{a};\underline{w})\mathfrak{A}_n(\underline{b};\underline{w}) \leq \mathfrak{A}_n(\underline{a}\,\underline{b};\underline{w}), \qquad (\check{C})$$

with equality if and only if $\underline{a}$, or $\underline{b}$ is constant.

COMMENTS (i) This famous inequality has been given many proofs; the most elementary depends on the identity:

$$n\sum_{i=1}^n a_i b_i - \sum_{i=1}^n a_i \sum_{i=1}^n b_i = \frac{1}{2}\sum_{i,j=1}^n (a_i - a_j)(b_i - b_j).$$

(ii) The relation $\mathfrak{A}_n(\underline{a}) + \mathfrak{A}_n(\underline{b}) = \mathfrak{A}_n(\underline{a}+\underline{b})$ is trivial and ($\check{C}$) is the non-trivial multiplicative analogue of this. Similarly $\mathfrak{G}_n(\underline{a})\mathfrak{G}_n(\underline{b}) = \mathfrak{G}_n(\underline{a}\,\underline{b})$, the case $r = 0$ of (1) below, is trivial; its additive analogue is given by **Power Mean Inequalities** (4), COMMENTS (iv).

(iii) While the conditions given are sufficient for ($\check{C}$) they are not necessary; a set of necessary and sufficient conditions has been given by Sasser & Slater.

EXTENSIONS (a) If $0 < r < \infty$ and $\underline{a}, \underline{b}$ are similarly ordered positive n-tuples then

$$\mathfrak{M}_n^{[r]}(\underline{a};\underline{w})\mathfrak{M}_n^{[r]}(\underline{b};\underline{w}) \leq \mathfrak{M}_n^{[r]}(\underline{a}\,\underline{b};\underline{w}), \qquad (1)$$

with equality if and only if either $r = 0$, or $r \neq 0$ and $\underline{a}$ or $\underline{b}$ is constant.
If $-\infty < r < 0$ and $\underline{a}, \underline{b}$ are oppositely ordered then (1) also holds.
If $0 < r < \infty, (-\infty < r < 0)$, and $\underline{a}, \underline{b}$ are oppositely, (similarly), ordered then (~ 1) holds.

(b) If the k non-negative n-tuples, $\underline{a}_i, 1 \leq i \leq k$, are similarly ordered

$$\prod_{i=1}^k \mathfrak{A}_n(\underline{a}_i,\underline{w}) \leq \mathfrak{A}\left(\prod_{i=1}^k \underline{a}_i;\underline{w}\right). \qquad (2)$$

(c) [ALZER] If $\underline{a}, \underline{w}$ are both strictly increasing then

$$\mathfrak{A}_n(\underline{a}\,\underline{b};\underline{w}) - \mathfrak{A}_n(\underline{a};\underline{w})\mathfrak{A}_n(\underline{b};\underline{w}) \geq K(\underline{w}) \min_{1\leq i,j \leq n-1}\{\Delta a_i, \Delta b_i\},$$

where

$$K(\underline{w}) = \frac{1}{W_n}\sum_{i=1}^n i^2 w_i - \left(\frac{1}{W_n}\sum_{i=1}^n i w_i\right)^2.$$

(d) [VASIĆ & DJORDJEVIĆ] If $\underline{a}, \underline{b}$ are non-negative increasing convex n-tuples and if $\underline{n} = \{0, 1, \ldots, n-1\}$ then

$$\mathfrak{A}_n(\underline{a};\underline{w})\mathfrak{A}_n(\underline{b};\underline{w}) \leq \frac{\mathfrak{A}_n(\underline{n};\underline{w})^2}{\mathfrak{A}_n(\underline{n}^2,\underline{w})}\mathfrak{A}_n(\underline{a}\,\underline{b};\underline{w}).$$

---

[7] П Л Чебышев. Also transliterated as Tchebyshev, Tchebichev, Chebyshev.

(e) [VASIĆ & PEČARIĆ] Under the hypothese of (b),

$$0 \leq \mathfrak{A}_2\left(\prod_{i=1}^{k} \underline{a}_i; \underline{w}\right) - \prod_{i=1}^{k} \mathfrak{A}_2(\underline{a}_i; \underline{w}) \leq \cdots \leq \mathfrak{A}_n\left(\prod_{i=1}^{k} \underline{a}_i; \underline{w}\right) - \prod_{i=1}^{k} \mathfrak{A}_2(\underline{a}_i; \underline{w}).$$

COMMENTS (iv) For an extension of (b) see **Mean Monotonic Sequence Inequalities**.
(v) Extension (d) of (Č) is an example of various results using n-convex sequences.
(vi) Better lower bounds for the difference between the right-hand side and left-hand side of (Č) have been obtained by Alzer.
(vii) Further extensions can be found in Pečarić, and extension to functions of index sets have been given by Vasić & Pečarić; see also [MI].

INTEGRAL ANALOGUES If $f, g \in \mathcal{L}([a,b])$ and are both monotonic in the same sense then

$$\int_a^b f \int_a^b g \leq (b-a) \int_a^b fg,$$

with equality if and only if one of the functions is constant almost everywhere.

COMMENTS (viii) There are extensions of this last result to weighted integral means and to products of more than two functions, see [AI]. In addition the concept of similarly ordered functions can be used, see [HLP].
(ix) For converse inequalities see **Grüsses' Inequalities** (a), **Karamata's Inequalities, Ostrowski's Inequalities** (b).
(x) A detailed review of the history and development of Čebišev's inequality has been given by Mitrinović & Vasić; see also [MFP].

REFERENCES [AI, pp. 36–41], [EM, vol.2, p. 119], [HLP, pp. 43–44, 168], [MI, pp. 230-234], [MPF, pp. 239–293, 351–358], [PPT, pp. 197–228]; [Alzer, 1992]

Čebišev[7] Polynomial Inequalities

The *Cebišev polynomial of degree n* is defined by

$$T_n(x) = \cos(n \arccos x), \qquad n \geq 1, \quad -1 \leq x \leq 1.$$

The *monic Cebišev Polynomial of degree n* is $\tilde{T}_n = 2^{1-n} T_n$.

If p_n is a monic polynomial of degree n that is not $\tilde{T}_n$ then

$$\|p_n\|_{\infty, [-1,1]} > \|\tilde{T}_n\|_{\infty, [-1,1]} = \frac{1}{2^{n-1}}.$$

COMMENTS (i) A monic polynomial has the term of highest degree with coefficient 1.
(ii) This inequality says that of all monic polynomials of degree n on the interval $[-1, 1]$ the one that deviates least from zero is the monic Cebišev polynomial of degree n
(iii) See also **Markov's Inequality** COMMENTS (i), and EXTENSIONS (b).

REFERENCES [EM, vol. 2, pp. 123–124]; [Inequalities, pp. 321–328].

---

[7] П Л Чебышев. Also transliterated as Tchebyshev, Tchebichev or Chebyshev.

Čebišev's[7] Probability Inequality

If X is a random variable with finite expectation then

$$P(|X - EX| \geq r) \leq \frac{\sigma^2 X}{r^2}.$$

COMMENTS (i) This is also known as the *Bienaymé-Čebišev* inequality.
(i) Refinements of this result can be found in **Bernšteĭn's Probability Inequality, Kolmogorov's Probability Inequality, Markov's Probability Inequality**.
(ii) See also **Markov's ProbabilityInequality** COMMENTS (ii).

REFERENCES [EM, vol. 2, pp. 119–120, vol. 5, pp. 295–296]; [Feller, vol. I, pp. 219–221], [Loève, pp. 234–236].

Chassan's Inequality

See **Ostrowski's Inequality**, COMMENTS (iii).

Chebyshev's Inequalities

See **Čebišev's Inequality, Čebišev Polynomial Inequalities, Čebišev's Probability Inequality**.

Chernoff Bounds

Let $(X, \mathcal{M}, \mu)$ be a measure space, $f : X^2 \to \mathbb{R}$, μ_n the product measure on X^n, and f_n the function $f_n(\underline{u}, \underline{v}) = \frac{1}{n} \sum_1^n f(u_i, v_i), \underline{u}, \underline{v} \in X^n$. Then for $r \in \mathbb{R}, s \geq 0$,

$$\mu_n\left(\{\underline{u}; f_n(\underline{u}, \underline{v}) < r\}\right) \leq \left(e^{sr} \sup_{v \in X} \int_X e^{-sf(u,v)} \, d\mu(u)\right)^n.$$

REFERENCES [General Inequalities, vol. 1, pp. 131–132].

Chong's Inequalities

(a) If $\underline{b}$ is a rearrangement of $\underline{a}$, both positive n-tuples, then

$$\sum_{i=1}^n \frac{b_i}{a_i} \geq n; \quad \prod_{i=1}^n a^{a_i} \geq \prod_{i=1}^n a^{b_i} \tag{1}$$

The inequalities are strict unless $\underline{a} = \underline{b}$.

(b) If $\underline{a}, \underline{b}$ are positive n-tuples and $\underline{a} \prec \underline{b}$ then

$$\prod_{i=1}^n a_i \leq \prod_{i=1}^n b_i. \tag{2}$$

(c) If $0 \leq x < y \leq 1$ and if the positive n-tuple $\underline{a}$ is not constant,

$$\mathfrak{A}_n\left(\mathfrak{G}_n^{1-x}(\underline{a}; \underline{w})\underline{a}^x; \underline{w}\right) < \mathfrak{A}_n\left(\mathfrak{G}_n^{1-y}(\underline{a}; \underline{w})\underline{a}^y; \underline{w}\right); \\ \mathfrak{G}_n\left(x\mathfrak{A}_n(\underline{a}; \underline{w}) + (1-x)\underline{a}; \underline{w}\right) < \mathfrak{G}_n\left(y\mathfrak{A}_n(\underline{a}; \underline{w}) + (1-y)\underline{a}; \underline{w}\right). \tag{3}$$

COMMENTS (i) The first inequality in (1) can be used to prove (GA).

(ii) (b) follows by first proving the case $\underline{a} = \lambda \underline{b} + (1-\lambda)\underline{b}'$ where $\underline{b}'$ is a permutation of $\underline{b}$. The general case follows from properties of the order relation $\prec$.
(iii) Applications of (2) occur in **Bernoulli's Inequality**, EXTENSIONS (a), and also in **Binomial Function Inequalities** COMMENTS (iv).
It can be used to prove the equal weight case of (GA) by noting that $\mathfrak{A}_n(\underline{a}) \prec \underline{a}$.
(iv) Both inequalities (3) reduce to (GA) if $x = 0$, $y = 1$.
A similar interpolation result for (C) is due to Callebaut, see **Cauchy's Inequality** EXTENSIONS (e).
REFERENCES   [MI, pp. 20, 81, 120], [MPF, pp. 37–40].

Circulant Matrix Inequalities
Given an n-tuple $\underline{a}$ then the *circulant matrix of $\underline{a}$* is

$$C(\underline{a}) = \begin{pmatrix} a_1 & a_2 & \cdots & a_n \\ a_n & a_1 & \cdots & a_{n-1} \\ \vdots & \vdots & \ddots & \vdots \\ a_2 & a_3 & \cdots & a_1 \end{pmatrix}.$$

[BECKENBACH & BELLMAN]   (a) If n is odd, and $\sum_{i=1}^{n} a_i \geq 0$ then

$$\det C(\underline{a}) \geq 0. \tag{1}$$

(b) If $n = 2m$ is even and if $\left|\sum_{k=1}^{m} a_{2k-1}\right| \geq \left|\sum_{k=1}^{m} a_{2k}\right|$ then (1) holds.

COMMENTS   (i) The cases of equality are discussed in the reference.
(ii) In addition the reference gives analogous results for the *skew-circulant matrix*, obtained by changing the sign of each element on one side of the diagonal of $C(\underline{a})$.
REFERENCES   [General Inequalities, vol. 1. pp. 39–48].

Clarkson's Inequalities
(a) If $p \geq 2$, q the conjugate index, $\underline{a}, \underline{b}$ complex sequences in ℓ_p, then

$$||\underline{a} + \underline{b}||_p^p + ||\underline{a} - \underline{b}||_p^p \leq 2^{p-1}(||\underline{a}||_p^p + ||\underline{b}||_p^p);$$
$$||\underline{a} + \underline{b}||_p^p + ||\underline{a} - \underline{b}||_p^p \leq 2(||\underline{a}||_p^q + ||\underline{b}||_p^q)^{p-1}.$$

(b) If $1 < p < 2$, q the conjugate index, $\underline{a}, \underline{b}$ complex sequences in ℓ_p, then

$$||\underline{a} + \underline{b}||_p^q + ||\underline{a} - \underline{b}||_p^q \leq 2(||\underline{a}||_p^p + ||\underline{b}||_p^p)^{q-1}.$$

COMMENTS   (i) These inequalities were used by Clarkson to prove that the spaces $\ell_p, \mathcal{L}^p$ are uniformly convex; see for instance [Hewitt & Stomberg].
(ii) These inequalities have easily stated integral analogues for functions in $\mathcal{L}^p$.
The proofs of the above results depend on two inequalities between complex numbers that are of some interest.

TWO COMPLEX NUMBER INEQUALITIES (a) If $z, w \in \mathbb{C}, p \geq 2$ then

$$|w + z|^p + |w - z|^p \leq 2^{p-1}(|w|^p + |z|^p). \tag{1}$$

(b) If $z, w \in \mathbb{C}, 1 < p \leq 2$, q the conjugate index, then

$$|w + z|^q + |w - z|^q \leq 2(|w|^p + |z|^p)^{q-1}.$$

COMMENTS (iii) This result is derived from the real case, the proof of which uses elementary calculus.
(iv) (1) is a converse of the **Beth & van der Corput Inequality**.
(v) These inequalities, and the following, have been extended to unitary spaces.

EXTENSIONS [KOSKELA] If $w, z \in \mathbb{C}$, $r, s > 0$, r' the conjugate index of r, and

$$t = \begin{cases} \min\{2, s\} & \text{if } r \leq 2, \\ \min\{r', s\} & \text{if } r > 2; \end{cases}$$

then

$$|w + z|^r + |w - z|^r \leq C (|w|^s + |z|^s)^{r/s}, \tag{2}$$

where

$$C = 2^{1 - \frac{r}{s} + \frac{r}{t}}.$$

There is equality in (2) : (i) for all w, z if $r = s = 2$; (ii) if $wz = 0$ and $0 < s < r' \leq 2$, or $0 < r < 2$ and $0 < s < 2$; (iii) $\Re(w\bar{z}) = 0$ and $0 < r < 2$; (iv) $w = \pm iz$ and $0 < r < 2 < s$; (v) $|w| = |z|$ and $r = 2 < s$; (vi) $w = \pm z$ and $2 < r$ and $r' < s$; (vii) $wz = 0$ or $w = \pm z$ and $2 < r = s'$.

COMMENTS (vi) This result follows using (H), or (r;s), when $0 < r \leq 2$, and (J) when $2 < r$ and $s \leq r'$.
(vii) The case of equality (i), $r = s = 2$, is just **Parallelogram Inequality**, COMMENTS (i).
(vii) Koskela used the above to extend the Clarkson inequalities.
(viii) See also **Hanner's Inequalities, von Neumann & Jordan Inequality**.

REFERENCES [MPF, pp. 534–558], [PPT, p. 135]; [Hewitt & Stromberg, pp. 225–227].

Clausius-Duhem Inequality
Let $\Omega_t \subset \mathbb{R}^3$ be a domain depending on time, t, with a smooth boundary, $\theta = \theta(\underline{x}, t)$ the temperature at $\underline{x} \in \Omega_t$, q the heat flux per unit area though $\partial \Omega_t$. $\rho = \rho(\underline{x}, t)$ the density at time t, and $\underline{x} \in \Omega_t$, $h = h(\underline{x}, t)$ the mass density of radiation heat, and $S(\overline{\Omega}_t)$ the total entropy of $\overline{\Omega}_t$ then

$$\frac{\partial S(\overline{\Omega}_t)}{\partial t} \geq \int_{\Omega_t} \frac{h}{\theta} \rho + \oint_{\partial \Omega_t} \frac{q}{\theta}.$$

COMMENTS For other entropy inequalities see **Entropy Inequalities, Shannon's Inequality**.
REFERENCES [EM, Supp., pp. 185–186].

Cohn-Vossen Inequality *If M is a non-compact Riemannian manifold with no boundary then*
$$\int_M K\,dS \leq 2\pi\chi,$$
where K is the Gaussian curvature, and χ the Euler characteristic.

COMMENTS   In case M is compact and closed, or with a smooth boundary, the inequality becomes the *Gauss-Bonnet theorem*:
$$\int_M K\,dS + \int_{\partial M} k_g\,d\ell = 2\pi\chi,$$
where k_g is the geodesic curvature.

REFERENCES   [EM, vol. 4, p. 196].

Complete Symmetric Function Inequalities   (a) *If $1 \leq r \leq n-1$ and $\underline{a}$ is a positive n-tuple, then*
$$q_n^{[r-1]}(\underline{a}) q_n^{[r+1]}(\underline{a}) \geq \left(q_n^{[r]}(\underline{a})\right)^2, \tag{1}$$
with equality if and only if $\underline{a}$ is constant.

(b) [MCLEOD, BASTON] *If $\underline{a}, \underline{b}$ are a positive n-tuples, r, s are integers, $1 \leq r \leq s \leq n$, and either $s = r$ or $s = r+1$ then*
$$\left(\frac{c_n^{[s]}(\underline{a}+\underline{b})}{c_n^{[s-r]}(\underline{a}+\underline{b})}\right)^{1/r} \leq \left(\frac{c_n^{[s]}(\underline{a})}{c_n^{[s-r]}(\underline{a})}\right)^{1/r} + \left(\frac{c_n^{[s]}(\underline{b})}{c_n^{[s-r]}(\underline{b})}\right)^{1/r}. \tag{2}$$

(c) [OZEKI] *If $\underline{a}$ is a positive log-convex sequence so is $q_n^{[r]}(\underline{a})$; that is,*
$$q_{n-1}^{[r]}(\underline{a}) q_{n+1}^{[r]}(\underline{a}) \geq \left(q_n^{[r]}(\underline{a})\right)^2, \quad 1 \leq r \leq n-1.$$

COMMENTS   (i) (1) is the analogue of **Elementary Symmetric Function Inequalities** (2).
(ii) Nothing is known about the other cases of (2), it would be of interest to complete the result and so obtain an analogue of **Marcus & Lopes's Inequality**.
(iii) $q_n^{[r]}$ is Schur concave, strictly if $r > 1$; for a definition see **Schur Convex Function Inequalities**.

REFERENCES   [MI, pp. 315, 333], [MO, pp. 81–82], [MPF, p. 165].

Complete Symmetric Mean Inequalities   (a) *If $\underline{a}$ is a positive n-tuple, and if $1 \leq r \leq n$ then*
$$\min \underline{a} \leq \mathfrak{Q}_n^{[r]}(\underline{a}) \leq \max \underline{a},$$
with equality if and only if $\underline{a}$ is constant.
(b) *If $\underline{a}$ is a positive n-tuple, $1 \leq r < s \leq n$ then*
$$\mathfrak{Q}_n^{[r]}(\underline{a}) \leq \mathfrak{Q}_n^{[s]}(\underline{a}),$$

with equality if and only if $\underline{a}$ is constant.

(c) If $\underline{a}, \underline{b}$ are positive n-tuples, r is an integer and $1 \leq r \leq n$ then

$$\mathfrak{Q}_n^{[r]}(\underline{a} + \underline{b}) \leq \mathfrak{Q}_n^{[r]}(\underline{a}) + \mathfrak{Q}_n^{[r]}(\underline{b}).$$

(d) If $\underline{a}$ is a positive n-tuple, r is an integer and $1 \leq r \leq n$ then

$$\mathfrak{P}_n^{[r]}(\underline{a}) \leq \mathfrak{Q}_n^{[r]}(\underline{a}),$$

with equality if and only if either $r = 1$ or $\underline{a}$ is constant.

COMMENTS (i) The inequality in (b) is the analogue of S(r;s).
(ii) The inequality in (c) follows from the case $s = r$ of **Complete Symmetric Function Inequalities** (2).
REFERENCES [MI, pp. 314–315, 333, 340], [MPF, pp. 16–17].

Complex Function Inequalities (a) If f is a continuous complex-valued function defined on $[a, b]$ then

$$\Re\left(e^{i\theta} \int_a^b f\right) \leq \int_a^b |f|, \tag{1}$$

and

$$\left|\int_a^b f\right| \leq \int_a^b |f|. \tag{2}$$

(b) If f is a complex-valued function continuous on the piecewise differentiable arc γ in $\mathbb{C}$ then

$$\left|\int_\gamma f(z)\,\mathrm{d}z\right| \leq \int_\gamma |f(z)|\,|\mathrm{d}z|.$$

COMMENTS (i) (1) follows from the right-hand side of the first inequality in **Complex Number Inequalities** (a). Then (2) follows from (1) by a suitable choice of θ.
(ii) A discrete analogue of (2) is **Complex Number Inequalities** EXTENSIONS (a). A converse van be found in **Wilf's Inequality** INTEGRAL ANALOGUES.
(iii) Most inequalities covered by this heading can be found elsewhere; in particular in the many entries for special functions, **Binomial, Factorial, Conjugate Harmonic, Hyperbolic, Laguerre, Logarithmic, Polynomial, Trigonometric**, etc. See also references in **Analytic Function Inequalities**.
REFERENCES [Ahlfors, 1966, pp. 101–104], .

Complex Number Inequalities (a) For all $z \in \mathbb{C}$,

$$-|z| \leq \Re z \leq |z|; \qquad -|z| \leq \Im z \leq |z|,$$

with equality on the right-hand side of first case if and only if $z \geq 0$.
(b) If z_1, z_2, z_3 are complex numbers then

$$|z_1 + z_2| \leq |z_1| + |z_2|, \tag{1}$$
$$|z_1 - z_2| \geq ||z_1| - |z_2||, \tag{2}$$
$$|z_1 - z_3| \leq |z_1 - z_2| + |z_2 - z_3|. \tag{3}$$

There is equality: in (1) if and only if $z_1 \bar{z}_2 \geq 0$, equivalently z_1, z_2 are on the same ray from the origin: in (2) if and only if $z_1 \bar{z}_2 \leq 0$: in (3) if and only if z_2 is between z_1 and z_3.
(c) If $|z|, |w| < 1$ then

$$\frac{|z| - |w|}{1 - |z||w|} \leq \left|\frac{z - w}{z\bar{w} - 1}\right| \leq \frac{|z| + |w|}{1 + |z||w|} < 1.$$

(d) If $\Re z \geq 1$ then

$$|z^{n+1} - 1| > |z|^n |z - 1|.$$

(e) If $0 < \theta < \pi/2$, $z = e^{i\theta} \cos \theta$ then

$$|1 - z| < |1 - z^n|.$$

(f) [BERGSTRÖM] If $p, q \in \mathbb{R}$, and if $pq(p+q) > 0$ then

$$\frac{|z+w|^2}{p+q} \leq \frac{|z|^2}{p} + \frac{|w|^2}{q}; \tag{4}$$

while if $pq(p+q) < 0$ then (~ 2) holds.
(g) [BOURBAKI-JANOUS] (i) If $\underline{z}$ is a complex n-tuple with $\sum_{k=1}^{n} |z_k| \leq h < 1$ then

$$\left|\prod_{k=1}^{n}(1 + z_k) - 1 - \sum_{k=1}^{n} z_k\right| \leq \frac{h^2}{1 - h}.$$

(ii) If $0 < \lambda \leq 1/(2\pi + 1)$ and for all $\mathcal{J} \subseteq \{1, \ldots, n\}$ we have that

$$\left|\prod_{k \in \mathcal{J}}(1 + z_k) - 1\right| \leq \lambda,$$

then

$$\sum_{k=1}^{n} |z_k| \leq \frac{2\pi \lambda}{\alpha(\lambda)}, \text{ where } \alpha(\lambda) = 1 + \sqrt{\frac{1 - \lambda(2\pi + 1)^2}{1 - \lambda}}.$$

(h) [REDHEFFER & SMITH] Let a, b be non-zero complex numbers, with $|a + b| = \sigma$, $|a - b| = \delta$, $\sigma \delta \neq 0$; then if $z \in \overline{D}$,

$$\max\{|az + b|, |a + bz|\} \geq \frac{\sigma \delta}{\sqrt{\sigma^2 + \delta^2}};$$

further given σ, δ there are a, b, z for which equality holds.

COMMENTS (i) (a) is immediate from the definitions.
(ii) For (1) just apply (a) to the identity

$$|z_1 \pm z_2|^2 = |z_1|^2 + |z_2|^2 \pm 2\Re z_1 \bar{z}_2.$$

(iii) Of course (3) is the complex number version of (T).
(iv) (f) follows from the identity

$$\frac{|z|^2}{p} + \frac{|w|^2}{q} - \frac{|z+w|^2}{p+q} = \frac{|qz - pw|^2}{pq(p+q)}.$$

It is generalized in **Norm Inequalities** (4).
(v) A slightly weaker particular case of (g)(ii), $\lambda = 1/81$, and $\frac{2\pi\lambda}{\alpha(\lambda)}$ replaced by $8/81$, was used by Bourbaki to prove that if a family of complex numbers was multipliable then it was summable.
(vi) The inequality in (h) is related to the **Goldberg-Straus Inequality**.

EXTENSIONS (a) If $\underline{z}$ is complex n-tuple,

$$\left| \sum_{k=1}^{n} z_k \right| \leq \sum_{k=1}^{m} |z_k|,$$

with equality if and only if for all non-zero terms $z_j/z_k > 0$.
(b) For all $z_1, z_2, z_3 \in \mathbb{C}$,

$$1 \leq |1 + z_1| + |z_1 + z_2| + |z_2 + z_3| + |z_3|$$

(c) [DE BRUYN] If $\underline{w}$ is a real n-tuple and $\underline{z}$ a complex n-tuple then

$$\left| \sum_{k=1}^{n} w_k z_k \right| \leq \frac{1}{\sqrt{2}} \left(\sum_{k=1}^{n} w_k^2 \right)^{1/2} \left(\sum_{k=1}^{n} |z_k|^2 + \left| \sum_{k=1}^{n} z_k^2 \right| \right)^{1/2},$$

with equality if and only if for some complex λ, $w_k = \Re \lambda z_k$, $1 \leq k \leq n$, and $\sum_{k=1}^{n} \lambda^2 z_k^2 \geq 0$.
(d) If $z, w \in \mathbb{C}, r \geq 0$ then

$$|z+w|^r \leq \begin{cases} |z|^r + |w|^r, & \text{if } r \leq 1, \\ 2^{r-1}(|z|^r + |w|^r), & \text{if } r > 1. \end{cases}$$

COMMENTS (vi) An integral analogue of (a) is **Complex Function Inequalities** (2).

CONVERSE INEQUALITIES   (a) If $\underline{z}$ is a complex n-tuple, $\underline{z} \neq \underline{0}$, then for some index set $\mathcal{I} \subseteq \{1, \ldots, n\}$,

$$\left|\sum_{k \in \mathcal{I}} z_k\right| > \frac{1}{\pi} \sum_{k \in \mathcal{I}} |z_k|.$$

(b) If $|\arg z - \arg w| \leq \theta \leq \pi$ and if $n \geq 1$ then

$$|z - w|^n \leq (|z|^n + |w|^n) \max\{1, 2^{n-1} \sin^n \theta/2\};$$

while if $w, z \neq 0$

$$|z - w| \geq \frac{1}{2}(|z| + |w|)\left|\frac{z}{|z|} - \frac{w}{|w|}\right|.$$

COMMENTS   (vii) The constant $1/\pi$ in (a) is best possible.
(viii) See also **Absolute Value Inequalities** (a), (4), COMMENTS (iii), **Beth & van der Corput's Inequality, Bohr's Inequality, Cauchy's Inequality,** EXTENSION (a), **Clarkson's Inequality** TWO COMPLEX NUMBER INEQUALITIES, EXTENSIONS, **Leindler's Inequality, Wilf's Inequality.**
REFERENCES   [AI, pp. 310–336], [MPF, pp. 78, 89–90, 499–500]; [General Inequalities, vol. 2, pp. 47–51]; [Janous].

Compound Mean Inequalities
See **Arithmetic-geometric Compound Mean Inequalities.**

Conjugate Convex Function Inequalities
If $f : \mathbb{R}^n \to \overline{\mathbb{R}}$ define $f^\# : \mathbb{R}^n \to \overline{\mathbb{R}}$ by

$$f^\#(\underline{a}) = \sup\{t; t = \underline{a}.\underline{b} - f(\underline{b}), \underline{b} \in \mathbb{R}^n\}.$$

If f is convex, possibly infinite, then the same is true of $f^\#$.
The domain can be generalized to any convex set.
$f^\#$ is called the *conjugate of f*. It is in a sense the best function for an inequality of type (1) below.

(a) If f, g have conjugates $f^\#, g^\#$ respectively, then

$$f \geq g \quad \Longrightarrow \quad f^\# \geq g^\#.$$

(b) [YOUNG'S INEQUALITY] If $f > -\infty$ and is convex on a convex domain, and if $f^\#$ is its conjugate

$$\underline{a}.\underline{b} \leq f(\underline{a}) + f^\#(\underline{b}). \qquad (1)$$

COMMENTS   (i) In particular if $f(x) = |x|^p/p, p > 1$ then $f^\#(x) = |x|^q/q, q$ the conjugate index. In this case (1) reduces to **Geometric-Arithmetic Mean Inequality** (2). If $f(x) = e^x$ then $f^\#(x) = x\log x - x, x > 0, = 0, x = 0, = -\infty, x < 0$; and then **Young's Inequalities** (1), and **Logarithmic Function Inequalities** (e) are particular cases of (1).
(ii) Inequality (1) is called *Fenchel's inequality* by Rockafellar.

REFERENCES [EM, vol. 2, pp. 336–337], [PPT, p. 241]; [Rockafellar, pp. 102–106], [Roberts & Varberg, pp. 21, 28–36, 110–111].

Conjugate Function Inequalities
See **Conjugate Harmonic Function Inequalities** COMMENTS (iv).

Conjugate Harmonic Function Inequalities
If f is analytic in D and if $f = u + iv$, the two real functions u, v are harmonic in D and v is called the *harmonic conjugate of u*.

[M.RIESZ] (a) If u is harmonic in $D, u \in \mathcal{H}_p(D), 1 < p < \infty$, and if v is the harmonic conjugate, then $v \in \mathcal{H}_p(D)$ and there is a constant C_p such that

$$||v||_p \leq C_p ||u||_p.$$

In the case $p = 1$,

$$||v||_1 \leq A \sup_{0<r<1} \frac{1}{2} \int_0^{2\pi} |u(re^{i\theta})| \log^+ |u(re^{i\theta})| \, d\theta.$$

(b) If Δ is any diameter of D and $u \in \mathcal{H}(D)$ then

$$\int_\Delta |v(z)| \, |dz| \leq \frac{1}{2}||u||.$$

COMMENTS (i) The norm is defined in **Analytic Function Inequalities** COMMENTS (i).
(ii) If p, q are conjugate indices then $C_p = C_q$. The exact value of the constant is known
(iii) An extension of (b) is given in **Féjer-Riesz Theorem**.
(iv) These results have real function analogues. In this case if $f \in \mathcal{L}(-\pi, \pi)$ the *conjugate function* is

$$\hat{f}(x) = -\frac{1}{\pi} \lim_{\epsilon \to 0+} \int_\epsilon^\pi \frac{f(x+t) - f(x-t)}{2 \tan t/2} \, dt.$$

With this definition of conjugate functions, the above theorem of M. Riesz holds.
This real result extends to higher dimensions.

REFERENCES [EM, vol. 2, pp. 336, 338–339; vol. 4, pp. 366–369; vol. 6, pp. 131–140]; [Hirschman, pp. 164–167], [Rudin, 1966, pp. 330–331, 345–347], [Zygmund, vol. I, pp. 51, 253–258].

Continued Fraction Inequalities
The *continued fraction*

$$a_0 + \cfrac{b_1}{a_1 + \cfrac{b_2}{a_2 + \cfrac{b_3}{a_3 + \cfrac{b_4}{a_4 + \cdots}}}} = a_0 + \frac{b_1|}{|a_1} + \cdots + \frac{b_n|}{|a_n} + \cdots$$

will be written $C = [a_0; a_1, a_2 \ldots : b_1, \ldots]$; and the finite continued fraction

$$C_n = [a_0; a_1, a_2 \ldots, a_n : b_1, \ldots, b_n] = a_0 + \frac{b_1|}{|a_1} + \cdots + \frac{b_n|}{|a_n}$$

is called the *n-th convergent* of the infinite continued fraction C.
In the case $b_n = 1, n \geq 1$ they are omitted from the bracket notation, and the continued fraction is said to be *regular* or *simple*.

(a) If $a_n > 0, b_n \geq 0, n \in \mathbb{N}$ then

$$C_{2n-1} > C_{2n+1} > C_{2n+2} > C_{2n}, n \geq 1.$$

(b) If C is regular continued fraction with a_n a positive integer, $n \geq 1$, then $C_n = P_n/Q_n$, with P_n, Q_n integers, $n \geq 1$, $C_n \to C$ and

$$C_{2n-1} > C_{2n+1} > C > C_{2n+2} > C_{2n}, \quad n \geq 1;$$
$$Q_n \geq 2^{(n-1)/2}, \quad n \geq 2;$$

further
$$|C - C_n| \leq \frac{1}{Q_n Q_{n+1}}. \tag{1}$$

COMMENTS   A converse of (1) can be found in **Mediant Inequalities** (1).
REFERENCES [EM, vol. 2, pp. 375–377], [MPḞ, pp. 661–665]; [Khinchin].

Convex Functions of Higher Order Inequalities
See **n-Convex Function Inequalities**.

Convex Function Inequalities
(a) If f is a convex function defined on the compact interval $[a, b]$ then for all $x, y \in [a, b]$ and $0 \leq \lambda \leq 1$,

$$f((1 - \lambda)x + \lambda y) \leq (1 - \lambda)f(x) + \lambda f(y); \tag{1}$$

if f is strictly convex then (1) is strict unless $x = y$, $\lambda = 0$ or $\lambda = 1$.
(b) If $a \leq x_1 < x_2 < x_3 \leq b$ then

$$f(x_2) \leq \frac{x_3 - x_2}{x_3 - x_1} f(x_1) + \frac{x_2 - x_1}{x_3 - x_1} f(x_3), \tag{2}$$

or equivalently

$$f(x_2)(x_3 - x_1) \leq f(x_3)(x_1 - x_2) + f(x_1)(x_2 - x_3),$$

$$0 \leq \begin{vmatrix} x_1 & f(x_1) & 1 \\ x_2 & f(x_2) & 1 \\ x_3 & f(x_3) & 1 \end{vmatrix}.$$

(c) With the same notation as in (b), but only requiring $x_1 < x_3$,

$$\frac{f(x_1) - f(x_2)}{x_1 - x_2} \leq \frac{f(x_2) - f(x_3)}{x_2 - x_3}. \tag{3}$$

(d) With the same notation as in (b), but no order restriction on x_1, x_2, x_3,

$$\frac{f(x_1)}{(x_1 - x_2)(x_1 - x_3)} + \frac{f(x_2)}{(x_2 - x_3)(x_2 - x_1)} + \frac{f(x_3)}{(x_3 - x_1)(x_3 - x_2)} \geq 0. \tag{4}$$

(e) If x_1, x_2, y_1, y_2 are points of $[a, b]$ with $x_1 \leq y_1$, $x_2 \leq y_2$, and $x_1 \neq x_2$, $y_1 \neq y_2$ then

$$\frac{f(x_2) - f(x_1)}{x_2 - x_1} \leq \frac{f(y_2) - f(y_1)}{y_2 - y_1}. \tag{5}$$

COMMENTS (i) Since (a) is the definition of convexity, and of strict convexity, no proof is required. In addition the definition of a concave function, strictly concave function, is one for which (~1) holds, strictly, under the conditions stated.

(ii) If (1) is only required for $\lambda = \frac{1}{2}$ then the function is said to be *Jensen*, *J*-, or *mid-point convex*. Such a function satisfies (1) with $\lambda \in \mathbb{Q}$. If f is bounded above then mid-point convexity implies convexity; the same is true if we only require that f is bounded above by a Lebesgue integrable function.

(iii) The geometric interpretation of (1) is: that

the graph of a convex f lies below its chords; if f strictly convex is equivalent to being convex and the graph containing no straight line segments.

(iv) (b)—(e) follow from (a) by rewriting and notation changes.

(v) The second equivalent form of (2) says that the area of the triangle $P_1 P_2 P_3$ is non-negative, where $P_i = (x_i, f(x_i))$, $1 \leq i \leq 3$.

(vi) The interpretation of (5) is that for convex functions the chords to the graph have slopes that increase to the right; see **Starshaped Function Inequalities** COMMENTS (ii).

(vii) It is important to note that if f has a second derivative then f is convex if and only if $f'' \geq 0$, and if $f'' > 0$, except possibly at a finite number of points, then f is strictly convex. So the exponential function is strictly convex, as is x^α if $\alpha > 1$ or $\alpha < 0$; while the logarithmic function is strictly concave, as is x^α if $0 < \alpha < 1$.
In any case, using (vi), if f is convex both $f'_\pm$ exist.

(viii) An important special case of (5) is

$$f(x + z) - f(x) \leq f(y + z) - f(y), \tag{6}$$

when $x \leq y$ and $x, y, y + z \in [a, b]$.
Inequality (6) is sometimes called the *property of equal increasing increments*. A function that has this property is said to be *Wright convex*.

DERIVATIVE INEQUALITIES   (a) If f is convex on $[a,b]$ and if $a \leq x \leq y < b$ then $f'_+(x) \leq f'_+(y)$; if f is strictly convex and $x \neq y$ this inequality is strict.
(b) If f is convex on $[a,b]$ and if $a < x \leq y \leq b$ then $f'_-(x) \leq f'_-(y)$; if f is strictly convex and $x \neq y$ this inequality is strict.
(c) If f is convex on $[a,b]$ then on $]a,b[$ we have $f'_- \leq f'_+$; this inequality is strict if f is strictly convex.
(d) If f is convex on $[a,b]$ and if x,y are points of $[a,b[$ then

$$f(y) \geq f(x) + (y-x)f'_+(x),$$

the inequality being strict if f is strictly convex.

The definition of a convex function of several variables, or even of a function on a vector space, is just (1) with the obvious vector interpretation of the notation; in addition now the domain must be a convex set. If f is twice differentiable then it is convex if for all $\underline{x} \in G$, and all real $u_i, u_j, 1 \leq i, j \leq n$,

$$\sum_{i=1}^{n}\sum_{j=1}^{n} \frac{\partial^2 f(\underline{x})}{\partial x_i \partial x_j} u_i u_j \geq 0.$$

HIGHER DIMENSIONS   (a) If f is continuously differentiable on the convex domain G in $\mathbb{R}^n$ and if $\underline{a}, \underline{b} \in G$ then

$$f(\underline{b}) - f(\underline{a}) \geq \sum_{i=1}^{n} \frac{\partial f(\underline{a})}{\partial a_i}(b_i - a_i).$$

(b) If f is convex on $\mathbb{R}^n$, symmetric and homogeneous of the first degree then

$$f(\underline{a}) \geq f(1,\ldots,1)\mathfrak{A}_n(\underline{a}).$$

COMMENTS   (viii) Definitions of the terms used in the last result are to be found in **Segre's Inequalities**.
(x) An important extension of (1), obtained by an induction argument, is (J).
(xi) A very important generalization of the whole concept of convexity is *higher order convexity*; see n-**Convex Function Inequalities**.
(xii) See also **Alzer's Inequalities** (a), **Arithmetic Mean Inequalities** INTEGRAL ANALOGUES, **Askey-Karlin Inequalities**, **Bernšteĭn Polynomial Inequalities** (b), **Bennett's Inequalities** COMMENTS (i), **Binomial Function Inequalities** COMMENTS (i), **Brunn-Minkowski Inequalities** (a), **Cauchy's Inequality** CONVERSE INTEGRAL ANALOGUES, **Conjugate Convex Function Inequalities**, **Convex Function Integral Inequalities**, **Convex Matrix Function Inequalities**, **Gigamma Function Inequalities** (a), **Favard's Inequalities**, **Hermite-Hadamard Inequality**, **Hua's Inequality** EXTENSION, **Jensen's Inequality**, **Jensen-Steffensen Inequality**, **Log-convex Function Inequalities**, **Order Inequalities** (b), **Petrović's Inequality**, **Power Mean Inequalities** INTEGRAL ANALOGUES (c), **Q-class Function Inequalities** COMMENTS (ii), **Quasi-arithmetic Mean Inequalities**

COMMENTS (i), **Quasi-convex Function Inequalities, Rearrangement Inequalities** EXTENSIONS (a), **Schur Convex Function Inequalities, Strongly Convex Function Inequalities, Szegö's Inequality, Thunsdorff's Inequality, Ting's Inequality**.

REFERENCES [AI, pp. 10–26], [EM, vol. 2, pp. 415–416], [HLP, pp. 70–81, 91–96], [MI, pp. 21–33], [MFP, pp. 1–19], [PPT, pp. 1–14], [MO, pp. 445–462]; [Roberts & Varberg].

Convex Function Integral Inequalities

[SKORDEV] If f is non-negative and convex on $[0,1]$ and $n \in \mathbb{N}$ then

$$\frac{1}{(n+1)(n+2)} \int_0^1 f \le \int_0^1 t^n f(t)\,dt \le \frac{2}{(n+2)} \int_0^1 f.$$

COMMENTS See also **Hermite-Hadamard Inequality, Petschke's Inequality, Rado's Inequality, Rahmail's Inequality, Thunsdorff's Inequality, Ting's Inequality**.

REFERENCES [Mitrinović & Pečarić].

Convex Matrix Function Inequalities

If I is a interval in $\mathbb{R}$ then a function $f : I \to \mathbb{R}$ is said to be a *convex matrix function of order n* if for all $n \times n$ Hermitian matrices A, B with eigenvalues in I, and all $\lambda, 0 \le \lambda \le 1$,

$$\overline{1-\lambda} f(A) + \lambda f(B) \ge f(\overline{1-\lambda}\,A + \lambda B);$$

where $A \ge B$ means that $A - B$ is positive semi-definite.

COMMENTS (i) $x^\alpha, 1 < \alpha \le 2$, is convex matrix function of all orders on $[0, \infty[$; and $-\sqrt{x}$, $x^\alpha, -1 \le \alpha < 0$, are convex matrix function of all orders on $]0, \infty[$.
(ii) A convex function of all orders on I is said to be *operator convex on I*.

[DAVIS] If f is a convex matrix function of order n on I and if A is an $n \times n$ Hermitian matrix with eigenvalues in I then

$$f(A)_{i_1,\ldots,i_m} \ge f(A_{i_1,\ldots,i_m}).$$

REFERENCES [MO, pp. 259–262]; [Roberts & Varberg, pp. 259–261]; [Chollet].

Convex Sequences of Higher Order Inequalities
See n-Convex Sequence Inequalities.

Convex Sequence Inequalities

(a) If $\underline{a}$ is a real convex sequence then

$$\Delta^2 \underline{a} \ge 0, \tag{1}$$

and $\Delta \underline{a}$ decreases.
(b) If $\underline{a}$ is a real bounded convex sequence then

$$\Delta \underline{a} \ge 0, \text{ and } \lim_{n\to\infty} n\Delta a_n = 0.$$

(c) If $\underline{a}$ is a real convex sequence and $\underline{b} = \{\frac{a_2}{1}, \frac{a_3}{2}, \ldots\}$ then

$$\Delta \underline{b} \geq 0.$$

(d) [OZEKI] If $\underline{a}$ is real convex sequence so is $\mathfrak{A}(\underline{a})$.
(e) If $\underline{a}, \underline{b}$ are positive convex sequences with $a_2 \geq a_1, b_2 \geq b_1$ then $\underline{c}$ is a convex where

$$c_n = \frac{1}{n} \sum_{i=1}^{n} a_i b_{n+1-i}.$$

COMMENTS (i) (1) is just the definition of a *convex sequence*.
(ii) If f is a mid-point convex function and $a_i = f(i)$, $i = 1, 2, \ldots$ then $\underline{a}$ is a convex sequence.
(iii) If (1) holds strictly we say that the sequence is *strictly convex*; while if ($\sim$1) hold, strictly, we say that the *sequence is concave, strictly concave*.
(iv) A definition of *higher order convex sequences* can be given; see **n-Convex Sequence Inequalities**, where a generalization of (d) can be found.
(v) See also **Barnes's Inequalities** (a), **Čebišev's Inequality** EXTENSIONS (d), **Log-convex Sequence Inequalities**, **Haber's Inequality** EXTENSIONS, **Nanson's Inequality**, **Thunsdorff's Inequality** DISCRETE ANALOGUE.

REFERENCES [AI, p. 202], [EM, vol. 2, pp. 419–420], [MI, pp. 8–9], [PPT, pp. 6, 277, 289]; [Pečarić, 1987, p. 165].

Convolution Inequalities
See **Young's Convolution Inequality**.

Copson's Inequality
If $\underline{a}$ is a non-negative sequence and if $p > 1$ then,

$$\sum_{n=1}^{\infty} \left(\sum_{k \geq n} \frac{a_k}{k} \right)^p \leq p^p \sum_{n=1}^{\infty} a_n^p. \tag{1}$$

There is equality if and only if $\underline{a} = \underline{0}$.
If $0 < p < 1$ then ($\sim$ 1) holds.

COMMENTS (i) This result can be deduced, using **Hölder's Inequality** OTHER FORMS (c) and **Hardy's Inequality** (1).

EXTENSIONS [BENNETT] (a) If $\underline{a}$ is a real sequence, and if $p > 1$ then

$$!\underline{a}!_p \leq \left(\sum_{n=1}^{\infty} \left(\sum_{k \geq n} \frac{a_k}{k} \right)^p \right)^{1/p} \leq p\,!\underline{a}!_p.$$

There is equality on the left if $\underline{a}$ has at most one non-zero entry, and on the right only when $\underline{a} = \underline{0}$.
(b) If $\underline{a}$ is a positive sequence, and if $\alpha > 0, 0 < p < 1$ then

$$\sum_{n\in\mathbb{N}} \left(\sum_{i=1}^{n} \frac{\binom{i+\alpha-1-n}{i-n} a_i}{\binom{i+\alpha}{i}} \right)^p \ge \left(\frac{\alpha!(p^{-1}-1)!}{(p^{-1}+\alpha-1)!} \right)^p \sum_{n\in\mathbb{N}} a_n^p.$$

COMMENTS   (ii) The notation in (a) is explained before **Hardy's Inequality** EXTENSIONS (c).
(iii) The inequality in (b) reduces to ($\sim$1) on putting $\alpha = 1$. The result should be compared with **Knopp's Inequalities** (a).
(iv) See also **Bennett's Inequalities** (d) **Hardy-Littlewood-Pólya Inequalities** DISCRETE ANALOGUES; and for other inequalities by Copson see **Hardy's Inequalities** COMMENTS (iv).
REFERENCES   [HLP, pp. 246–247]; [Bennett, 1996, pp. 25–28, 38–39], [Grosse-Erdmann].

Copula Inequalities   If $c : S = [0,1] \times [0,1] \to [0,1]$ is a copula then:
(a) if $\underline{x}_i = (x_i, y_i) \in S, i = 1, 2$ with $0 \le x_1 \le x_2 \le 1, 0 \le y_1 \le y_2 \le 1$,

$$c(x_1, y_1) - c(x_1, y_2) - c(x_2, y_1) + c(x_2, y_2) \ge 0; \tag{1}$$

(b) for all $\underline{x}_i \in S, i = 1, 2$,

$$|c(\underline{x}_1) - c(\underline{x}_2)| \le |x_1 - x_2| - |y_1 - y_2|;$$

(c) for all $\underline{x} = (x, y) \in S$,

$$\max\{x + y - 1, 0\} \le c(\underline{x}) \le \min\{x, y\}.$$

COMMENTS   (i) A *copula* is a function such as c that satisfies (1) and

$$c(0, a) = c(a, 0), \qquad c(1, a) = c(a, 1) = a, \qquad 0 \le a \le 1. \tag{2}$$

(ii) The inequalities in (b) and (c) are easily deduced from (1) and (2).
(iii) For a multiplicative analogue of this see **Totally Positive Function Inequalities**.
REFERENCES   [EM, Supp., pp. 199–201]; [General Inequalities, vol. 1, pp. 133–149; vol. 4, p. 397].

Cordes's Inequality   If A, B are positive bounded linear operators on the Hilbert space X and if $0 \le \alpha \le 1$ then

$$\|A^\alpha B^\alpha\| \le \|AB\|^\alpha.$$

COMMENTS   This inequality is equivalent to the Heinz-Kato Inequality, and to the **Löwner-Heinz Inequality**.

REFERENCES [EM, Supp., p. 289].

Correlation Inequalities
See **Ahlswede-Daykin Inequality, FKG Inequality, Holley's Inequality**.

Counter Harmonic Mean Inequalities
If $-\infty < p < \infty$ then p counter-harmonic mean, of order n, of the positive sequence $\underline{a}$ with positive weight $\underline{w}$ is

$$\mathfrak{H}_n^{[p]}(\underline{a};\underline{w}) = \frac{\sum_{i=1}^n w_i a_i^p}{\sum_{i=1}^n w_i a_i^{p-1}}.$$

This definition is completed by defining

$$\mathfrak{H}_n^{[-\infty]}(\underline{a};\underline{w}) = \min \underline{a}; \quad \mathfrak{H}_n^{[\infty]}(\underline{a};\underline{w}) = \max \underline{a}.$$

These are justified as limits of the previous definition.
In particular,

$$\mathfrak{H}_n^{[0]}(\underline{a};\underline{w}) = \mathfrak{H}_n(\underline{a};\underline{w}); \quad \mathfrak{H}_n^{[1]}(\underline{a};\underline{w}) = \mathfrak{A}_n(\underline{a};\underline{w}).$$

(a) If $1 \leq r \leq \infty$ then

$$\mathfrak{H}_n^{[r]}(\underline{a};\underline{w}) \geq \mathfrak{M}_n^{[r]}(\underline{a};\underline{w}), \quad (1)$$

while if $-\infty \leq r \leq 1$ inequality ($\sim$1) holds. The inequalities are strict unless $r = \pm\infty, 1$ or $\underline{a}$ is constant.
(b) If $-\infty \leq r \leq 0$ then

$$\mathfrak{H}_n^{[r]}(\underline{a};\underline{w}) \leq \mathfrak{M}_n^{[r+1]}(\underline{a};\underline{w}). \quad (2)$$

Inequality (2) is strict unless $r = -\infty$ or $\underline{a}$ is constant.
(c) If $-\infty \leq r < s \leq \infty$ then

$$\mathfrak{H}_n^{[r]}(\underline{a};\underline{w}) \geq \mathfrak{H}_n^{[s]}(\underline{a};\underline{w}), \quad (3)$$

with equality only if $\underline{a}$ is constant.
(d) [BECKENBACH] If $1 \leq r \leq 2$ then

$$\mathfrak{H}_n^{[r]}(\underline{a}+\underline{b};\underline{w}) \leq \mathfrak{H}_n^{[r]}(\underline{a};\underline{w}) + \mathfrak{H}_n^{[r]}(\underline{b};\underline{w}). \quad (4)$$

If $0 \leq r \leq 1$ inequality ($\sim$4) holds. Inequality (4) is strict unless $r = 1$ or $\underline{a} \sim \underline{b}$.

COMMENTS (i) In (a) and (b) the extreme values of the parameter r are trivial and in the other cases follow by an application of simple algebra and (r;s).
(ii) In (c) the inequalities for the extreme values of the parameters are trivial; the other cases follow from the convexity properties of the power means, see **Power Mean Inequalities** (d), (e).
(iii) (d) follows by an application of **Radon's Inequality** and (M).
(iv) The counter harmonic means have been generalized by both Gini and Bonferroni, see **Gini-Dresher Mean Inequalities**.

REFERENCES [BB, pp. 27–28], [MI, pp. 185–190], [MPF, pp. 156–163], [PPT, pp. 122–124].

Cyclic Inequalities (a) [DAYKIN] If the positive n-tuple $\underline{a}$ is extended to a sequence by defining $a_{n+r} = a_r, r \in \mathbb{N}$, then

$$\sum_{i=1}^{n} \frac{a_i + a_{i+2}}{a_i + a_{i+1}} \geq n\gamma(n),$$

where

$$\gamma(n) > \frac{n+1}{2n}.$$

(b) [ALZER] If $\underline{a}$ is a positive n-tuple and if $a_{n+1} = a_1$ then

$$\frac{\mathfrak{A}_n(\underline{a})}{\mathfrak{G}_n(\underline{a})} \geq \frac{1}{n}\sum_{i=1}^{n} \frac{a_i}{a_{i+1}} \geq \left(\sum_{i=1}^{n} \frac{a_i}{a_{i+1}} - (n-1)\right)^{1/n} \geq 1,$$

with equality if and only if $a_1 = \cdots = a_n$.

COMMENTS   (i) It has been shown by Elbert that $\lim_{n\to\infty} \gamma(n)$ exists, γ, say; and if $n \geq 3$,

$$\gamma(n) \geq \gamma = .978912\ldots.$$

EXTENSIONS [KOVAĆEC]   Let $s : I^2 \to \mathbb{R}$, where $I \subseteq \mathbb{R}$ is an interval, have the following properties: (i) $\sigma(x) = s(x,x)$ is increasing on I, (ii) $\tau(x) = s(a,x) + s(x,b)$ is increasing on $[\max\{a,b\},\infty[\cap I$. If then $\underline{a}$ is an n-tuple with elements in I, and if $a = \min \underline{a}$, then

$$\sum_{i=1}^{n} s(a_i, a_{i+1}) \geq ns(a,a),$$

where $a_{n+1} = a_1$ and

COMMENTS   (ii) A particular case of this last result, $I =]0, \infty[$, $s(x,y) = x/y$, is **Korovkin's Inequality**.
(iii) See also **Elementary Symmetric Function Inequalities** COMMENTS (vi), **Schur's Inequality**, **Shapiro's Inequality**. In addition many cyclic inequalities are given in the second reference.

REFERENCES   [AI, pp. 132–138], [MFP, pp. 407–471]; [General Inequalities, vol. 4, pp. 327–41]; [Alzer, 1990].

- end of C -

D

Davies & Petersen's Inequality If $\underline{a}$ is a non-negative sequence, and $p \geq 1$ then
$$A_n^p \leq p \sum_{k=1}^{n} a_k A_k^{p-1}.$$

COMMENTS   See also **Pachpatte's Series Inequalities**.
REFERENCES   [Davies & Petersen].

Davis's Inequality
See **Martingale Inequalities** COMMENTS (ii).

Derivative Inequalities   (a) If u, v are differentiable on $]a, b[$ and such that for any $c, a < c < b, u(c) = v(c)$ implies $u'(c) < v'(c)$, and if $u(a+) < v(a+)$ then
$$u < v.$$
(b) If f is differentiable on $[a, b]$ and if for all points $x \in A, A \subseteq [a, b], |f'(x)| \leq M$ then
$$|f[A]| \leq M|A|.$$

COMMENTS   (i) (a) has many extensions and important applications; see [Walter]. In particular we can drop the condition of differentiability, and replace the first inequality condition by either $u'_+(c) < v'_+(c)$, or $u'_-(c) < v'_-(c)$.
(ii) The result in (b) can be generalized by dropping the differentiability condition and replacing "for all points $x \in A, A \subseteq [a, b], |f'(x)| \leq M$" by "for all points $x \in A, A \subseteq [a, b], |\overline{f}'_+(x)| \leq M$, and $|\underline{f}'_-(x)| \geq -M$". A completely different generalization is in [Zygmund].

EXTENSIONS   If u, v are continuous on $[a, b]$, differentiable on $]a, b[$, and such that for some function f,
$$u'(x) - f(x, (u(x)) < v'(x) - f(x, v(x)), \quad a < x < b \quad \text{and if} \quad u(a) < v(a)$$
then
$$u(x) < v(x), \quad a < x < b.$$

COMMENTS   (iii) There are many other inequalities involving derivatives; see in particular **Alzer's Inequalities** (a), EXTENSIONS, **Hardy-Littlewood-Landau Derivative Inequalities**, **Integral Inequalities** (c), (d), **Mean Value Theorem of Differential Calculus**.
REFERENCES   [Saks, pp. 226–227], [Walter, pp. 54–57], [Zygmund, vol. II, pp. 88–89].

Descartes' Rule of Signs If r is the number of positive roots of the polynomial $p(x) = a_0 + \cdots a_n x^n$, where the coefficients are all real, and if v the number of sign variations in $\{a_n, \ldots, a_0\}$ then
$$0 \leq r \leq v;$$
more precisely $v - r$ is always an even non-negative integer.

COMMENTS In calculating v all zeros in the coefficients are ignored, and in calculating r multiple roots are counted according to their muliplicity.

REFERENCES [EM, vol. 3, p. 59].

Determinant Inequalities (a) [KY FAN] If A, B are real positive definite matrices and $0 \leq \lambda \leq 1$, then
$$\det\big((1-\lambda)A + \lambda B\big) \geq (\det A)^{1-\lambda} \det B^\lambda.$$

(b) [MINKOWSKI] If A, B are non-negative $n \times n$ Hermitian matrices then
$$(\det A)^{1/n} + (\det B)^{1/n} \leq (\det(A+B))^{1/n}; \tag{1}$$
and hence,
$$\det A + \det B \leq \det(A+B).$$

(c) [BERGSTRÖM] If A, B are real positive definite matrices then
$$\frac{\det(A+B)}{\det(A'_i + B'_i)} \geq \frac{\det(A)}{\det(A'_i)} + \frac{\det(B)}{\det(B'_i)}.$$

(d) If A is a real positive definite $n \times n$ matrix then
$$|\det A| \leq |\det A_{1,2,\ldots k}||\det A_{k+1,\ldots,n}|.$$

In particular
$$|\det A| \leq \left|\prod_{i=1}^{n} a_{ii}\right|.$$

EXTENSIONS (a) [KY FAN] (i) If A, B are positive definite $n \times n$-matrices and if $|A|_k$ is the product of the first k smallest eigenvalues and if $0 \leq \lambda \leq 1$ then
$$|\lambda A + (1-\lambda)B|_k \leq |A|_k^\lambda |B|_k^{1-\lambda}.$$

(ii) If A, B are real symmetric $n \times n$-matrices and if $_k|A|$ is the sum of the first k largest eigenvalues and if $0 \leq \lambda \leq 1$ then
$$_k|\lambda A + (1-\lambda)B| \geq {}_k|A|^\lambda {}_k|B|^{1-\lambda}.$$

(b) [OPPENHEIM] *Under same assumptions as in (a),*

$$|A+B|_k^{1/k} \leq |A|_k^{1/k} + |B|_k^{1/k}.$$

COMMENTS (i) Many variants and extensions can be found in [MPF].
(ii) See also **Circulant Matrix Inequalities, Gram Determinant Inequalities, Hadamard's Determinant Inequality, Permanent Inequalities**.

REFERENCES [BB, pp. 63–64, 74–75], [EM, vol.6, p. 248], [HLP, p. 16], [MPF, pp. 211–238]; [Marcus & Minc, pp. 115,117].

Difference Means of Gini

If $m > 1$, and $\underline{a}$ is a positive m-tuple, the *Gini difference mean* of $\underline{a}$ is,

$$\mathfrak{D}_m(\underline{a}) = \frac{2}{m(m-1)} \sum_{\substack{i,j=1 \\ 1 \leq i < j \leq m}}^m |a_i - a_j|.$$

If then $\underline{b}$ is a positive n-tuple, $n > 1$, the *Gini mixed difference mean* of $\underline{a}$ and $\underline{b}$ is,

$$\mathfrak{D}_{m,n}(\underline{a},\underline{b}) = \frac{1}{mn} \sum_{i=1}^m \sum_{j=1}^n |a_i - b_j|.$$

ANAND-ZAGIER INEQUALITY *If $\underline{a}$ is a positive m-tuple, $\underline{b}$ is a positive n-tuple, with $m,n > 1$, then,*

$$\frac{\mathfrak{D}_{m,n}(\underline{a},\underline{b})}{\mathfrak{A}_m(\underline{a})\mathfrak{A}_n(\underline{b})} \geq \frac{1}{4}\left(\frac{(m-1)\mathfrak{D}_m(\underline{a})}{m\mathfrak{A}_m^2(\underline{a})} + \frac{(n-1)\mathfrak{D}_n(\underline{b})}{n\mathfrak{A}_n^2(\underline{b})}\right). \quad (1)$$

COMMENTS (i) This inequality was conjectured by Anand and proved by Zagier.
(ii) The quantity $\phi_{11}(\underline{a}) = (n-1)\mathfrak{D}_n(\underline{a})/4n\mathfrak{A}(\underline{a})$ was suggested by Gini as a measure of inequality, the *Gini coefficient*.
(iii) It is easy to see that

$$\phi_{11}(\underline{a}) = 1 - \frac{\sum_{i,j=1}^n \min\{a_i,a_j\}}{n^2\mathfrak{A}_n(\underline{a})} = 1 + \frac{1}{n} - \frac{2\sum_{i=1}^n ia_{[i]}}{n^2\mathfrak{A}_n(\underline{a})}.$$

(iv) As a result of the previous comment it is seen that (1) is equivalent to

$$2\frac{\sum_{i,j=1}^n \min\{a_i,b_j\}}{\mathfrak{A}_n(\underline{a})\mathfrak{A}_n(\underline{b})} \leq \frac{\sum_{i,j=1}^n \min\{a_i,a_j\}}{\mathfrak{A}_n(\underline{a})} + \frac{\sum_{i,j=1}^n \min\{b_i,b_j\}}{\mathfrak{A}_n(\underline{b})}.$$

COMMENTS An integral analogue for these means can be found in **Mulholland's Inequality**.

REFERENCES [EM, vol. 4, p. 277], [MO, p. 411], [MPF, pp. 584–586].

Digamma Function Inequalities

The *digamma* or *psi function* is defined as:

$$\mathcal{F}(z) = \psi(z+1) = (\log z!)'.$$

The derivatives of this function are called the the *multigamma functions*; in particular the *trigamma function*, *tetragamma function*, etc.

(a) The reciprocal of the trigamma function is convex, or equivalently
$$\psi''^2 \geq \frac{\psi'\psi'''}{2}.$$

(b)
$$\sum_{k \in \mathbb{N}} \frac{1}{(z+k)^3} > \frac{\sqrt{3}}{2} \left(\sum_{k \in \mathbb{N}} \frac{1}{(z+k)^2} \right)^{1/2} \left(\sum_{k \in \mathbb{N}} \frac{1}{(z+k)^4} \right)^{1/2}.$$

COMMENTS The second inequality is a deduction from the first. It is form of ($\sim$C).

REFERENCES [Abramowitz & Stegun, pp. 258-260], [Jeffreys & Jeffreys, pp. 465–466]; [Trimble et al].

Dirichlet Kernel Inequalities

The quantity
$$D_n(x) = \frac{1}{2} + \sum_{k=1}^{n} \cos kx = \frac{1}{2} \sum_{k=-n}^{n} e^{ikx} = \frac{\sin(n+1/2)x}{2\sin x/2}, \quad n \in \mathbb{N},$$
is called the *Dirichlet kernel*, or sometimes the *Dirichlet summation kernel*; and
$$D_n^*(x) = D_n(x) - \frac{1}{2}\cos nx = \frac{\sin nx}{\tan x/2}, \quad n \in \mathbb{N},$$
the *modified Dirichlet kernel*.

(a) If $n \in \mathbb{N}$ then
$$|D_n^*(x)| \leq n.$$

(b) If $n \geq 1$ and $0 < t \leq \pi$ then
$$|D_n^*(x)| \leq \frac{1}{t}.$$

(c) If $0 \leq x \leq \pi$ then
$$|D_n^{(r)}(x)| \leq \frac{Cn^{r+1}}{1+nx}.$$

(d) [MAKAI] If $n \geq 1$ and $k, m \in \mathbb{N}$ then
$$\sum_{i=0}^{m} D_k\left(\frac{2\pi}{n}i\right) > 0;$$
in particular if $0 \leq k < n, 1 \leq m < n$,
$$\sum_{i=1}^{m} D_k\left(\frac{2\pi}{n}i\right) < \min\left\{\frac{n}{2}, \frac{n-k+m}{2}\right\}.$$

COMMENTS The Dirichlet kernel has analogues in more general forms of convergence. These more general kernels have inequalities that are extensions of the inequalities above. For the case $(C, 1)$-summability see **Fejér Kernel Inequalities**, and for Abel summability see **Poisson Kernel Inequalities**.

REFERENCES [AI, p. 248], [MPF, p. 584]; [Zygmund, vol. I, pp. 49–51, 94–95; vol. II, p. 60].

Distortion Theorems [KOEBE] If f is univalent in D, with $f(0) = 0, f'(0) = 1$, then

$$\frac{|z|}{(1+|z|)^2} \leq |f(z)| \leq \frac{|z|}{(1-|z|)^2};$$

$$\frac{1-|z|}{(1+|z|)^3} \leq |f'(z)| \leq \frac{1+|z|}{(1-|z|)^3}.$$

There is equality if and only if f is a rotation of the Koebe function.

COMMENTS (i) The *Koebe function* is $f(z) = \sum_{n=1}^{\infty} nz^n = z/(1-z)^2$, and if $\alpha \in \mathbb{R}$, $a(z) = z/(1 - e^{i\alpha}z)^2$ is a *rotation of the Koebe function*.
(ii) These results give estimates for the distortion at a point under a conformal map.

EXTENSIONS If K is a compact subset of the region $\Omega \subseteq \mathbb{C}$, then there is a constant M, depending on K, such that for all f, univalent on Ω, and all $z, w \in K$,

$$\frac{1}{M} \leq \frac{|f'(z)|}{|f'(w)|} \leq M.$$

COMMENTS (iii) See also **Rotation Theorems**.

REFERENCES [EM, vol. 3, pp. 269–271]; [Ahlfors, 1973, pp. 84–85], [Conway, vol. II, pp. 65–70].

Dočev's Inequality
See **Geometric-Arithmetic Mean Inequality**, CONVERSE INEQUALITIES (b).

Doob's Upcrossing Inequality
If $\mathcal{X} = (X_n, \mathcal{F}_n, n \in \mathbb{N})$ is a sub-martingale, and if $\beta_m(a, b)$ is the number of upcrossings of $[a, b]$ by $\mathcal{X}$ in m steps then

$$E\beta_m(a, b) \leq \frac{E|X_m| + |a|}{b - a}.$$

COMMENTS For another inequality by Doob see **Martingale Inequalities** (a).
REFERENCES [EM, vol. 6, p. 110]; [Loève, p. 532].

Doubly Stochastic Inequalities
See **Order Inequalities** (a), **van der Waerden's Conjecture**.

Dresher's Inequality
See **Gini-Dresher Mean Inequalities**.

Duff's Inequality
If $p > 0$ and f differentiable almost everywhere on $[0, a]$ then

$$\int_0^a |(f')^*|^p \leq \int_0^a |f'|^p.$$

COMMENTS For the definition of decreasing rearrangement of a function see **Notations 5**.

EXTENSIONS [MITRINOVIĆ PEČARIĆ] If $p > 0$ and f differentiable almost everywhere on $[0, a]$ and H a non-decreasing function then

$$\int_0^a H \circ |(f')^*| \le \int_0^a |H \circ |f'|.$$

REFERENCES [Duff], [Mitrinović & Pečarić, 1989].

Dunkl & Williams' Inequality
If X is a unitary space, $x \ne 0, y \ne 0$ then

$$||x - y|| \ge \left(\frac{||x|| + ||y||}{2}\right) \left|\left|\frac{x}{||x||} - \frac{y}{||y||}\right|\right|,$$

with equality if and only if either $||x|| = ||y||$, or $||x|| + ||y|| = ||x - y||$.

COMMENTS (i) For the definition of a unitary space see **Inner Product Space Inequalities** COMMENTS (i).
(ii) This is related to (T).
(iii) It appears to be an open question as to whether this inequality characterizes inner product spaces.

REFERENCES [MPF, pp. 515–519].

- end of D -

E

Efron-Stein Inequality If $X_i, 1 \leq i \leq n+1$, are independent identically distributed random variables and S a symmetric function then,

$$\sigma^2 S(X_1, \ldots, X_n) \leq EQ;$$

where, writing $\tilde{S} = \frac{1}{n+1} \sum_{i=1}^{n+1} S(X, \ldots, X_i, X_{i+1}, \ldots, X_{n+1})$,
$Q = \sum_{i=1}^{n+1} \left(S(X, \ldots, X_i, X_{i+1}, \ldots, X_{n+1}) - \tilde{S} \right)^2$.

COMMENTS   A definition of symmetric functions is given in **Segre's Inequalities**.

REFERENCES   [Tong, ed., pp. 112–114].

Eigenvalue Inequalities   (a) [SCHUR] If $A = \{a_{ij}\}$ is an $n \times n$ complex matrix then

$$\sum_{i=1}^{n} |\lambda_i(A)|^2 \leq \sum_{i,j=1}^{n} |a_{ij}|^2, \tag{1}$$

with equality if and only if A is normal.

(b) If A, B are two non-negative $n \times n$ matrices, such that $B - A$ is also non-negative then,

$$|\lambda_{[1]}(A)| \leq |\lambda_{[1]}(B)|.$$

(c) [CAUCHY] If A is a Hermitian $n \times n$ matrix then for $1 \leq s \leq n - r$,

$$\lambda_{[s+r]}(A) \leq \lambda_{[s]}(A_{i_1,\ldots,i_{n-r}}) \leq \lambda_{[s]}(A).$$

(d) If A, B are non-negative $n \times n$ matrices and if C is their Hadamard product then

$$\lambda_{[1]}(C) \leq \lambda_{[1]}(A)\lambda_{[1]}(B).$$

(e) If A is an $n \times n$ complex matrix then

$$|\lambda_{[1]}(I + A)| \leq 1 + \lambda_{[1]}(A);$$

there is equality if A is non-negative.

(f) [KOMAROFF] If A, B are Hermitian matrices, A positive semi-definite, then for each $m, 1 \leq m \leq n$

$$\sum_{k=1}^{m} \lambda_{(n-k+1)}(AB) \leq \sum_{k=1}^{m} \lambda_{(k)}(A)\lambda_{(k)}(B);$$

$$\sum_{k=1}^{m} \lambda_{(k)}(A)\lambda_{(n-k+1)}(B) \leq \sum_{k=1}^{m} \lambda_{(k)}(AB).$$

COMMENTS (i) Schur's result has been used to prove (GA). A is a *normal matrix* if it is Hermitian and $AA^* = A^*A$.
(ii) $\rho(A) = |\lambda_{[1]}(A)|$ is called the *spectral radius* of A.
(iii) The *Hadamard product* of the matrices A, B is the matrix $C = (a_{ij}b_{ij})_{\substack{1 \le i \le n \\ 1 \le j \le n}}$.
(iv) See also **Determinant Inequalities** EXTENSIONS, **Hirsch's Inequalities**, **Rayleigh-Ritz Ratio**, **Weyl's Inequality**.

REFERENCES [MI, p. 78]; [Marcus & Minc, pp. 119, 126, 142], [Horn & Johnson, pp. 491,507]; [Komaroff].

Elementary Symmetric Function Inequalities (a) If $1 \le r \le n-1$ then

$$e_n^{[r-1]}(\underline{a})e_n^{[r+1]}(\underline{a}) < \left(e_n^{[r]}(\underline{a})\right)^2; \tag{1}$$

$$p_n^{[r-1]}(\underline{a})p_n^{[r+1]}(\underline{a}) \le \left(p_n^{[r]}(\underline{a})\right)^2; \tag{2}$$

with equality in (2) if and only if $\underline{a}$ is a constant.
(b) If $1 \le r < s \le n$ then,

$$e_n^{[r-1]}(\underline{a})e_n^{[s]}(\underline{a}) < e_n^{[r]}(\underline{a})e_n^{[s-1]}(\underline{a});$$

$$p_n^{[r-1]}(\underline{a})p_n^{[s]}(\underline{a}) \le p_n^{[r]}(\underline{a})p_n^{[s-1]}(\underline{a});$$

with equality in the second if and only if $\underline{a}$ is constant.
(c) If $1 \le r \le n-1$ then,

$$e_n^{[r-1]}(\underline{a}) > e_n^{[r]}(\underline{a}) \quad \Longrightarrow \quad e_n^{[r]}(\underline{a}) > e_n^{[r+1]}(\underline{a});$$

$$p_n^{[r-1]}(\underline{a}) > p_n^{[r]}(\underline{a}) \quad \Longrightarrow \quad p_n^{[r]}(\underline{a}) > p_n^{[r+1]}(\underline{a}).$$

(d) If $1 \le r+s \le n$ then,

$$p_n^{[r+s]}(\underline{a}) \le p_n^{[r]}(\underline{a})p_n^{[s]}(\underline{a}),$$

with equality if and only if $\underline{a}$ is constant.

COMMENTS (i) Inequality (1) is an easy consequence of (2). In fact (2) gives more,

$$e_n^{[r-1]}(\underline{a})e_n^{[r+1]}(\underline{a}) \le \frac{r(n-r)}{(r+1)(n-r+1)}\left(e_n^{[r]}(\underline{a})\right)^2.$$

A similar result to this last inequality is EXTENSION (e) below.
An inductive proof of (1) can be given; both (1) and (2) follow from **Newton's Inequalities** (1). Note that (2) in the case $n=2$ is just (GA).
(ii) (b), (c) are immediate consequences of (1) and (2), and (d) follows from (2).
(iii) Extensions of these "quadratic" inequalities to "cubic" inequalities have been given by Rosset.

EXTENSIONS   (a) [LOG-CONCAVITY] If $1 \leq t < r < s \leq n$ then

$$p_n^{[r]}(\underline{a}) \geq \left(p_n^{[t]}(\underline{a})\right)^{(s-r)/(s-t)} \left(p_n^{[s]}(\underline{a})\right)^{(r-t)/(s-t)}.$$

There is equality if and only if $\underline{a}$ is constant.

(b) [POPOVICIU-TYPE] If $1 \leq r < s \leq n$, and if $0 < p \leq q$ then

$$\left(\frac{e_n^{[r]}(\underline{a})}{e_{n+1}^{[r]}(\underline{a})}\right)^p \geq \left(\frac{e_n^{[s]}(\underline{a})}{e_{n+1}^{[s]}(\underline{a})}\right)^q.$$

(c) [OBRECHKOFF]

$$\frac{\left(3e_n^{[3]}(\underline{a}) - e_n^{[1]}(\underline{a})e_n^{[2]}(\underline{a})\right)^2}{2\left((e_n^{[1]}(\underline{a}))^2 - 2e_n^{[2]}(\underline{a})\right)\left(2(e_n^{[2]}(\underline{a}))^2 - 3e_n^{[1]}(\underline{a})e_n^{[3]}(\underline{a})\right)} \leq \frac{n-1}{n},$$

with equality if and only if $\underline{a}$ is constant.

(d)

$$e_n^{[r]} > 0, 1 \leq r \leq n \iff a_r > 0, 1 \leq r \leq n.$$

(e) [JECKLIN] If $1 \leq 2r \leq n$ then

$$\sum_{i=1}^{n}(-1)^{i+1}e_{n-i}^{[r-1]}(\underline{a})e_{n+i}^{[r+1]}(\underline{a}) \leq \frac{\binom{n}{r}-1}{2\binom{n}{r}}\left(e_n^{[r]}(\underline{a})\right)^2.$$

(f) [OZEKI] If $\underline{a}$ is a positive log-convex sequence then so is $p_n^{[r]}(\underline{a})$, that is

$$p_{n-1}^{[r]}(\underline{a})p_{n+1}^{[r]}(\underline{a}) \geq \left(p_n^{[r]}(\underline{a})\right)^2, 1 \leq k \leq n-1.$$

(g) [LIN & TRUDINGER]

$$e_n^{[r-1]}(\underline{a}_i') \leq C_{n,r}e_n^{[r-1]}(\underline{a}).$$

COMMENTS   (iv) The log-concavity result is an easy induction from (2); see [Ku et al].

(v) A series of elementary inequalities in the case $n = 3$ have been given by Klamkin and can be found in [MPF].

(vi) Elementary symmetric function inequalities can be regarded as examples of **Cyclic Inequalities**.

(vii) $e_n^{[r]}$ is Schur concave, strictly so if $r > 1$; for the definition see **Schur Convex Function Inequalities**.

(viii) The constant in the inequality in (g) is estimated in the reference.

(ix) See also **Marcus & Lopez's Inequality, Symmetric Mean Inequalities**.

REFERENCES   [AI, pp. 95–107, 211–212, 336–337], [HLP, pp. 51–55, 104–105], [MI, pp. 283–290, 303–305], [MO, pp. 78–81], [MPF, pp. 163–164]; [Ku, Ku & Zhang], [Lin & Trudinger], [Rosset].

Entire Function Inequalities (a) [HADAMARD] If f is an entire function of finite order λ, and if γ is the genus of f, then,

$$\gamma \leq \lambda \leq \gamma + 1.$$

(b) If f is an entire function of order at most σ then

$$\sup_{x \in \mathbb{R}} |f^{(r)}(x)| \leq \sigma^r \sup_{x \in \mathbb{R}} |f(x)|, \quad r \in \mathbb{N}.$$

(c) If f is an entire function of finite order λ and if $n(r)$ is the number of zeros of f in $\{z;\ |z| \leq r\}$ then for all $\epsilon > 0$ there is a constant K such that

$$n(r) < K r^{\lambda + \epsilon}.$$

COMMENTS (i) The order, λ, of an entire function f is the infimum of the μ such that

$$\max_{|z| \leq r} |f(z)| < e^{r^\mu}, \quad r > r_0.$$

(ii) The *genus* of an entire function f is the smallest integer γ such that f can be represented in the form

$$f(z) = z^m e^{g(z)} \prod_n \left(1 - \frac{z}{a_n}\right) e^{z/a_n + (1/2)(z/a_n)^2 + \cdots + (1/\gamma)(z/a_n)^\gamma},$$

where g is a polynomial of degree at most γ; if there is no such representation then $\gamma = \infty$.
(iii) A special case of (b) is **Bernšteĭn's Inequality** (a).

REFERENCES [EM, vol. 1, p. 366; vol. 3, p. 385–387]; [Ahlfors,1966, pp. 194, 206–210], [Titchmarsh, 1939, p. 249].

Entropy Inequalities If $(X, \mathcal{A}, P)$ is a finite probability space with $\mathcal{B} \subseteq \mathcal{A}$ and H the entropy function, then

$$H(\mathcal{A}|\mathcal{B}) \leq H(\mathcal{A}); \tag{1}$$
$$H(\mathcal{A}\mathcal{B}) \leq H(\mathcal{A}) + H(\mathcal{B}).$$

COMMENTS (i) The entropy function H is defined in **Shannon's Inequality** COMMENTS (i). In the present context

$$H(\mathcal{A}) = -\sum_{A \in \mathcal{A}} p(A) \log_2 p(A).$$

(ii) The main results in this area are due to Hinčin[1], who called inequality (1) *Shannon's fundamental inequality*.
(iii) See also **Shannon's Inequality**.

REFERENCES [EM, vol. 3, pp. 387–388], [MPF, pp. 646–648]; [General Inequalities, vol. 2, pp. 435–445; vol. 5, pp. 411–417], [Tong, ed., pp. 68–77].

---

[1] А Я Хинчин. Also transliterated as Khintchine.

Enveloping Series Inequalities

If $A \in \mathbb{C}$ and if the series $\sum_{n \in \mathbb{N}} a_n$ envelops A, then for all $n \in \mathbb{N}$

$$\left| A - \sum_{k=0}^{n} a_k \right| < |a_{n+1}|.$$

COMMENTS (i) This is just the definition of a series that envelops a number. In particular this is the case in the real situation if for all $n \in \mathbb{N}$, $A - \sum_{k=0}^{n} a_k = \theta_n a_{n+1}$, for some $\theta_n, 0 < \theta_n < 1$.
(ii) **Gerber's Inequality** shows that the Taylor series of the function $(1+x)^\alpha$ envelops the function; and the **Trigonometric Function Inequalities** (n) show the same for sine and cosine functions.
(iii) The first references gives many such results for various special functions.

REFERENCES [General Inequalities, vol. 2, pp. 161–175]; [Pólya & Szegö, 1972, pp. 32–36].

Equimeasurable Function Inequalities

See Duff's Inequality, Hardy-Littlewood Maximal Inequalities, Modulus of Continuity Inequalities, Spherical Rearrangement Inequalities, Variation Inequalities.

Equivalent Inequalities

Many very different pairs of inequalities are equivalent in the sense that given the one inequality the other can be proved and conversely. This has been pointed out in various entries; see for instance **Bernoulli's Inequality** COMMENTS (viii), **Bilinear Form Inequalities, Cauchy's Inequality** COMMENTS (i), **Cordes's Inequality** COMMENTS, **Geometric-Arithmetic Mean Inequality** COMMENTS (iii), **Geometric Mean Inequalities** COMMENTS (i), **Harmonic Mean Inequalities** COMMENTS (i), (iv), **Hausdorff-Young Inequalities** COMMENTS (i), **Heinz-Kato-Furuta Inequality** COMMENTS (iv), **Hölder's Inequality** COMMENTS (ii), **Isoperimetric Inequalities** COMMENTS (i), **Kantorovič's Inequality** COMMENTS (i), **Löwner-Heinz Inequality** COMMENTS (iii), **Paley's Inequalities** COMMENTS (iii), **Polyá & Szegö's Inequality** COMMENTS (i), (iv), **Popoviciu's Geometric-Arithmetic Mean Inequality Extension** COMMENTS (ii), **Power Mean Inequalities** COMMENTS (ii), **Q-class Function Inequalities** COMMENTS (i), **Rado's Geometric-Arithmetic Mean Inequality Extension** COMMENTS (i), **Rennie's Inequality** COMMENTS (i), **Sobolev's Inequalities** COMMENTS (ii), **Steffensen's Inequality** COMMENTS (i), **Walsh's Inequality** COMMENTS (ii).

A detailed discussion can be found in the references

REFERENCES [MI, pp. 97, 141, 170, 220–224], [MO, pp. 457–462], [MPF, pp. 191–209].

Erdös's Inequality

Let p be a polynomial of degree n with real roots all of which lie outside the interval $]-1, 1[$ then

$$\|p'\|_{\infty,[-1,1]} \leq \frac{en}{2} \|p\|_{\infty,[-1,1]}.$$

COMMENTS This should be compared with **Markov's Inequality**.

EXTENSIONS [MÁTÉ] *Let p be a polynomial of degree n with n − k real roots all of which lie outside the interval* $]-1,1[$ *then*

$$||p'||_{\infty,[-1,1]} \leq 6ne^{\pi\sqrt{k}}||p||_{\infty,[-1,1]}.$$

REFERENCES [Máté].

Erdös & Grünwald's Inequality *If p is a polynomial with real roots, $p(0) = p(1) = 0$, then*

$$\frac{2}{3}\frac{2p'(1)p'(-1)}{p'(1) - p'(-1)} \leq \int_{-1}^{1} p \leq \frac{4}{3}||p||_{\infty,[-1,1]}.$$

COMMENTS (i) This can be interpreted as:

the area under the graph of the polynomial p lies between two-thirds of the area of the containing rectangle and two thirds of the area of the tangential triangle.

(ii) This result has been extended to higher dimensions in the reference.

REFERENCES [René].

Ergodic Inequality *If $T : \mathcal{L}_\mu(X) \to \mathcal{L}_\mu(X)$ is a linear operator with $|T|_1 \leq 1$, $|T|_\infty \leq 1$, and if $f \in \mathcal{L}_\mu^p(X)$ write*

$$\tilde{f} = \sup_{n \geq 1} \left| \frac{1}{n} \sum_{k=0}^{n-1} T^k f \right|.$$

Then

$$||\tilde{f}||_{p,\mu} \leq \begin{cases} 2q^{1/p}||f||_{p,\mu}, \ q \ \text{the conjugate index}, & \text{if } 1 < p < \infty, \\ 2\left(\mu(X) + \int_X |f|\log^+ |f|\, d\mu\right) & \text{if } p = 1. \end{cases}$$

COMMENTS $|T|_p = \sup ||Tf||_{p,\mu}$ where the supremum is taken over all $f \in \mathcal{L}_\mu(X)$ with $||f||_{p,\mu} < \infty$.

REFERENCES [Dunford & Schwarz, pp. 669, 678–679], [Dictionary, vol. I, pp. 531–533].

Error Function Inequalities (a) [CHU]

$$\sqrt{1 - e^{-ax^2}} \leq \mathfrak{a}(x) \leq \sqrt{1 - e^{-bx^2}} \iff 0 \leq a \leq 1/2, \text{ and } b \geq 2/\pi.$$

(b) [GORDON] If $x > 0$,

$$\frac{x}{x^2 + 1} \leq \mathfrak{r}(x) \leq \frac{1}{x}.$$

COMMENTS The lower bound in (b) has been improved by Birnbaum to $(\sqrt{4+x^2}-x)/2$. Other extensions and improvements can be found in the reference.

REFERENCES [AI, pp. 177–181, 385].

Euler's Constant Inequalities

The real number

$$\gamma = \lim_{n\to\infty}\left(\sum_{i=1}^{n} 1/i - \log n\right) = -(x!)'_{x=0} \approx 0.57721566490\cdots,$$

is called *Euler's constant*. It is not known whether it is irrational or not.

(a) [RAO]

$$\frac{1}{2n} - \frac{1}{8n^2} < \left(\sum_{i=1}^{n} 1/i - \log n\right) - \gamma < \frac{1}{2n}.$$

(b) [SANDHAM]

$$\gamma < \sum_{i=1}^{p}\frac{1}{i} + \sum_{i=1}^{q}\frac{1}{i} - \sum_{i=1}^{pq}\frac{1}{i} \leq 1.$$

REFERENCES [AI, pp. 187–188], [EM, vol.3, p. 424]; [Apostol et al, Part I, pp. 389–390].

Exponential Function Inequalities

REAL INEQUALITIES (a) If $a \geq b > 0$ then

$$a^x \geq b^x,$$

with equality if and only if either $a = b$ or $x = 0$.
(b) If $n \neq 3$ then $\qquad n^{1/n} < 3^{1/3}.$
(c) [WANG] If $x \neq 0$,

$$e^x > \left(1+\frac{x}{n}\right)^n > e^x\left(1+\frac{x}{n}\right)^{-x} \qquad n=1,2,\ldots;\qquad(1)$$

and if $0 < x \leq 1$,

$$e^x < \left(1+\frac{x}{n}\right)^{n+1}, \qquad n=1,2,\ldots.\qquad(2)$$

In particular, if $x \neq 0$,

$$e^x > 1+x.\qquad(3)$$

(d) If $x \neq e$

$$e^x > x^e.\qquad(4)$$

(e) If $x > 0$ then

$$x^x \geq e^{x-1}.$$

(f) If $0 < x < e$ then
$$(e+x)^{e-x} > (e-x)^{e+x}.$$

(g) If $x, y > 0$ then
$$x^y + y^x > 1.$$

(h) [TSIBULIS[2]] If $1 < y < x$ then
$$yx^y\left(y^x - (y-1)^x\right) > xy^x\left(x^y - (x-1)^y\right).$$

(j) If $a, x > 0$ then
$$e^x > \left(\frac{ex}{a}\right)^a.$$

(k) If $x < 1, x \neq 0$ then
$$e^x < \frac{1}{1-x}.$$

COMMENTS (i) Inequalities (1), (2) are basic in the theory of the exponential function, the right-hand sides tending to the left-hand side as $n \to \infty$; see **Binomial Function Inequalities** COMMENTS (iv).
If $x > 0$ the right-hand side of (1) can be replaced by $e^x(1 + x/n)^{-x/2}$.
(ii) Inequality (3) is an immediate consequence of Taylor's Theorem. In a certain sense (3) characterizes the number e in that if $a^x > 1 + x, x \neq 0$ then $a = e$; see [Halmos].
(iii) (d) follows from the strict concavity of the logarithmic function at $x = e$.
As we see from (d), the graphs of e^x and x^e touch at (e, e^e). However the graphs of a^x and x^a, $a > 1, a \neq e$ cross at $x = a$, and if $a > 1$ at another point; call these points α, β with $\alpha < \beta$. [If $a < 1$ consider $\beta = \infty$; of course one of α, β is always a.] Then $a^x < x^a$ if $\alpha < x < \beta$, while the opposite inequality holds if either $x < \alpha$ or $x > \beta$.
(iv) The inequality in (j) is important for large values of a.

COMPLEX INEQUALITIES (a) If $z \in \mathbb{C}$ with $0 < |z| < 1$ then
$$\frac{|z|}{4} < |e^z - 1| < \frac{7|z|}{4}.$$

(b) If $z \in \mathbb{C}$
$$|e^z - 1| \leq e^{|z|} - 1 \leq |z|e^{|z|}.$$

(c) [KLOOSTERMAN] If $z \in \mathbb{C}$ then
$$\left|e^z - \left(1 + \frac{z}{n}\right)^n\right| < \left|e^{|z|} - \left(1 + \frac{|z|}{n}\right)^n\right| < \frac{|z|^2}{2n}e^{|z|}.$$

[INEQUALITIES INVOLVING THE REMAINDER OF THE TAYLOR SERIES] Let the function I_n denote the $(n+1)$-th remainder of the Taylor series for e^x; that is,
$$I_n(x) = e^x - \sum_{i=0}^{n} \frac{x^i}{i!} = \sum_{i=n+1}^{\infty} \frac{x^i}{i!}, \quad x \in \mathbb{R}, \text{ or } \mathbb{C}.$$

---

[2] Also spelt Cibulis.

(a) [SEWELL] If $f\, x \geq 0$ then
$$I_n(x) \leq \frac{xe^x}{n}.$$

(b) If $|z| \leq 1$ then
$$|I_n(z)| \leq \frac{1}{(n+1)!}\left(1 + \frac{2}{n+1}\right).$$

(c) [ALZER] if $n \geq 1, x > 0$ then
$$I_n(x)I_{n+1}(x) > \frac{n+1}{n+2}I_n^2(x).$$

The constant is best possible.

COMMENTS (v) This last result has been extended in [Alzer, Brenner & Ruehr].

EXTENSIONS (a) If $x \neq 0$ and if n is odd then
$$e^x > 1 + x + \cdots + \frac{x^n}{n!}.$$
If n is even this inequality holds if $x > 0$, while the opposite holds if $x < 0$.

(b) [NANJUNDIAH]
$$\frac{2n+2}{2n+1}\left(1 + \frac{1}{n}\right)^n < e < \left(1 + \frac{1}{n}\right)^{n+\frac{1}{2}}.$$

(c) If $x, n > 0$ then
$$e^x < \left(1 + \frac{x}{n}\right)^{n+x/2}.$$
In particular if $n \geq 1$, $0 \leq x \leq n$,
$$e^x < \left(1 + \frac{x}{n}\right)^n + \frac{x^2 e^x}{n}.$$

(d)
$$\frac{e}{2n+2} < e - \left(1 + \frac{1}{n}\right)^n < \frac{e}{2n+1}.$$

COMMENTS (vi) Inequality in (b) follows from the particular case of the **Logarithmic Mean Inequalities** (1), $\mathcal{L}_1(a,b) > \mathcal{L}_0(a,b) > \mathcal{L}_{-2}(a,b)$.

(ix) See also **Binomial Function Inequalities** COMMENTS (iii) and (iv), **Series Inequalities** (d).

REFERENCES [AI, pp. 266–269, 279–281, 323–324], [HLP, pp. 103–104, 106], [MI, p. 130]; [Abramowicz & Stegun, 1965, p. 70], [Apostol et al, Part II, pp. 445–452], [Borwein & Borwein, p. 317], [Dienes, 1957, p. 135], [Halmos, p. 19], [Melzak, 1976, pp. 64–67]; [Alzer, Brenner & Ruehr], [Tsibulis], [Wang].

Extended Mean Inequalities

See **Logarithmic Mean Inequalities**.

- end of E -

F

Factorial Function Inequalities (a) [GAUTSCHI] (i) If $0 \leq x \leq 1$ then

$$n^{1-x} \leq \frac{n!}{(n-1+x)!} \leq (n+1)^{1-x}.$$

(ii)
$$x!\left(\frac{1}{x}\right)! \geq 1.$$

(b) [WATSON] If $x \geq 0$ then

$$\sqrt{x+\frac{1}{4}} \leq \frac{x!}{(x-\frac{1}{2})!} \leq \sqrt{x+\frac{1}{\pi}}.$$

(c) [GURLAND] If $n \in \mathbb{N}$

$$\sqrt{n+\frac{1}{4}} \leq \frac{n!}{(n-\frac{1}{2})!} \leq \sqrt{n+\frac{1}{4}+\frac{1}{16n+12}}.$$

(d) [KHINTCHINE] If $n_i \in \mathbb{N}, 1 \leq i \leq k, n = n_1 + \cdots n_k$ then

$$\frac{n_1! \cdots n_k!}{(2n_i)! \cdots (2n_k)!} \leq \frac{1}{2^n}.$$

(e) If $n \geq 2$ then

$$n^{n/2} < n! < \left(\frac{n+1}{2}\right)^n.$$

(f) Writing $(2n)!! = \prod_{i=1}^{n}(2i), (2n-1)!! = \prod_{i=1}^{n}(2i-1)$ we have if $n \geq 2$ that

$$n^n > (2n-1)!!; \qquad (n+1)^n > 2n!! > ((n+1)!)^n.$$

(g) If $n \geq 2$ then

$$(n!)^4 < \left(\frac{n(n+1)^3}{8}\right)^n.$$

(h) If $n \geq 3$ then

$$n! < n^n < (n!)! < n^{n^n} < ((n!)!)!.$$

(j) [OSTROWSKI] If $n, p \in \mathbb{N}$, $0 \leq x \leq n$ then
$$n! \geq \left| x^{(p-1)/p} \prod_{i=1}^{n} (x - i) \right|.$$

(k) [WHITTAKER & WATSON] If $\underline{a}$ is a positive n-tuple
$$(\mathfrak{A}_n(\underline{a}))! \leq \mathfrak{G}_n(\underline{a}!).$$

(ℓ) [ELIEZER] If $x, y > 0$,
$$\mathfrak{G}_2\left(\frac{x!}{x^{x+1}}, \frac{y!}{y^{y+1}}\right) \geq \frac{((x+y)/2)!}{((x+y)/2)^{1/2(x+y)+1}}.$$

(m) [MINC & SATHRE] If $n = 1, 2, \ldots$,
$$\frac{n}{n+1} \leq \frac{(n!)^{1/n}}{((n+1)!)^{1/(n+1)}}.$$

COMMENTS (i) Inequalities (a)–(d) and other similar inequalities have been discussed by Slavić.
(ii) The inequalities in (k), (ℓ) use **Log-convex Function Inequalities** COMMENTS (ii).
(iii) See also **Alzer's Inequalities** (b), **Beta Function Inequalities, Binomial Coefficient Inequalities, Digamma Function Inequalities, Stirling's Formula, Wallis's Inequality**.

REFERENCES [AI, pp. 192–194, 285–286], [MO, pp. 75–76]; [General Inequalities, vol. 3, pp. 277–280], [Hájos, Hungarian Problem Book II, pp. 13,55], [Jeffreys & Jeffreys, pp. 462–473]; [Alzer, 1993[(3)], 1995[(2)]], [Merkle], [Slavić].

Faĭziev's[1] Inequality If $\underline{a}$ is a non-negative n-tuple then
$$\left(a_1^{n+1} + \cdots + a_n^{n+1}\right) \geq (a_1 \cdots a_n)(a_1 + \cdots + a_n).$$

REFERENCES [Faĭziev].

Fan-Taussky-Todd Inequality
See **Ky Fan-Taussky-Todd Inequalities**.

Farwig & Zwick's Lemma
If f is n-convex and has a continuous n-th derivative on $[a, b]$; and if for some points $a \leq y_0 \leq \cdots y_{n-1} \leq b$, $[y_k, \ldots, y_{n-1}; f] \geq 0, 0 \leq k \leq n-1$, then
$$f^k(b) \geq 0, \quad 0 \leq k \leq n-1.$$

COMMENTS The strong condition on the n-th derivative can be relaxed.

REFERENCES [PPT, pp. 30–32].

---

[1] Р Ф Файзиев.

Fatou's Lemma If $f_n, n \in \mathbb{N}$ is a sequence of non-negative measurable functions defined on the measurable set $E, E \subseteq \mathbb{R}$, then

$$\int_E \liminf_{n \to \infty} f_n \leq \liminf_{n \to \infty} \int_E f_n.$$

COMMENTS (i) Strict inequality is possible in this basic result of measure theory; consider for instance $E = \mathbb{R}, f_n = 1_{[n,n+1]}, n \in \mathbb{N}$.
(ii) The result extends to general measure spaces.
REFERENCES [Hewitt & Stromberg, p. 172], [Rudin, p. 246].

Favard's Inequalities (a) Let f be a non-negative, continuous, concave function not identically zero on $[a, b]$ and put $\overline{f} = \mathfrak{A}_{[a,b]}(f)$. If g is convex on $[0, 2\overline{f}]$ then

$$\mathfrak{A}_{[a,b]}(g \circ f) \leq \mathfrak{A}_{[0,2\overline{f}]}(g).$$

(b) If f is a non-negative concave function on $[0, 1]$, and $p \geq 1$ then

$$\int_0^1 f \geq \frac{(p+1)^{1/p}}{2} \|f\|_p.$$

COMMENTS (i) The inequality in (b) is connected with **Grüss-Barnes's Inequality**, and is a particular case of **Thunsdorff's Inequality**. See also **Arithmetic Mean Inequalities** INTEGRAL ANALOGUES, **Geometric-Arithmetic Mean Inequalities** INTEGRAL ANALOGUES (b).

EXTENSIONS [BERWALD] Let f be a non-negative, continuous, concave function not identically zero on $[a, b]$ and let h be continuous and strictly monotonic on $[0, \ell]$, where $z = \ell$ is the unique positive solution of the equation

$$\mathfrak{A}_{[0,z]}(h) = \mathfrak{A}_{[a,b]}(h \circ f).$$

If g is convex with respect to h on $[0, \ell]$ then

$$\mathfrak{A}_{[a,b]}(g \circ f) \leq \mathfrak{A}_{[0,\ell]}(g).$$

COMMENTS (iii) For a definition of "g is convex with respect to h" see **Quasi-arithmetic Mean Inequalities** COMMENTS (i).
REFERENCES [BB, pp. 43–44], [PPT, pp. 212–217]; [Brenner & Alzer], [Maligranda, Pečarić & Persson, 1994].

Fejér-Jackson-Gronwall Inequality
See **Fejér-Jackson Inequality** COMMENTS (i).

Fejér-Jackson Inequality If $0 < x < \pi$ then

$$0 < \sum_{i=1}^{n} \frac{\sin ix}{i} < \pi - x. \tag{1}$$

COMMENTS   (i) It is the left-hand side of this that is known as the *Fejér-Jackson inequality*, or sometimes the *Fejér-Jackson-Gronwall inequality*.

EXTENSIONS   (a) If $0 < x < \pi$, $-1 \le \epsilon \le 1$, $p - q = 0$ or 1 then

$$\sum_{i=1}^{p} \frac{\sin(2i-1)x}{2i-1} > \epsilon \sum_{i=1}^{q} \frac{\sin 2ix}{2i}; \tag{2}$$

in particular

$$0 < \sum_{i=1}^{n} (-1)^{n-1} \frac{\sin ix}{i}.$$

(b) [ASKEY-STEINIG] If $0 \le x < y \le \pi$ then

$$\frac{1}{\sin x/2} \sum_{i=1}^{n} \frac{\sin ix}{i} > \frac{1}{\sin y/2} \sum_{i=1}^{n} \frac{\sin iy}{i}.$$

(c) [TUDOR] If $0 < x < \pi/(2m-1)$ then

$$\sum_{i=m}^{m+n-1} \frac{\sin ix}{i} > 0, \quad 0 < x < \pi.$$

(d) [SZEGÖ, SCHWEITZER] If $0 \le x \le 2\pi/3$ then

$$\sum_{i=1}^{n} \binom{n+2-i}{2} i \sin ix > 0.$$

COMMENTS   (ii) If we put $\epsilon = -1$ in (2) we get (1). The case $\epsilon = 0$ gives an analogous result, due to Fejér:

$$\sum_{i=1}^{n} \frac{\sin(2i-1)x}{2i-1} > 0.$$

(iii) See also **Trigonometric Polynomial Inequalities** (b), (c), **Young's Inequalities** (d).

REFERENCES   [AI, pp. 249, 251, 255], [MPF, pp. 611–627]; [Gluchoff & Hartmann], [Tudor]

Fejér Kernel Inequalities
The quantity

$$K_k(x) = \frac{1}{k+1} \sum_{i=0}^{k} \frac{\sin(i+1/2)x}{2\sin x/2} = \frac{2}{k+1} \left(\frac{\sin(k+1/2)x}{2\sin x/2} \right)^2,$$

is called the *Fejér kernel*.
In terms of the Dirichlet kernel

$$K_k(x) = \frac{1}{k+1} \sum_{i=0}^{k} D_i(x).$$

(a) If $k \in \mathbb{N}$ then

$$k+1 > K_k(x) \geq 0.$$

(b) If $k \geq 1, 0 < x \leq \pi$ then for some constant C,

$$K_k(x) \leq \frac{C}{(k+1)x^2}.$$

(c) If $k \geq 1, t \leq x \leq \pi$ then

$$K_k(x) \leq \frac{1}{2(k+1)\sin^2 t/2}.$$

COMMENTS (i) Extensions of these results to other kernels can be found in the references.
(ii) See also **Dirichlet Kernel Inequalities, Poisson Kernel Inequalities**.
REFERENCES [Hewitt & Stromberg, p. 292], [Zygmund, vol. I, pp. 88–89, 94–95; vol. II, p. 60].

Fejér-Riesz Theorem
If f is analytic in $\overline{D}$ and if $p > 0$, then for all θ,

$$\int_{-1}^{1} |f(re^{i\theta})|^p \, dr \leq \frac{1}{2} \int_{-\pi}^{\pi} |f(e^{it})|^p \, dt.$$

The constant is best possible.

COMMENTS The case $p = 1$ is given in **Conjugate Harmonic Function Inequalities** (b).
REFERENCES [AI, p. 334]; [Heins, pp. 122, 347], [Zygmund, vol. I, p. 258].

Fenchel's Inequality
See **Conjugate Convex Function Inequalities** COMMENTS (ii).

Fischer's[2] Inequalities If $\underline{p}, \underline{q}$ are positive n-tuples, $n \geq 3$, with $P_n = Q_n = 1$ and if $c \leq -1$ or $c \geq 0$ then

$$\sum_{i=1}^{n} p_i \left(\frac{p_i}{q_i}\right)^c \geq 1. \tag{1}$$

If $0 < c < 1$ then (~ 1) holds.

COMMENTS The power function is essentially the only function for which an inequality of this type holds. Any function $f :]0, 1[\to]0, \infty[$ for which

$$\sum_{i=1}^{n} p_i \frac{f(p_i)}{f(q_i)} \geq 1, n \geq 3,$$

is internal on $]0, 1/2]$; for a definition of internal function see **Internal function Inequalities**.

REFERENCES [MPF, pp. 641–643].

FKG Inequality If X is a distributive lattice, let $\ell : X \to [0, \infty[$ satisfy

$$\ell(a)\ell(b) \leq \ell(a \vee b)\ell(a \wedge b), \quad \text{for all } a, b \in X.$$

If $f, g : X \to \mathbb{R}$ are increasing then

$$\sum_{x \in X} \ell(x)f(x) \sum_{x \in X} \ell(x)g(x) \leq \sum_{x \in X} \ell(x) \sum_{x \in X} \ell(x)f(x)g(x).$$

COMMENTS (i) This is also known as *the Fortuin-Kasteleyn-Ginibre inequality*; it is an example of a *correlation inequality*.
(ii) A function such as ℓ is called *log supermodular*.

REFERENCES [EM, Supp., pp. 201–202, 253–254].

Fortuin-Kasteleyn-Ginibre Inequality
See **FKG Inequality**.

Four Function Inequality
See **Ahlswede-Daykin Inequality**.

Fourier Transform Inequalities
If $f \in \mathcal{L}(\mathbb{R})$ the Fourier transform of f is

$$\hat{f}(t) = \frac{1}{\sqrt{2\pi}} \int_{\mathbb{R}} f(x) e^{-ixt} \, dx.$$

In the case of other assumptions on f the existence of the transform is more delicate; see the references.

---

[2] This is P Fischer.

We can also define the *Fourier sine, and Fourier cosine transforms,*

$$\sqrt{\frac{2}{\pi}} \int_0^\infty f(t) \sin t \, dt, \qquad \sqrt{\frac{2}{\pi}} \int_0^\infty f(t) \cos t \, dt.$$

(a) If $f \in \mathcal{L}([0,\infty[)$ with $\lim_{t \to \infty} f(t) = 0$ then

$$\sum_{i=1}^\infty (-1)^k f(k\pi) < \int_0^\infty f(t) \cos t \, dt < \sum_{i=0}^\infty (-1)^k f(k\pi);$$

$$\sum_{i=0}^\infty (-1)^k f\left((k+\frac{1}{2})\pi\right) < \int_0^\infty f(t) \sin t \, dt < f(0) + \sum_{i=0}^\infty (-1)^k f\left((k+\frac{1}{2})\pi\right).$$

(b) [TITCHMARSH'S THEOREM] If $f \in \mathcal{L}^p(\mathbb{R})$, $1 < p \leq 2$, and if q is the conjugate index, then $\hat{f} \in \mathcal{L}^q(\mathbb{R})$ and

$$\left(\frac{1}{\sqrt{2\pi}} \int_\mathbb{R} |\hat{f}|^q\right)^{1/q} \leq \frac{p^{1/p}}{q^{1/q}} \left(\frac{1}{\sqrt{2\pi}} \int_\mathbb{R} |f|^p\right)^{1/p}, \qquad (1)$$

with equality if and only if $f(x) = ae^{-\alpha x^2}$, $\alpha > 0$, almost everywhere.

COMMENTS (i) (a) is a deduction from **Steffensen's Inequality**.
(ii) Titchmarsh obtained (1) with the constant on the right-hand side equal to one as an integral analogue of **Hausdorff-Young Inequalities** (a). The correct value of the constant and the cases of equality were given by Beckner. The similarity of the constant to that in **Young's Convolution Inequality** is more than a coincidence; see the references. In addition the result can be extended to $\mathbb{R}^n$, $n \geq 2$.

EXTENSIONS (a) [PALEY-TITCHMARSH] If $|f(x)|^q x^{q-2} \in \mathcal{L}(\mathbb{R})$, $q > 2$, then $\hat{f} \in \mathcal{L}^q(\mathbb{R})$ and

$$\int_\mathbb{R} |\hat{f}|^q \leq C \int_\mathbb{R} |f(x)|^q |x|^{q-2} \, dx,$$

where C depends only on q.
If $f \in \mathcal{L}^p(\mathbb{R})$, $1 < p < 2$, then $|\hat{f}(x)|^p x^{p-2} \in \mathcal{L}^p(\mathbb{R})$ and

$$\int_\mathbb{R} |\hat{f}(x)|^p |x|^{p-2} \, dx \leq C \int_\mathbb{R} |f|^p,$$

where C depends only on p.
(b) [RIEMANN-LEBESGUE LEMMA] If $f \in \mathcal{L}(\mathbb{R})$ then $\hat{f} \in C_0(\mathbb{R})$ and

$$\|\hat{f}(x)\|_\infty \leq \|f\|.$$

COMMENTS (iii) Extension (a) is an integral analogue of **Paley's Inequalities**. These results have been further extended to allow weight functions.

REFERENCES [AI, pp. 109–110], [EM, vol.4, pp. 80–83]; [General Inequalities, vol. 5, pp. 217–232], [Lieb & Loss, pp. 120–122], [Rudin, 1966, pp. 184–185; 1973, p. 169], [Titchmarsh, 1948, pp. 3–4, 96–113], [Zygmund, vol. II, pp. 246–248, 254–255].

Frank-Pick Inequality
See **Thunsdorff's Inequality** COMMENTS (ii).

Frequency Inequalities If A is a plane membrane and Λ is its principal frequency then

$$j_{0_1}\sqrt{\frac{K}{2|A|}} \geq \Lambda \geq \frac{j_{0_1}\pi}{\sqrt{|A|}},$$

where j_{0_1} is the smallest positive root of the Bessel function of the first kind J_0, and K is a constant depending on A. Equality occurs when the membrane is circular.

COMMENTS (i) The *principal frequency* of a plane membrane A can be defined as

$$\Lambda^2 = \inf \frac{\int_A |\nabla f|^2}{\int_A |f|^2},$$

where the inf is over all f zero on ∂A.
(ii) Precisely: let $h(x, s)$ denote the distance of $\underline{x}$, $\underline{x} \in \overset{\circ}{A}$, from the tangent to ∂A at the point s, where s is the arc length on ∂A, $0 \leq s \leq \ell$; then $K = \inf\{\underline{x} \in \overset{\circ}{A};\ \oint_{\partial A} \frac{1}{h(\underline{x},s)} ds\}$.
(iii) $j_{0_1} \approx 2.4048$; [Abramowicz & Stegun, p. 409].
(iv) This is an example of **Symmetrization Inequalities**.

REFERENCES [Pólya & Szegö, 1951, pp. 2, 8, 9, 87–94, 195–196].

Friederichs's Inequality If $\Omega \subset \mathbb{R}^n$ is a simply connected bounded domain with a locally Lipschitz boundary and if $f \in \mathcal{W}^{1,2}(\Omega)$ then

$$\int_\Omega f^2 \leq C \left(\int_\Omega |\nabla f|^2 + \oint_{\partial \Omega} f^2 \right)$$

where C is a constant that only depends on Ω and n.

COMMENTS (i) A definition of $\mathcal{W}^{1,2}(\Omega)$ is given in **Sobolev's Inequalities**.
(ii) This inequality has been the subject of much study and generalization; see the references. It is sometimes called a *Poincaré inequality*, or a *Nirenberg inequality*; see **Poincaré's Inequalities**. The case $n = 2$, $f \in C^2(\overline{\Omega})$ is due to Friederichs.
(iii) If the condition $f = 0$ holds on ∂R this can be regarded as a generalization of **Wirtinger's Inequality**.

REFERENCES [EM, vol.4, p. 115]; [Opic & Kufner, p. 2]; [Dostanić].

Function Inequalities (a) [BOAS] If $A = [0, 1[$ or $[0, 1]$ and f, g are strictly starshaped functions, with $f \in C(A)$, $f(0) = 0$, $1 < f(1) \leq \infty$, $0 < x < 1$, $f(x) \neq x$; and $g \in C(f[A])$, $g(1) \leq 1$, then

$$f \circ g \leq g \circ f.$$

In particular

$$\arcsin\left(\sinh \frac{x}{2}\right) \leq \sinh\left(\frac{1}{2} \arcsin x\right).$$

(b) [KUBO[3]] If $f, g \in \mathcal{C}([a,b])$, f piecewise monotonic with $g[I] \subseteq I$ if f is monotonic on the closed interval I, and if f is increasing, decreasing, on I then $g(x) \geq x$, $g(x) \leq x$, respectively; then
$$f \circ g \geq f.$$

COMMENTS (i) For a definition of strictly starshaped see **Starshaped Function Inequalities** COMMENTS (i).
(ii) For a particular case of (b) see **Trigonometric Function Inequalities** (d).
(ii) See also under particular functions: **Absolutely and Completely Monotonic, Almost Symmetric, Analytic, Bernšteĭn Polynomials, Bessel, Beta, Binomial, Bounded Variation, Čebišev Polynomials, Complete Symmetric, Conjugate Convex, Conjugate Harmonic, Convex, Complex, Digamma, Elementary Symmetric, Entire, Equimeasurable, Error, Exponential, Factorial, Harmonic, Hyperbolic, Increasing, Internal, Laguerre, Legendre, Lipschitz, Logarithmic, Maximal, Monotonic, n-Convex, N-Function, Polynomial, Q-class, Quasi-conformal, Schur Convex, Subharmonic, Starshaped, Subadditive, Symmetric, Totally Positive, Trigonometric, Trigonometric Polynomial Ultraspherical Polynomials, Wright Convex, Univalent, Zeta.**

REFERENCES [*Boas, 1979*], [*Kubo, T*].

- end of F -

---
[3] This is T Kubo.

G

Gabriel's Problem   If Λ is a closed convex curve lying inside the closed convex curve Γ and if f is analytic inside Γ then:

$$\int_\Lambda |f(z)|\,|dz| < A \int_\Gamma |f(z)|\,|dz|,$$

where A is some absolute constant.

COMMENTS   It is conjectured that $A = 2$; the best known value is $A = 3.6$.

REFERENCES   [AI, p. 336].

Gagliardo-Nirenberg Inequality
See **Sobolev's Inequalities** (b).

Gale's Inequality   If d is the diameter of a bounded domain in $\mathbb{R}^n$, $n \geq 2$ and ℓ the edge of the smallest circumscribing regular simplex then:

$$\ell \leq d\sqrt{\frac{n(n+1)}{2}}.$$

COMMENTS   See also **Bonnensens Inequality, Isoperimetric Inequalities, Young's Inequalities** (e).

REFERENCES   [EM, vol. 5, p. 204].

Gamma Function Inequalities
See **Factorial Function Inequalities**.

Gaussian Error Function Inequalities
See **Error Function Inequalities**.

Gaussian Probability Function Inequalities
See **Error Function Inequalities**.

Gauss's Inequality If $f :]0, \infty[\to \mathbb{R}$ is decreasing and if $x > 0$ then:

$$x^2 \int_x^\infty f \leq \frac{4}{9} \int_0^\infty t^2 f(t)\,dt.$$

EXTENSIONS [VOLKOV[1]] If $f, G : [0, \infty[\to \mathbb{R}$ with f decreasing, G increasing and differentiable with $G(t) \geq t$, $t > 0$ then:

$$\int_{G(0)}^\infty f \leq \int_0^\infty fG'.$$

COMMENTS (i) Taking $G(t) = x + 4t^3/27x^2$ gives the original result.

CONVERSE INEQUALITY [ALZER] If $f :]0, \infty[\to [0, \infty[$ and if $x > 0$ then,

$$x^2 \int_x^\infty f \geq 3 \int_0^x t^2 f(t+x)\,dt.$$

The constant on the right-hand side is best possible.

COMMENTS (ii) See also **Moment Inequalities, Steffensen's Inequalities** (a).

REFERENCES [AI, p. 300], [MPF, pp. 313, 319]; [Alzer, 1991].

Gauss-Winkler Inequality If $v_r, r \geq 0$ denotes the r-th absolute moment of a probability distribution P that has a continuous decreasing derivative, and if $r \leq s$ then:

$$\big((r+1)v_r\big)^{1/r} \leq \big((s+1)v_s\big)^{1/s}.$$

COMMENTS (i) Without the constant factors the inequality is weaker, and is a particular case of (r;s).
(ii) The original Gauss result was the case $r = 2$, $s = 4$ of this:

if $f \geq 0$ is decreasing, then provided the integrals exist,

$$\left(\int_0^\infty x^2 f(x)\,dx \right)^2 \leq \frac{5}{9} \int_0^\infty f(x)\,dx \int_0^\infty x^4 f(x)\,dx.$$

EXTENSIONS [BEESACK] If, with the above notation, $(-1)^k P^{(k)}$ is positive, continuous and decreasing, $1 \leq k \leq n_0$, x then if $r \leq s$,

$$\left(\binom{r+k}{k} v_r \right)^{1/r} \leq \left(\binom{s+k}{k} v_s \right)^{1/s}, \quad 1 \leq k \leq n_0;$$

and if $m < n < r$,

$$\binom{n+k}{k} v_n \leq \left(\binom{m+k}{k} v_m \right)^{(r-n)/(r-m)} \left(\binom{r+k}{k} v_r \right)^{(n-m)/(r-m)}.$$

COMMENTS (ii) Compare these results with **Klamkin & Newman Inequalities**.

REFERENCES [MPF, pp. 53–56], [PPT, pp. 217–228]; [General Inequalities, vol. 7, pp. 27–37];[Pečarić & Varošanec].

Gegenbauer Polynomial Inequalities
See **Ultraspherical Polynomial Inequalities**.

---

[1] В Н Волков.

Geometric-Arithmetic Mean Inequality

If $\underline{a}$ is a positive n-tuple, $n \geq 1$, then:

$$\mathfrak{G}_n(\underline{a};\underline{w}) \leq \mathfrak{A}_n(\underline{a};\underline{w}), \qquad (GA)$$

with equality if and only if $\underline{a}$ is constant.

COMMENTS (i) The most direct proof is first to obtain the case $n = 2$, and then make an induction on n. However there are many proofs available, a total of 52 are given in [MI]; and this does not count the indirect proofs where (GA) is a special case of a more general result. It is surprising that attacking (GA) via the equal weight case is more bothersome than a direct attack; for, although the $n = 2$ case is now trivial, the induction is more difficult, and the move to general weights needs some care.

SOME SPECIAL CASES (a) [EQUAL WEIGHT CASE] If $\underline{a}$ is a positive n-tuple

$$(a_1 \cdots a_n)^{1/n} \leq \frac{a_1 + \cdots + a_n}{n}, \qquad (1)$$

with equality if and only if $\underline{a}$ is constant.
(b) [CASE $n = 2$] If x, y, p are positive, $x = u^p, y = v^q$, q the conjugate index of p, then

$$x^{1/p} y^{1/q} \leq \frac{x}{p} + \frac{y}{q}, \quad \text{or} \quad uv \leq \frac{u^p}{p} + \frac{v^q}{q}, \qquad (2)$$

with equality if and only if $x = y$, or $u = v$.
(c) [EQUAL WEIGHT CASE; $n = 2$] If a, b are positive

$$\sqrt{ab} \leq \frac{a+b}{2}, \qquad (3)$$

with equality if and only if $a = b$.

COMMENTS (ii) (1) has many proofs, for instance it is a particular case of **Newton's Inequality** (1), (2), see, **Newton's Inequality** COMMENTS (iv); see also [MI].
(iii) (2) is just **Bernoulli's Inequality** (2). In fact (GA) is equivalent to (B); see **Bernoulli's Inequality**, COMMENTS (viii).
(iv) (3) is easily seen to be equivalent to $(c - d)^2 \geq 0$; put $c = \sqrt{a}, d = \sqrt{b}$.
(v) Inequalities (1), (2) have some simple geometric interpretations; see **Geometric Inequalities** (a).
There are many extensions that will be discussed in their own rights; see **Chong's Inequalities** (b), **Cyclic Inequalities** (b), **Geometric Mean Inequalities** (d), **Kober & Dianada's Inequalities**, **Nanjundiah's Mixed Mean Inequalities**, **O'Shea's Inequality**, **Popoviciu's Geometric-Arithmetic Mean Inequality Extension**, **Rado's Geometric-Arithmetic Mean Inequality Extension**, **Redheffer's Inequalities**, **Sierpinski's Inequalities**. Here we will mention a few of the others.

EXTENSIONS OF GA (a) [DIFFERENT WEIGHTS]

$$\mathfrak{G}_n(\underline{a};\underline{u}) \leq \frac{V_n \mathfrak{G}_n(\underline{u};\underline{u})}{U_n \mathfrak{G}_n(\underline{v};\underline{u})} \mathfrak{A}_n(\underline{a};\underline{v}),$$

with equality if and only if $a_1v_1u_i^{-1} = \cdots = a_nv_nu_n^{-1}$.
(b) [LUPAŞ & MITROVIĆ]

$$1 + \mathfrak{G}_n(\underline{a}) \leq \mathfrak{G}_n(1+\underline{a}) \leq 1 + \mathfrak{A}_n(\underline{a}),$$

with equality if and only if $\underline{a}$ is constant.

(c) [SIEGEL] If $n \geq 2$ and $\underline{a}$ has no two terms equal then:

$$\mathfrak{G}_n(\underline{a})\mathfrak{G}_n(\underline{b}) \leq \mathfrak{A}_n(\underline{a}),$$

where $b_i = 1 + (n-i)/(t+i-1)$, $i = 1, \ldots, n$, and t is the unique positive root of the equation

$$\frac{1}{n!} \prod_{i=0}^{n-2} \left(\frac{t+i}{t-i}\right)^{n+i-1} \prod_{1 \leq i < j \leq n} (a_i - a_j)^2 = \left(\prod_{i=1}^n a_i\right)^{n-1}.$$

(d) [HUNTER] If $n \geq 2$ and $\underline{a}$ has no two terms equal then:

$$\mathfrak{G}_n(\underline{a}) \leq \left\{(1+(n-1)t)(1-t)^{n-1}\right\}^{1/n} \mathfrak{A}_n(\underline{a}),$$

where t, $0 < t < 1$, is a root of the equation

$$\sum_{1 \leq i < j \leq n} (a_i - a_j)^2 = t^2(n-1)A_n^2.$$

(e) [FINK & JODHEIT] If $0 < a_1 \leq \cdots \leq a_n$ then:

$$\mathfrak{G}_n(\underline{a}) \leq \mathfrak{A}_n(\underline{a}; \underline{w}),$$

where $\underline{w}$ is defined by

$$w_i = \frac{W_n}{n} \prod_{k<i} \left\{1 - \left(1 - \left(\frac{a_k}{a_i}\right)^{1/n}\right)^n\right\}, \quad 1 \leq i \leq n.$$

(f) [MIJALKOVIĆ & KELLER]

$$\mathfrak{A}_n(\underline{a}; \underline{w}) \leq \mathfrak{G}_n(\underline{a}; \underline{a}\,\underline{w}).$$

(g) [ZACIU]

$$\left(\frac{\mathfrak{A}_n(\underline{a})}{\mathfrak{A}_n(\underline{b})}\right)^{\mathfrak{A}_n(\underline{a})} \leq \left(\frac{\mathfrak{G}_n(\underline{a}^{\underline{a}})}{\mathfrak{G}_n(\underline{b}^{\underline{b}})}\right).$$

(h) [ÅKERBERG]

$$\frac{1}{n}\sum_{i=1}^n \frac{(i!)^{1/i}}{i+1}\mathfrak{G}_i(\underline{a}) \leq \mathfrak{A}_n(\underline{a}).$$

(j) [ALZER]

$$\frac{3}{e}(\mathfrak{A}_n(\underline{a};\underline{w}) - \mathfrak{G}_n(\underline{a};\underline{w})) \geq \mathfrak{A}_n(\underline{a};\underline{w})e^{-\mathfrak{G}_n/\mathfrak{A}_n} - \mathfrak{G}_n(\underline{a};\underline{w})e^{-\mathfrak{A}_n/\mathfrak{G}_n} \geq 0,$$

with equality if and only if $\underline{a}$ is constant.

(k) [SCHAMBERGER]

$$\frac{(b-a)^2}{8\mathfrak{A}_2(a,b)} \leq \mathfrak{A}_2(a,b) - \mathfrak{G}_2(a,b) \leq \frac{(b-a)^2}{8\mathfrak{G}_2(a,b)}.$$

(ℓ) If $u, p > 0, v > -1/p$ then:

$$uv \leq u\frac{u^p - 1}{p} + \left(\frac{1+pv}{1+p}\right)^{(1+p)/p}.$$

(m) [(GA) WITH GENERAL WEIGHTS] If $0 < a_1 \leq \cdots \leq a_n$ and $\underline{w}$ is a real n-tuple with

$$W_n \neq 0, \quad 0 \leq \frac{W_i}{W_n}, 1 \leq i \leq n, \quad (4)$$

then (GA) holds, with equality if and only if $\underline{a}$ is constant.

COMMENTS (vi) (b) is related to **Weierstrass's Inequalities**; see also **Geometric Mean Inequalities** (f), **Harmonic Mean Inequalities** (d). The left hand inequality follows using (GA) and **Schur Convex Function Inequalities** (b).
(vii) The constant in (j), cannot be improved.
(viii) Inequality (ℓ) is obtained by a simple change of variable in (2).
(ix) (m) can be considered as a special case of the **Jensen-Steffensen Inequality** or it can be proved directly by the method used to prove that result. A particular case of (m) is **Bernoulli's Inequality** (1).
The restriction (4) on weights is in a way quite natural in that we have the following:
if $0 < a_1 \leq \cdots \leq a_n$ and $\underline{w}$ is a real n-tuple then:

$$\min \underline{a} \leq \mathfrak{A}_n(\underline{a};\underline{w}) \leq \max \underline{a} \quad (5)$$

if and only if (1) holds.

There are many converse forms for (GA) of which we give a few of the most important.

CONVERSE INEQUALITIES (a) [MOND & SHISHA] Assume that $W_n = 1$, and that $0 < m \leq \underline{a} \leq M$ then:

$$0 \leq \mathfrak{A}_n(\underline{a};\underline{w}) - \mathfrak{G}_n(\underline{a};\underline{w}) \leq \mathfrak{A}_2(m, M; 1-\theta, \theta) - \mathfrak{G}_2(m, M; 1-\theta, \theta), \quad (6)$$

where if $\mu = M/m$

$$\theta = \frac{\log(\mu - 1) - \log \log \mu}{\log \mu}.$$

There is equality on the right of (6) if and only if for some subset J of $1, \ldots, n$ we have
$$\sum_{i \in J} w_i = \theta \qquad a_i = M,\ i \in J, \qquad a_i = m,\ i \notin J.$$

(b) [DOČEV] If $\underline{a}$ and μ are as in the previous result then:
$$1 \leq \frac{\mathfrak{A}_n(\underline{a}; \underline{w})}{\mathfrak{G}_n(\underline{a}; \underline{w})} \leq \frac{(\mu - 1)\mu^{1/\mu - 1}}{e \log \mu}.$$

Further, the right-hand side of the last expression is increasing as a function of M, and decreasing as a function of m.

COMMENTS (x) A proof of the result of Dočev can be obtained using the method given in [MI, Theorem 12, pp.28–29]. Further results of this form have been given by Clausing; see [General Inequalities].

INTEGRAL ANALOGUES (a) If $f \in \mathcal{L}_\mu([a,b])$, $f \geq 0$ then:
$$\mathfrak{G}_{[a,b]}(f; \mu) \leq \mathfrak{A}_{[a,b]}(f; \mu).$$

(b) [BERWALD] If $f \geq 0$ on $[a,b]$ is concave then:
$$\mathfrak{G}_{[a,b]}(f; \mu) \geq \frac{2}{e} \mathfrak{A}_{[a,b]}(f; \mu).$$

COMMENTS (xi) Inequality (b) is a particular case of Berwald's inequality, see **Favard's Inequalities** EXTENSIONS.
(xii) See also **Conjugate Convex Function Inequalities** COMMENTS (i), **Factorial Function Inequalities** (k), **Henrici's Inequality, Kalajdžić Inequality** COMMENTS (ii), **Ky Fan's Inequality, Levinson's Inequality, Log-convex Functions Inequalities** (2), **Matrix Inequalities** (c), **Mixed Mean Inequalities** SPECIAL CASES (b), **Muirhead Symmetric Function and Mean Inequalities, Order Inequalities** COMMENTS (v), **OShea's Inequality** COMMENTS, **Power mean Inequalities, Rearrangement Inequalities, Sierpinski's Inequalities, Symmetric Mean Inequalities** COMMENTS (ii), **Weierstrass's Inequalities** RELATED INEQUALITIES (b).

REFERENCES   All books on inequalities and many others contain discussions of (GA). [AI, pp. 84–85, 208, 344], [BB, pp. 3–15, 44], [HLP, pp. 17–21, 61, 136–139], [MI, pp. 43–94, 117–118, 122–126], [MPF, pp. 36–40], [MO, pp. 72–73]; [General Inequalities, vol. 3, pp. 61–62], [Hájos et al, pp. 70–72], [Pólya & Szegö, 1972, pp. 62–65, 67]; [Alzer, 1990], [Bullen, 1998].

Geometric Inequalities

(a) DERIVED FROM (GA). The equal weight case of (GA) and of (HG) have some simple geometric interpretations; in this section we will call these special cases (GA) and (HG) respectively.
(i) *The case $n = 2$.* Let ABCD be a trapezium with parallel sides AB, DC, with $AB = b > DC = a$. Let IJ and GH be parallel to the AB with the first bisecting AD and the second

dividing the trapezium into two similar trapezia. If K is the point of intersection of AC and BD, let the line EKF be parallel to AB. Then since $EF = \mathfrak{H}_2(a,b)$, $GH = \mathfrak{G}_2(a,b)$, $IJ = \mathfrak{A}_2(a,b)$ it follows from (HG) and (GA) that EF is nearer to DC than GH, and GH is nearer to DC than IJ.

(ii) *The case* $n = 3$. A simple deduction from (GA) in this case is:

of all the triangles of given perimeter the equilateral has greatest area.

(iii) It is easy to see that (GA) is equivalent to either of the following:

if $\prod_{i=1}^{n} a_i = 1$ then $\sum_{i=1}^{n} a_i \geq n$;

if $\sum_{i=1}^{n} a_i = 1$ then $\prod_{i=1}^{n} a_i \leq n^{-n}$;

with equality in either case if and only if $\underline{a}$ is constant.

These forms give (GA) simple geometric interpretations:

Of all n-parallelepipeds of given volume the one of least perimeter is the n-cube;

of all n-parallelepipeds of given perimeter the one of greatest volume is the n-cube;

of all the partitions of the interval $[0,1]$ *into* n *sub-intervals, the partition into equal sub-intervals is the one for which the product of the sub-interval lengths is the greatest.*

(b) DERIVED FROM S(r;s). If a_1, a_2, a_3, are the sides of a parallelepiped then $\mathfrak{P}_3^{[1]}(a_1, a_2, a_3)$, $\mathfrak{P}_3^{[2]}(a_1, a_2, a_3)$ and $\mathfrak{P}_3^{[3]}(a_1, a_2, a_3)$ are, respectively the side of the cube of the same perimeter, P say, the side of the cube of the same area, A say, and the side of the cube of the same volume, V say. The inequality S(r;s) says that unless $a_1 = a_2 = a_3$,

$$P > A > V.$$

(c) DERIVED FROM (C). If a, b, c are the lengths of the semi-axes of an ellipsoid of surface area S then

$$\frac{4\pi}{3}(ab + bc + ca) \leq S \leq \frac{4\pi}{3}(a^2 + b^2 + c^2),$$

and

$$S \geq 4\pi(abc)^{2/3}.$$

There is equality if and only if $a = b = c$.

(d) DERIVED FROM ORDER INEQUALITIES (b). If $ABC, A'B'C'$ are two triangles with the angles of the second lying between the smallest and largest angle of the first then if f is continuous and convex,

$$f(A') + f(B') + f(C') \leq f(A) + f(B) + f(C);$$

further if f is strictly convex then there is equality if and only if $A = A', B = B', C = C'$. In particular if $0 < kA \le \pi$ then

$$\sin kA + \sin kB + \sin kC \le \sin kA' + \sin kB' + \sin kC'.$$

COMMENTS   See also **Adamović's Inequality** COMMENTS (ii), **Bessel Function Inequalities** COMMENTS (ii), **Beth & van der Corput Inequality** COMMENTS (iii), **Blaschke-Santaló Inequality**, **Bonnenesen's Inequality**, **Brunn-Minkowski inequalities**, **Gale's Inequality**, **Hadamard's Determinant Inequality** COMMENTS (ii), **Isodiametric Inequality**, **Isoperimetric Inequalities**, **Jensen's Inequality** COMMENTS (ii), (viii), **Mahler's Inequalities**, **Mixed-Volume Inequalities**, **n-Simplex Inequality**, **Parallelogram Inequality** COMMENTS (ii), **Schwarz's Lemma** COMMENTS (ii), **Symmetrization Inequalities**, **Triangle Inequality** COMMENTS (iv), **Young's Inequalities** (e).
REFERENCES   [HLP, pp. 36–37], [MI, pp. 38–39, 53–55, 294–295], [MO, pp. 192–214]; [Pólya & Szegö, 1972, pp. 72–73]; [Oppenheim].

Geometric Mean Inequalities
(a) If $\underline{a}$ is a positive n-tuple,

$$\min \underline{a} \le \mathfrak{G}_n(\underline{a}; \underline{w}) \le \max \underline{a},$$

with equality if and only if $\underline{a}$ is constant.
(b) If $\underline{a}, \underline{b}$ are positive n-tuples,

$$\mathfrak{G}_n(\underline{a}; \underline{w}) + \mathfrak{G}_n(\underline{b}; \underline{w}) \le \mathfrak{G}_n(\underline{a} + \underline{b}; \underline{w}) \tag{1}$$

with equality if and only if $\underline{a} \sim \underline{b}$.
(c) [DAYKIN & SCHMEICHEL] If $\underline{a}$ is a positive n-tuple, define $a_{n+k} = a_k$, and the n-tuple $\underline{b}$ by $b_k = A_{k+n} - A_k$, $1 \le k \le n$, then:

$$n\mathfrak{G}_n(\underline{a}) \le \mathfrak{G}_n(\underline{b})$$

(d) If $\underline{a}$ is a positive n-tuple and the n-tuples $\underline{b}, \underline{c}$ are defined by $b_k = \mathfrak{A}_n(\underline{a}) - (n-1)a_k \ge 0$, $c_k = (\mathfrak{A}_n(\underline{a}) - a_k)/(n-1) \ge 0$, $1 \le k \le n$, then:

$$\mathfrak{G}_n(\underline{b}) \le \mathfrak{G}_n(\underline{a}) \le \mathfrak{G}_n(\underline{c}).$$

(e) [REDHEFFER & VOIGT] Let $\underline{a}$ be a positive sequence, with $\sum_{i=1}^{\infty} a_i < \infty$, then:

$$\sum_{i=1}^{\infty} (-1)^{i-1} i \mathfrak{G}(\underline{a}) \le \sum_{i=1}^{\infty} a_{2i-1},$$

whenever the left-hand side converges. If $a_{2i-1} \ne o(1/i)$ the inequality is strict, but if $a_{2i-1} = o(1/i)$ there is exactly one choice of $\{a_{2i}\}$ for which equality holds.
(f) [KLAMKIN & NEWMAN] If $\underline{a}$ is a non-negative n-tuple with $\sum_{i=1}^{n} a_i = 1$,

$$\mathfrak{G}_n(1 \pm \underline{a}) \ge (n \pm 1)\mathfrak{G}_n(\underline{a}),$$

and
$$\frac{\mathfrak{G}_n(1+\underline{a})}{\mathfrak{G}_n(1-\underline{a})} \geq \frac{n+1}{n-1}. \tag{2}$$

COMMENTS (i) Inequality (1) is sometimes called *Hölder's Inequality* and is just **Power Mean Inequalities** (4). It is easily seen to be equivalent to (H).

(ii) The inequality in (e) can be deduced from **Rado's Geometric-Arithmetic Mean Inequality** EXTENSION (2); various extensions are given in [Redheffer & Voigt].

(iii) The inequalities in (f) are related to **Weierstrass's Inequalities**, and to **Geometric-Arithmetic Mean Inequalities** EXTENSIONS (b), In particular (2) is the equal weight case of **Weierstrass's Inequalities** RELATED INEQUALITIES (b). In addition they can be deduced using Schur convexity.

(iv) See also **Alzer's Inequalities** (b), **Bessel Function Inequalities** (a), **Binomila Function Inequalities** (j), **Harmonic Mean Inequalities** (b) and (d), **Hölder's Inequality** SPECIAL CASES (a), **Kaluza-Szegö Inequality**, **Levinson's Inequality** SPECIAL CASES, **Logarithmic Mean Inequalities** (b). (c), (d), COROLLARIES, EXTENSIONS, **Mitrinović & Djoković's Inequality**, **Mixed Mean Inequalities** SPECIAL CASES (b), **Muirhead Symmetric Function and Mean Inequalities** COMMENTS (ii).

INTEGRAL ANALOGUES If $f, g \in \mathcal{L}[a,b]$,

$$\mathfrak{G}_{[a,b]}(f) + \mathfrak{G}_{[a,b]}(g) \leq \mathfrak{G}_{[a,b]}(f+g),$$

with equality if and only if the right-hand side is zero or $Af = Bg$ almost everywhere, where the real A, B are not both zero.

REFERENCES [HLP, pp. 21–24, 138], [MI, pp. 93–94, 170–171], [MO, p. 78]; [Pólya & Szegö, 1972, pp. 69–70]; [Redheffer & Voigt].

Gerber's Inequality
If $\alpha \in \mathbb{R}$, $n \in \mathbb{N}$, and $x > -1$,

$$\binom{\alpha}{n+1} x^{n+1} \begin{cases} \geq 0 \\ = 0 \\ \leq 0 \end{cases} \implies (1+x)^\alpha \begin{cases} \geq \sum_{i=0}^{n} \binom{\alpha}{i} x^i, \\ = \sum_{i=0}^{n} \binom{\alpha}{i} x^i, \\ \leq \sum_{i=0}^{n} \binom{\alpha}{i} x^i. \end{cases} \tag{1}$$

COMMENTS (i) This is a simple consequence of Taylor's Theorem for the function $a(x) = (1+x)^\alpha$.

(ii) (B) is the case $n = 1$ of (1).

(iii) See also **Kaczmarz & Steinhaus's Inequalities**.

EXTENSIONS If $a < 0 < b$ and $f :]a,b[\to \mathbb{R}$ has derivatives of order $(n+1)$ on $]a,b[$, let $T_n(x)$ be the n-th order Taylor polynomial of f centred at 0, that is

$$T_n(x) = \sum_{i=0}^{n} \frac{x^i}{i!} f^{(i)}(x).$$

Then:

$$x^{n+1} f^{(n+1)}(x) \begin{cases} \geq 0, \\ = 0, \\ \leq 0, \end{cases} a < x < b, \implies f(x) \begin{cases} \geq T_n(x), \\ = T_n(x), \\ \leq T_n(x). \end{cases}$$

Further if $x^{n+1} f^{(n+1)}(x)$ has only a finite number of zeros in $]a,b[$ the first and last inequalities are strict.

COMMENTS (iv) This is an immediate consequence of Taylor's theorem; for another such inequality see the last reference.

(v) See also **Enveloping Series Inequalities** COMMENTS (ii), **Trigonometric Function Inequalities** COMMENTS (vi).

REFERENCES [AI, pp. 35–36], [MPF, p. 66]; [General Inequalities, vol. 2, pp. 143–144].

Gini-Dresher Mean Inequalities

If $\underline{a}$ is a positive n-tuple, and $p, q \in \mathbb{R}$, with $p \geq q$, define

$$\mathfrak{B}_n^{p,q}(\underline{a}; \underline{w}) = \begin{cases} \left(\dfrac{\sum_{i=1}^{n} w_i a_i^p}{\sum_{i=1}^{n} w_i a_i^q} \right)^{1/(p-q)}, & \text{if } p \neq q, \\ \left(\prod_{i=1}^{n} a_i^{w_i a_i^p} \right)^{1/(\sum_{i=1}^{n} w_i a_i^p)}, & \text{if } p = q. \end{cases}$$

These quantities are known as *Gini* or *Gini-Dresher Means* of $\underline{a}$, and have been studied and generalized by various authors. The definition of $\mathfrak{B}_n^{p,p}(\underline{a}, \underline{w})$ is justified by limits.

Particular cases are other well known means:

$$\mathfrak{B}_n^{p,0}(\underline{a}; \underline{w}) = \mathfrak{M}_n^{[p]}(\underline{a}; \underline{w}), \ p \geq 0; \mathfrak{B}_n^{0,q}(\underline{a}; \underline{w}) = \mathfrak{M}_n^{[q]}(\underline{a}; \underline{w}), \ q \leq 0;$$

$$\mathfrak{B}_n^{p,p-1}(\underline{a}; \underline{w}) = \mathfrak{H}_n^{[p]}(\underline{a}; \underline{w}).$$

A special case of these means is the *Lehmer family of means*: if a, b are positive and r real

$$\mathfrak{Le}^r(a,b) = \frac{a^{r+1} + b^{r+1}}{a^r + b^r}.$$

Clearly:

$$\mathfrak{Le}^r(a,b) = \mathfrak{B}_2^{r+1,r}(a,b);$$

and in particular

$$\mathfrak{Le}^0(a,b) = \mathfrak{A}_2(a,b), \ \mathfrak{Le}^{-1/2}(a,b) = \mathfrak{G}_2(a,b), \ \mathfrak{Le}^{-1}(a,b) = \mathfrak{H}_2(a,b).$$

COMMENTS (i) These means have been extended to what are known as *Biplanar means*; see [MI]. In addition it is easy to define integral analogues.

(a) Unless $\underline{a}$ is constant

$$\min \underline{a} < \mathfrak{B}_n^{p,q}(\underline{a}; \underline{w}) < \max \underline{a}. \qquad (1)$$

(b) If $p_1 \leq p_2$, $q_1 \leq q_2$ then

$$\mathfrak{B}_n^{p_1,q_1}(\underline{a}; \underline{w}) \leq \mathfrak{B}_n^{p_2,q_2}(\underline{a}; \underline{w}). \qquad (2)$$

(c) [LOSONCZI] If $\underline{a}, \underline{b}, \underline{w}$ are positive n-tuples and if $p \geq 1 \geq q \geq 0$ then

$$\mathfrak{B}_n^{p,q}(\underline{a} + \underline{b}; \underline{w}) \leq \mathfrak{B}_n^{p,q}(\underline{a}; \underline{w}) + \mathfrak{B}_n^{p,q}(\underline{b}; \underline{w}).$$

COMMENTS   (ii) This last inequality, (c), generalizes **Counter Harmonic Mean Inequalities** (4). The conditions on p, q are necessary as well as sufficient.
(iii) See also **Liapunov's Inequality** COMMENTS (ii) for a particular case of (2);
(iv) For another kind of mean due to Gini see **Difference Means of Gini**.

REFERENCES   [MI, pp. 189–190, 316–317], [MPF, pp. 156–163, 491], [PPT, pp. 119–121]; [General Inequalities, vol. 3, pp. 107–122]; [Alzer 1988], [Losonczi & Páles, 1988], [Páles, 1988].

Goldberg-Straus Inequality   If $z_j \in \mathbb{C}, 1 \leq j \leq n$, with $\sigma = \left|\sum_{j=1}^n z_j\right|$, $\delta = \max_{1 \leq j,k \leq n} |z_j - z_k|$, and $\sigma\delta > 0$, then for all m-tuples $\underline{a}_j, 1 \leq j \leq n$,

$$\max \left|\sum_{j=1}^n z_j \underline{a}_{k_j}\right| \geq \frac{\sigma\delta}{2\sigma + \delta} \max_{1 \leq j \leq n} |\underline{a}_j|,$$

where the maximum on the left-hand side is over all permutations of the indices, $k_1, \ldots, k_n$.

COMMENTS   (i) The value of the constant on the right-hand side is not optimal for any n, but is the best that can be chosen independently of n.
(ii) A related inequality is **Complex Number Inequalities** (h).

REFERENCES   [General Inequalities, vol. 2, pp. 37–51; vol.3, pp. 195–204].

Goldstein's Inequality   If $\underline{a}_j, |\underline{a}_j| = 1, 1 \leq j \leq m$, are real n-tuples that do not lie in the same half-space then:

$$\left|\sum_{j=1}^m \underline{a}_j\right| < m - 2.$$

COMMENTS   (i) This is due to Klamkin & Newman; the case $n = 2$ is due to Goldstein.
(ii) The result is best possible in the sense that we can get as close to $m - 2$ by considering a sequence of convex polytopes that converge to a line segment.

REFERENCES   [Klamkin & Newman].

Gorny's Inequality If $f : [a,b] \to \mathbb{R}$ has bounded derivatives of all orders k, $0 \leq k \leq n$, define

$$M_k = \|f^{(k)}\|_{\infty,[a,b]}, \quad M_n' = \max\{M_n, M_0 n!(b-a)^{-n}\},$$

then

$$M_k \leq 4e^{2k} \binom{n}{k}^k M_0^{1-k/n} M_n^{k/n}, \quad 0 < k < n;$$

and

$$f^{(k)}((a+b)/2) \leq 16(2e)^k M_0^{1-k/n} M_n'^{k/n}, \quad 0 < k < n.$$

COMMENTS These are extensions of the **Hardy-Littlewood-Landau Inequalities**.
REFERENCES [AI, pp. 138–139].

Gram Determinant Inequalities (a) If $\underline{a}_j$, $1 \leq j \leq m$, are real n-tuples define

$$G = G(\underline{a}_1, \ldots, \underline{a}_m) = \det \begin{pmatrix} \underline{a}_1 \cdot \underline{a}_1 & \cdots & \underline{a}_1 \cdot \underline{a}_m \\ \vdots & \ddots & \vdots \\ \underline{a}_m \cdot \underline{a}_1 & \cdots & \underline{a}_m \cdot \underline{a}_m \end{pmatrix};$$

then,

$$\prod_{j=1}^{m} |\underline{a}_j|^2 \geq G(\underline{a}_1, \ldots, \underline{a}_m) \geq 0, \tag{1}$$

with equality on the right-hand side if and only if the $\underline{a}_j, 1 \leq j \leq m$, are linearly dependent, and equality on the left-hand side if and only if the $\underline{a}_j, 1 \leq j \leq m$, are mutually orthogonal.
(b) With the above notation,

$$\frac{G(\underline{a}_1, \ldots, \underline{a}_m)}{G(\underline{a}_1, \ldots, \underline{a}_k)} \leq \frac{G(\underline{a}_2, \ldots, \underline{a}_m)}{G(\underline{a}_2, \ldots, \underline{a}_k)} \leq \cdots \leq G(\underline{a}_{k+1}, \ldots, \underline{a}_m). \tag{2}$$

In particular

$$G(\underline{a}_1, \ldots, \underline{a}_m) \leq G(\underline{a}_1, \ldots, \underline{a}_k) G(\underline{a}_{k+1}, \ldots, \underline{a}_m).$$

(c) If $\underline{a}_j, \underline{b}_j, 1 \leq k \leq m$, are real n-tuples then

$$G(\underline{a}_1, \ldots, \underline{a}_m) G(\underline{b}_1, \ldots, \underline{b}_m) \geq \left| \det \begin{pmatrix} \underline{a}_1 \cdot \underline{b}_1 & \cdots & \underline{a}_1 \cdot \underline{b}_m \\ \vdots & \ddots & \vdots \\ \underline{a}_m \cdot \underline{b}_1 & \cdots & \underline{a}_m \cdot \underline{b}_m \end{pmatrix} \right|^2, \tag{3}$$

with equality if and only if the vectors $\underline{a}_j, 1 \leq j \leq m$, and $\underline{b}_j, 1 \leq j \leq m$, span the same space.

COMMENTS (i) This result holds for m vectors in a unitary space.

(ii) The determinant G is called *the Gram determinant*, and the inequality on the right in (1) is called *Gram's Inequality*.
(iii) The inequality (3) can be regarded as a generalization of (C), to which it reduces when $m = 1$.
(iv) For an integral analogue see **Mitrinović & Pečarić's Inequality**.

REFERENCES   [AI, pp. 45–48], [EM, vol.4, p. 293], [MPF, pp. 595–609], [PPT, p. 201]; [Courant & Hilbert, pp. 34–36], [Mitrinovic & Pečarić, 1991$^{(2)}$, pp. 46–51]; [Dragomir & Mond], [Sinnadurai, 1963.]

Greub & Rheinboldt's Inequality
See **Polyá & Szegö's Inequalities** COMMENTS (iii).

Gronwall-Bellman Inequalities
See **Gronwall's Inequality** COMMENTS (iii).

Gronwall's Inequality   Let f, u, v be continuous on $[a, b]$ with $v \geq 0$ then

$$f(x) \leq u(x) + \int_a^x vf, \quad a \leq x \leq b,$$

implies that

$$f(x) \leq u(x) + \int_a^x u(y) v(y) e^{\int_y^x v} \, dy, \quad a \leq x \leq b. \tag{1}$$

COMMENTS   (i) The significance of this result is that the right-hand side of (1) no longer involves f.
(ii) If u is absolutely continuous the right-hand side of (1) can be written

$$e^{\int_a^x v} \left(u(a) + \int_a^x u'(y) e^{\int_y^a v} \, dy \right).$$

(iii) This inequality is of importance in the theory of differential equations, as a result there have been many extensions both in the conditions under which it applies, and to similar inequalities for other differential equations. Related results are due to Bellman and Bihari; so that the reference is made to inequalities of *Gronwall-Bellman*, *Bellman-Bihari*, and *Bihari-Bellman type*.

REFERENCES   [AI, pp 374–375], [BB. pp. 131–136], [EM, vol. 3, pp. 170–171]; [Inequalities III, pp. 333–340], [Walter, pp. 11–41].

Grothendieck's Inequality   If $n \geq 2$ and $(a_{ij})_{1 \leq i,j \leq n}$ is a real or complex square matrix with the property that

$$\left| \sum_{i,j=1}^n a_{ij} x_i y_j \right| \leq \sup_{1 \leq i,j \leq n} |x_i| |y_j|,$$

then there is a constant $K_G(n)$ such that for any elements $\underline{u}, \underline{v}$ of an n-dimensional Hilbert space,

$$\left| \sum_{i,j=1}^{n} a_{ij} \langle \underline{u}_i, \underline{v}_j \rangle \right| \leq K_G(n) \sup_{1 \leq i,j \leq n} \|\underline{u}_i\| \|\underline{v}_j\|.$$

COMMENTS (i) The constant $K_G(n)$ increases with n and the limit, as $n \to \infty$, K_G say, is called *Grothendieck's constant*.
(ii) In the real case $\pi/2 < K_G \leq \pi/2 \log(1 + \sqrt{2})$.
(iii) The value in the complex case is smaller since then $K_G < \pi/2$.
REFERENCES [MPF, pp. 569–571]; [General Inequalities, vol. 6, pp. 201–206, 481].

Grunsky's Inequalities
If f is univalent in D with $f(0) = f'(0)$ then for all complex n-tuples $\underline{z}$ with $z_r \in D$, $1 \leq r \leq n$, and all complex n-tuples $\underline{w}$

$$\sum_{r,s=1}^{n} w_r \overline{w}_s \log \frac{1}{1 - z_r \overline{z}_s} \geq \left| \sum_{r,s=1}^{n} w_r \overline{w}_s \log \left\{ \frac{z_r z_s}{f(z_r) f(z_s)} \frac{f(z_r) - f(z_s)}{z_r - z_s} \right\} \right|.$$

COMMENTS (i) Whenever $z_r = z_s$ the last term on the right-hand side is interpreted as $f'(z_s)$
(ii) Conversely if f is analytic on D with $f(0) = f'(0)$ and the inequality holds for all such $\underline{z}, \underline{w}$ then f is univalent.
(iii) This inequality can be written as $\underline{w}^* A \underline{w} \geq |\underline{w}^T B \underline{w}|$, where $\underline{w}^T$ is the transpose of $\underline{w}$ and A, B are the matrices with entries the log term on the left-hand side, right-hand side, respectively.
(iv) For other forms of this inequality see [EM].
REFERENCES [EM, vol. 1, p. 245; vol.3, p. 269; vol. 9, p. 338]; [Horn & Johnson, p. 202].

Grüss-Barnes's Inequality
If $f, g : [0,1] \to [0, \infty[$ are concave, and if $p, q \geq 1$ then

$$\int_0^1 fg \geq \frac{(p+1)^{1/p}(q+1)^{1/q}}{6} \|f\|_p \|g\|_q. \tag{1}$$

If $0 < p, q \leq 1$ (~1) holds with 6 replaced by 3.

COMMENTS (i) The case $p = q = 1$ is due to Grüss, the case $p, q \geq 1$ and conjugate indices is by Bellman and the general result is by Barnes.
(ii) The general case follows from Grüss's result and **Favard's Inequalities** (b).

EXTENSIONS [BORELL] (a) Under the same conditions as above

$$\int_0^1 fg \geq \frac{(p+1)^{1/p}(q+1)^{1/q}}{6} \|f\|_p \|g\|_q + \frac{f(0)g(0) + f(1)g(1)}{6}.$$

(b) If $f_i, 1 \leq i \leq n$ satisfy the conditions above then

$$\int_0^1 \prod_{i=1}^n f_i \geq \frac{[(n+1)/2]![n/2]!}{(n+1)!} \prod_{i=1}^n (p_i+1)^{1/p_1} \|f\|_{p_i}.$$

COMMENTS (iii) (b) was published independently by Godunova & Levin[2].

REFERENCES [AI, p. 73], [PPT, pp. 223–224]; [Maligranda, Pečarić & Persson, 1994].

Grüsses'[3] Inequalities
(a) If f, g are integrable on $[0,1]$, with $\alpha \leq f \leq A$, and $\beta \leq g \leq B$, then

$$\left|\mathfrak{A}_{[0,1]}(fg) - \mathfrak{A}_{[0,1]}(f)\mathfrak{A}_{[0,1]}(g)\right| \leq \frac{1}{4}(A-\alpha)(B-\beta).$$

(b) [G.GRÜSS] If $m, n \in \mathbb{N}$ then

$$\frac{1}{m+n+1} - \frac{1}{(m+1)(n+1)} \leq \frac{4}{45}.$$

COMMENTS (i) (a) is a converse of the integral analogue of (Č). The result was conjectured by H Grüss and proved by G Grüss. For another such result see **Karamata's Inequalities**.
(ii) The constant on the right-hand side of (a) is best possible as can be seen by taking $f(x) = g(x) = \operatorname{sgn}(x - (a+b)/2)$.

DISCRETE ANALOGUES [BIERNACKI, PIDEK & RYLL-NARDZEWSKI] If $\underline{a}, \underline{b}$ are real n-tuples with $\alpha \leq \underline{a} \leq A$, and $\beta \leq \underline{b} \leq B$, then

$$|\mathfrak{A}(\underline{a}\,\underline{b}) - \mathfrak{A}(\underline{a})\mathfrak{A}(\underline{b})| \leq \frac{1}{n}\left[\frac{n}{2}\right]\left(1 - \frac{1}{n}\left[\frac{n}{2}\right]\right)(A-\alpha)(B-\beta).$$

(iii) There are many generalizations in the references; see also **Ostrowski's Inequalities** (b).

REFERENCES [AI, pp.70–74, 190–191], [MPF, pp.295–310], [PPT, 206–212].

Guha's Inequality
If $p \geq q > 0$, $x \geq y > 0$ then

$$(px + y + a)(x + qy + a) \geq [(p+1)x + a][(q+1)y + a].$$

COMMENTS (i) A proof of this inequality follows by noting that the difference between the left-hand side and right-hand side is just $(px - qy)(x - y)$.
(ii) This inequality is the basis of a proof of (GA).

REFERENCES [MI, p. 77].

- end of G -

---

[2] Е К Годунова, В И Левин.

[3] This refers to H & G Grüss.

H

Haber's Inequality If $a + b > 0$, $a \neq b$ then

$$\left(\frac{a+b}{2}\right)^n < \frac{a^n + a^{n-1}b + \cdots + b^n}{n+1}, \quad n = 2, 3, \ldots; \tag{1}$$

if $a + b < 0$ then (1) holds if n is odd, but if n is even the ($\sim$ 1) holds.

COMMENTS (i) This can be proved by induction.
(ii) A special case of (1) is **Polynomial Inequalities** (4).

EXTENSIONS [MERCER] If $\underline{a} = \{a_0, a_1, \ldots\}$ is convex then

$$\frac{1}{2^n} \sum_{i=0}^{n} \binom{n}{i} a_i \leq \mathfrak{A}_{n+1}(\underline{a}).$$

REFERENCES [General Inequalities, vol. 2, pp. 143–144]; [Mercer].

Hadamard's Determinant Inequality If A is an $m \times n$ complex matrix with rows the complex n-tuples $\underline{r}_k$, $1 \leq k \leq m$, then:

$$\det(AA^*) \leq \prod_{k=1}^{m} |\underline{r}_k|^2.$$

There is equality if one of $\underline{r}_k$ is zero, or, if no $\underline{r}_k$ is zero and the $\underline{r}_k$, $1 \leq k \leq m$, are orthogonal.

COMMENTS (i) This inequality is often referred to as *Hadamard's inequality*; as is Corollaries (b) below.
(ii) In the real case the geometric meaning of the Hadamard inequality is:

in $\mathbb{R}^n$ the volume of a polyhedron formed by n vectors is largest when the vectors are orthogonal.

COROLLARIES (a) If A is an $n \times n$ complex matrix with rows the complex n-tuples $\underline{r}_k$, $1 \leq k \leq n$, and columns the complex n-tuples $\underline{c}_k$, $1 \leq k \leq n$, then

$$|\det A| \leq \min\left\{\prod_{k=1}^{n} |\underline{r}_k|, \prod_{k=1}^{n} |\underline{c}_k|\right\}.$$

(b) If A is a positive definite $n \times n$ matrix then

$$\det A \leq \prod_{i=1}^{n} a_{ii},$$

with equality if and only if A is diagonal.

COMMENTS (iii) The cases of equality for the corollaries follow easily from those in the main result.

REFERENCES [BB, p. 64], [EM, vol.4, pp. 350–351], [HLP, p. 34–36], [MO, pp. 223–224], [MPF, p. 213]; [Courant & Hilbert, pp. 36–37], [Horn & Johnson, pp. 476–482], [Marcus & Minc, p. 114], [Price, pp. 607–610].

Hadamard's Integral Inequality
See **Hermite-Hadamard Inequality**.

Hadamard's Three Circles Theorem
If f is analytic on the closed annulus $a \leq |z| \leq b$ then:

$$M_\infty(f;r) \leq M_\infty(f;a)^{\log(b/r)/\log(b/a)} M_\infty(f;b)^{\log(r/a)/\log(b/a)}, \quad a \leq r \leq b.$$

COMMENTS (i) This result says that $M_\infty(f;r)$ is a log-convex function of $\log r$.
(ii) A similar result holds for harmonic functions, and can be extended to the solutions of more general elliptic partial differential equations; see **Harnack's Inequalities** COMMENTS.
(iii) See also **Hardy's Analytic Function Inequality** and **Phragmen-Lindelöf Inequality**.

REFERENCES [AI, p. 19], [EM, vol. 4, p. 351]; [Ahlfors,1966, p. 165], [Conway, vol.I, p. 133], [Pólya & Szegö, 1972, pp. 164–166], [Protter & Weinberger, pp. 128–137].

Hajela's Inequality
Let μ be a probability measure on X, $1 \leq \sqrt[3]{3}p < q < \infty$, $f \in \mathcal{L}^p_\mu(X)$, not constant μ-almost everywhere, then

$$0 < \frac{q-p}{q}\left(\|f\|^p_{q,\mu} - \|f\|^p_{p,\mu} - \|f\|^p_{p,\mu}\log\|f\|^p_{q,\mu} + \int_X |f|^p \log|f|^p \, d\mu\right)$$
$$\leq \|f\|^p_{q,\mu} - \|f\|^p_{p,\mu}.$$

COMMENTS The left inequality holds under the wider condition $1 \leq p < q$.

REFERENCE [Hajela].

Halmos's Inequality *If f is absolutely continuous on $[0,1]$ with $f(0) = f(1) = 1$ and $\int_0^1 |f'| = 1$ then $|f| \leq 1/2$.*

REFERENCES   [Halmos, p. 62].

Hamy Mean Inequalities
If $\underline{a}$ is a positive n-tuple, and if $1 \leq r \leq n, r \in \mathbb{N}$, define

$$\mathfrak{S}_n^{[r]}(\underline{a}) = \frac{1}{r!\binom{n}{r}} \sideset{}{'}\sum_{r} \left(\prod_{j=1}^r a_{i_j} \right)^{1/r}.$$

This quantity is called the *Hamy Mean of $\underline{a}$*.
It reduces to well known means in special cases:

$$\mathfrak{S}_n^{[n]}(\underline{a}) = \mathfrak{P}_n^{[n]}(\underline{a}) = \mathfrak{G}_n(\underline{a}), \text{ and } \mathfrak{S}_n^{[1]}(\underline{a}) = \mathfrak{P}_n^{[1]}(\underline{a}) = \mathfrak{A}_n(\underline{a}).$$

(a) With the above notation, if $r, s \in \mathbb{N}, 1 \leq r < s \leq n$, then:

$$\mathfrak{S}_n^{[s]}(\underline{a}) \leq \mathfrak{S}_n^{[r]}(\underline{a}),$$

with equality if and only if $\underline{a}$ is constant.
(b) If $1 \leq r \leq n$,

$$\mathfrak{S}_n^{[r]}(\underline{a}) \leq \mathfrak{P}_n^{[r]}(\underline{a}),$$

with equality if and only if either $r = 1$, $r = n$ or $\underline{a}$ is a constant.

COMMENTS   (b) is an easy consequence of (r;s).
REFERENCES   [MI, pp. 311–313].

Hanner's Inequalities *If $f, g \in \mathcal{L}_\mu^p(X)$ and $2 \leq p < \infty$ then*

$$\|f+g\|_p^p + \|f-g\|_p^p \leq \left| \|f\|_p - \|g\|_p \right| + (\|f\|_p + \|g\|_p)^p; \quad (1)$$

$$2^p(\|f\|_p^p + \|g\|_p^p) \leq (\|f+g\|_p + \|f-g\|_p)^p + \left| \|f+g\|_p - \|f-g\|_p \right|^p. \quad (2)$$

If $1 \leq p \leq 2$ then $(\sim 1), (\sim 2)$ hold.

COMMENTS   These inequalities are connected with the uniform convexity of the $\mathcal{L}^p$ spaces. See also **Clarkson's Inequalities**
REFERENCES   [Lieb & Loss, pp. 42–44, 69].

Hardy's Analytic Function Inequality *If f is analytic in the open annulus $a < |z| < b$ and $1 \leq p < \infty$ then, $M_p(f;r)$ is a log-convex function of $\log r, a < r < b$.*

COMMENTS   This result, which also holds for harmonic functions, should be compared to **Hadamard's Three Circle Theorem**, and the **Phragmen-Lindelöf Inequality**.
REFERENCES   [AI, p. 19]; [Conway, vol. I, p. 134], [Inequalities III, pp. 9–21].

Hardy's Inequality If $p > 1$ and $\underline{a}$ is a non-negative, non-null ℓ_p sequence,

$$\sum_{i=1}^{\infty} \mathfrak{A}_i^p(\underline{a}) < \left(\frac{p}{p-1}\right)^p \sum_{i=1}^{\infty} a_i^p; \qquad (1)$$

or

$$||\mathfrak{A}||_p < \frac{p}{p-1}||\underline{a}||_p.$$

The constant is best possible.

COMMENTS (i) Both inequality (1) and its integral analogue, (5) below, are known as Hardy's inequality, or sometimes as the Hardy-Landau inequality, as it was Landau who determined the value of the constant.
(ii) Putting a_i for $a_i^p, i \geq 1$, in (1) and letting $p \to \infty$, gives a weak form of **Carleman's Inequality**.

EXTENSIONS (a) [REDHEFFER] With the hypotheses of (a) above,

$$\sum_{i=1}^{n} \mathfrak{A}_i^p(\underline{a}) + \frac{np}{p-1}\mathfrak{A}_n^p(\underline{a}) < \left(\frac{p}{p-1}\right)^p \sum_{i=1}^{n} a_i^p. \qquad (2)$$

(b) [COPSON] If $p > 1, a_i \geq 0, w_i > 0, i = 1, \ldots,$ and not all the a_i are zero then

$$\sum_{i=1}^{\infty} w_n \mathfrak{A}_i^p(\underline{a}; \underline{w}) < \left(\frac{p}{p-1}\right)^p \sum_{i=1}^{\infty} w_n a_i^p. \qquad (3)$$

COMMENTS (iii) Inequality (2) is an example of a recurrent inequality, see **Recurrent Inequalities**.
(iv) Inequality (3) can be deduced from (1) and **Hölder's Inequality** OTHER FORMS (c). It is known as Copson's inequality, it is the weighted form of (1). Another inequality of the same name is discussed in **Copson's Inequality**.

A converse and a further important extension of Hardy's Inequality, (1), needs some extra notation.
If $p > 1$ and $\underline{a}$ is a real sequence such that $\sum_{i=1}^{n}|a_i|^p = O(n)$ we say that $\underline{a} \in \mathfrak{g}(p)$. The collection of all such sequences is a Banach space with norm defined by

$$||\underline{a}||_{\mathfrak{g}(p)} = \sup_n \left(\frac{1}{n}\sum_{i=1}^{n}|a_i|^p\right)^{1/p}.$$

Consider all factorizations of a real sequence $\underline{a}$ as $\underline{a} = \underline{u}\,\underline{v}, \underline{u} \in \ell_p, \underline{v} \in \mathfrak{g}(q)$, q the conjugate index, and define

$$!\underline{a}!_p = \inf ||\underline{u}||_p ||\underline{v}||_{\mathfrak{g}(q)},$$

where the inf is taken over all such factorizations of $\underline{a}$.

EXTENSIONS (c) [BENNETT] If $p > 1$,

$$(p-1)^{-1/p} !\underline{a}!_p \leq ||\mathfrak{A}_i^p(\underline{a})||_p \leq q !\underline{a}!_p, \qquad (4)$$

q being the conjugate index. The constants are best possible, and the right-hand inequality is strict unless $\underline{a} = \underline{0}$.

COMMENTS (v) Bennett's result actually gives more: the centre term in the inequalities is finite if and only if the sequence admits a factorization of the type above.
(vi) (1) follows from (4) by taking $\underline{u} = \underline{a}, \underline{v} = \underline{e}$.
(vii) Other converse inequalities have been studied by Neugebauer.

INTEGRAL ANALOGUES If $p > 1$ and, $f \geq 0$,

$$\int_0^\infty \left(\frac{1}{x}\int_0^x f\right)^p dx \leq \left(\frac{p}{p-1}\right)^p \int_0^\infty f^p; \qquad (5)$$

$$\int_0^\infty \left(\int_x^\infty f\right)^p dx \leq p^p \int_0^\infty x^p f^p(x) \, dx. \qquad (6)$$

The inequalities are strict unless $f = 0$ almost everywhere, and the constants are best possible.

COMMENTS (viii) Inequalities (5) and (6) can be rewritten by putting $F' = f$ to give an inequality between a function and its derivative. In this and the original form they have been the subject of much study and generalization. These extensions include introducing weights, varying the exponents on each side, using fractional integrals, replacing the Cesàro matrix on the left-hand side by a more general matrix, and going to higher dimensions, amongst many others. See **Bennett's Inequalities** (b), **HELP Inequalities**, **Sobolev's Inequalities**.

EXTENSIONS (a) If $f \geq 0$, and $p > 1, m > 1$ or $p < 0, m < 1$, then:

$$\int_0^\infty x^{-m} \left(\int_0^x f\right)^p dx \leq \left|\frac{p}{m-1}\right|^p \int_0^\infty x^{p-m} f^p(x) \, dx. \qquad (7)$$

If $0 < p < 1, m \neq 1$ then (~ 7) holds.
(b) [IZUMI & IZUMI] If $p > 1, m > 1, f > 0$ then

$$\int_0^\pi x^{-m} \left(\int_{x/2}^x f\right)^p dx < \left(\frac{p}{m-1}\right)^p \int_0^\pi x^{-m} |f(x/2) - f(x)|^p \, dx.$$

REFERENCES [AI, p. 131], [EM, vol.4, p. 369], [HLP, pp. 239–249], [PPT, pp. 229–239]; [Bennett, 1996, pp. 13–18], [General Inequalities, vol. 2, pp. 55–80; vol. 3, pp. 205–218; vol. 4, pp. 47–57; vol. 6, pp. 33–58; vol. 7, pp. 3–16], [Grosse-Erdmann], [Opic & Kufner], [Zwillinger, p. 206]; [Leindler], [Mohupatra & Vajravelu], [Neugebauer], [Pachpatte, 1988], [Pachpatte & Love], [Ross].

Hardy-Landau Inequality
See **Hardy's Inequality** COMMENTS (i).

Hardy-Littlewood-Landau Derivative Inequalities

If $f : [0,\infty[\to \mathbb{R}$ has bounded derivatives of orders $k, 0 \leq k \leq n$, let $M_k = \|f^{(k)}\|_\infty$ then:

$$M_k^n \leq C_{n,k}^n M_0^{n-k} M_n^k, \quad 0 < k < n; \tag{1}$$

where

$$C_{n,1} \leq 2^{n-1}. \tag{2}$$

In particular:0

$$C_{2,1}^2 = 2,\ C_{3,1}^3 = 9/8,\ C_{3,2}^3 = 3,\ C_{4,1}^4 = 512/375,\ C_{4,2}^4 = 36/25,\ C_{5/2}^5 = 125/72.$$

COMMENTS (i) Inequality (2), and the value of $C_{2,1}$ are due to Hadamard, and the other values were obtained by Šilov[1], and the values are best possible.
(ii) The study of such inequalities was emphasised by Hardy, Littlewood and Landau.
(iii) It is natural to try to extend (1) to other $\mathcal{L}^p$ norms; see **Hardy-Littlewood-Pólya Inequalities (a)**, **Kolmogorov's Inequalities**. Another find of inequality between derivatives can be found in **Sobolev's Inequalities**.
(iv) For another extension see **Gorny's Inequality**, and for ones where ordinary derivatives are replaced by by Sturm-Liouville expressions see **HELP Inequalities**.
(v) The above inequalities have also been studied in their discrete forms; see [General Inequalities].
(vi) See also **Halmos's Inequality, Opial's Inequalities, Wirtinger's Inequality**.

REFERENCES [AI, pp. 138–140], [EM, vol.5, p. 295], [HLP, pp. 187–193]; [General Inequalities, vol. 5, pp. 367–379; vol. 6, pp. 459–462]; [Benyon, Brown & Evans].

Hardy-Littlewood Maximal Inequalities

(a) If f is a non-negative measurable function on $]0, a[$ that is finite almost everywhere and if $E \subseteq [0, a]$ is measurable then

$$\int_E f \leq \int_0^{|E|} f^*.$$

(b) [THE MAXIMAL THEOREM] If $f \in \mathcal{L}(0, a), f \geq 0$ define

$$\theta(x) = \theta_f(x) = \sup_{0 \leq y < x} \frac{1}{x-y} \int_y^x f, \quad 0 < x \leq a.$$

If then $\chi \geq 0$ is an increasing function,

$$\int_0^a \chi \circ \theta_f \leq \int_0^a \chi \circ \theta_{f^*}.$$

COMMENTS (i) Since f^* is decreasing the function θ_{f^*} is given by

$$\theta_{f^*}(x) = \frac{1}{x} \int_0^x f^*.$$

---

[1] Г Е Шилов.

(ii) If we replace f^* by f_* and θ_f by

$$\theta^{\blacktriangle}(x) = \theta_f^{\blacktriangle}(x) = \sup_{x<y\leq a} \frac{1}{y-x} \int_x^y f, \quad 0 \leq x < a.$$

the above result remains valid.
(iii) Hence if

$$\Theta_f(x) = \max\{\theta_f(x), \theta_f^{\blacktriangle}(x)\} = \sup_{\substack{0\leq y\leq a \\ y\neq x}} \frac{1}{y-x} \int_x^y f, \quad 0 \leq x \leq a.$$

we get that

$$\int_0^a \chi \circ \Theta_f \leq 2 \int_0^a \chi \circ \theta_{f^*}.$$

Taking particular functions χ we get the following important results.

COROLLARIES (a) If $f \in \mathcal{L}^p(0,a), p > 1, f \geq 0$ then $\Theta_f \in \mathcal{L}_p(0,a)$ and

$$\int_0^a \Theta_f^p \leq 2\left(\frac{p}{p-1}\right)^p \int_0^a f^p.$$

(b) If $f \in \mathcal{L}(0,a), f \geq 0$ then $\Theta_f \in \mathcal{L}^p(0,a), 0 < p < 1$, and

$$\int_0^a \Theta_f^p \leq \frac{2a^{1-p}}{1-p} \int_0^a f^p.$$

(c) If $f\log^+ f \in \mathcal{L}(0,a), f \geq 0$ then $\Theta_f \in \mathcal{L}(0,a)$ and

$$\int_0^a \Theta_f \leq 4 \int_0^a f\log^+ f \, + \, C.$$

COMMENTS   (iv) The above can easily be extended to functions non-negative, measurable, finite almost everywhere on any bounded interval $[a,b]$.

If f is measurable, finite almost everywhere and periodic, with period 2π, define

$$M(x) = M_f(x) = \sup_{0<|t|<\pi} \frac{1}{t} \int_x^{x+t} |f|;$$

and let $\theta_{|f|}$ be defined using the interval $[-2\pi, 2\pi]$, then:

$$M_f \leq \theta_{|f|}.$$

COMMENTS   (v) The above results imply integral inequalities for the function M_f, the Hardy-Littlewood maximal function.
(vi) See also **Kakeya's Maximal Function Inequality, Rearrangement Inequalities, Spherical Rearrangement Inequalities**.

REFERENCES   [EM, vol. 4, pp. 370–371]; [Hewitt& Stromberg, pp. 422–429], [Zygmund, vol. I, pp. 30–33].

Hardy-Littlewood-Pólya Inequalities
(a) If $f, f'' \in \mathcal{L}^2([0, \infty[)$ then:

$$\|f'\|_{2,[0,\infty[}^2 < 2\|f\|_{2,[0,\infty[}\|f''\|_{2,[0,\infty[}, \tag{1}$$

unless $f(x) = Ae^{-BX/2}\sin(Bx\sqrt{3}/2 - \pi/3)$ almost everywhere.
If $f, f'' \in \mathcal{L}^2(\mathbb{R})$ then

$$\|f'\|_{2,\mathbb{R}}^2 < \|f\|_{2,\mathbb{R}}\|f''\|_{2,\mathbb{R}}, \tag{2}$$

unless $f(x) = 0$ almost everywhere.
(b) If $\underline{a}, \underline{b}$ are non-negative, non-nul sequences, p, q conjugate indices then:

$$\sum_{i=1}^{\infty}\sum_{j=1}^{\infty} \frac{a_i b_j}{\max\{i,j\}} < pq \left(\sum_{i=1}^{\infty} a_i^p\right)^{1/p} \left(\sum_{j=1}^{\infty} b_j^q\right)^{1/q}. \tag{3}$$

The constant is best possible.

COMMENTS (i) (1) is often called the *Hardy-Littlewood-Pólya inequality* and has been the object of much research; the same name is also given to the much easier inequality (2). It is natural to see what can be said for the other $\mathcal{L}^p$ classes; see **Hardy-Littlewood-Landau Derivative Inequalities**, **Kolmogorov's Inequalities**.
In addition the possibility of weights, and more general derivative operators has been explored; see [General Inequalities], **HELP Inequalities**.
(ii) An integral analogue of (3) is easily stated.

DISCRETE ANALOGUES [COPSON]   If $\underline{a}$ is a real sequence,

$$\left(\sum_{n=0}^{\infty}(\Delta a_n)^2\right)^2 \leq 4 \sum_{n=0}^{\infty} a_n^2 \sum_{n=0}^{\infty}(\Delta^2 a_n)^2;$$

$$\left(\sum_{n=-\infty}^{\infty}(\Delta a_n)^2\right)^2 \leq \sum_{n=-\infty}^{\infty} a_n^2 \sum_{n=-\infty}^{\infty}(\Delta^2 a_n)^2;$$

with equality if and only if $\underline{a} = \underline{0}$.

COMMENTS (iii) As with the integral originals these discrete inequalities have been extended, to higher order differences for instance.
(iii) See also **Multilinear Form Inequalities** and **Sobolev's Inequalities**.
REFERENCES [HLP, pp. 187–193, 254], [PPT, p. 234]; [General Inequalities, vol. 5, pp. 29–63]; [Borogovac & Arslanagić].

Hardy-Littlewood-Pólya-Schur Inequalities
Let $p > 1$, q the conjugate index, and $K : [0, \infty[\times[0, \infty[\to [0, \infty[$, homogeneous of degree -1 with

$$\int_0^{\infty} \frac{K(x,1)}{x^{1/p}}\,dx = \int_0^{\infty} \frac{K(1,y)}{y^{1/q}}\,dx = C, \tag{1}$$

where both of the integrands are strictly decreasing then:

$$\sum_{m,n\in\mathbb{N}} K(m,n)a_m b_n \leq C||\underline{a}||_p ||\underline{b}||_q, \qquad (2)$$

with equality if and only if $\underline{a} = \underline{b} = \underline{0}$;
and

$$\left(\sum_{n\in\mathbb{N}}\left(\sum_{m\in\mathbb{N}} K(m,n)a_m\right)^p\right)^{1/p} \leq C||\underline{a}||_p, \quad \left(\sum_{m\in\mathbb{N}}\left(\sum_{n\in\mathbb{N}} K(m,n)b_n\right)^q\right)^{1/q} \leq C||\underline{b}||_q,$$

with equality if and only if $\underline{a} = \underline{0}$ in the first case and $\underline{b} = \underline{0}$ in the second.
In all cases the constant in the right-hand side is best possible.

COMMENTS (i) The case $p = 2$ is due to Schur, and the result has been the object of much study and generalization. Homogeneous functions are defined in **Segre's Inequalities**.
(ii) A particular case of (2) is **Hilbert's Inequalities** (2).

INTEGRAL ANALOGUES Let $p > 1$, q the conjugate index, and $K : [0,\infty[^2 \to [0,\infty[$, homogeneous of degree -1 with (1) holding, then:

$$\int_0^\infty \int_0^\infty K(x,y)f(x)g(y)\,dx\,dy \leq C||f||_{p,[0,\infty[}||g||_{q,[0,\infty[}; \qquad (3)$$

$$\left(\int_0^\infty \left(\int_0^\infty K(x,y)f(x)\,dx\right)^p dy\right)^{1/p} \leq C||f||_{p,[0,\infty[}; \qquad (4)$$

$$\left(\int_0^\infty \left(\int_0^\infty K(x,y)g(y)\,dy\right)^q dx\right)^{1/q} \leq C||g||_{q,[0,\infty[}. \qquad (5)$$

If K is positive there is equality in (3) if and only if $f = 0$ almost everywhere, and $g = 0$ almost everywhere, in (4) if and only if $f = 0$ almost everywhere, and in (5) if and only if $g = 0$ almost everywhere.

COMMENTS (iii) A particular case of (3) is **Hilbert's Inequalities** (3).
(iv) The various forms of these inequalities are equivalent; see **Bilinear Form Inequalities**.
REFERENCES [HLP, pp. 227–232, 243–246]; [General Inequalities, vol. 2, pp. 277–286, 458–461; vol. 3, p. 207].

Hardy-Littlewood-Sobolev Inequalities

(a) If $f_1,\ldots,f_m, h$ are non-negative,

$$\int_\mathbb{R} (f_1 \star \cdots \star f_m)h \leq \int_\mathbb{R} (f_1^{(*)} \star \cdots \star f_m^{(*)})h^{(*)}.$$

(b) If $f \in \mathcal{L}^p(\mathbb{R}^n), g \in \mathcal{L}^q(\mathbb{R}^n)$, where $p, q > 1$ and $1/p + 1/q + \lambda/n = 2, 0 < \lambda < n$, then

$$\left|\int_{\mathbb{R}^n}\int_{\mathbb{R}^n} f(x)|x-y|^{-\lambda} h(y)\,dx\,dy\right| \leq C||f||_p ||g||_q,$$

where C depends on n, λ, p only.

COMMENTS (i) The definition of convolution given in **Notations 4** is easily extended to the convolution of more than two functions.
(ii) The case $m = 2$ of (a) is by F Riesz, based on an argument of Hardy-Littlewood; the extension to general m is due to Sobolev.
(iii) Inequality (a) has been further generalized in the reference.
(iv) In the case $p = q$ of (b) the exact value of the constant as well as the cases of equality are known. The inequality is sometimes called the *weak Young inequality*.
REFERENCES   [HLP, pp. 279–288]; [Lieb & Loss, pp. 98–102].

Harmonic Function Inequalities
See **Analytic Function Inequalities** COMMENTS (ii), **Conjugate Harmonic Function Inequalities, Hadamard's Three Circles Theorem** COMMENTS (ii), **Hardy's Analytic Function Inequality** COMMENTS, **Harnack's Inequalities, Maximum-Modulus Principle** COMMENTS (i).

Harmonic Mean Inequalities
Results concerning this mean are usually subsumed under the results involving the arithmetic mean, to which it is closely related; see **Notations 3**. However there are some results that are worth noting.

If $\underline{a}$ is a positive n-tuple:
(a)
$$\min \underline{a} \leq \mathfrak{H}_n(\underline{a}; \underline{w}) \leq \max \underline{a},$$
with equality if and only if $\underline{a}$ is constant.
(b) [HARMONIC -GEOMETRIC MEAN INEQUALITY]
$$\mathfrak{H}_n(\underline{a}; \underline{w}) \leq \mathfrak{G}_n(\underline{a}; \underline{w}), \qquad (HG)$$
with equality if and only if $\underline{a}$ is constant.
(c) [KLAMKIN] If $n \geq 2$ then
$$W_n \sqrt{\frac{\mathfrak{A}_n(\underline{a}; \underline{w})}{\mathfrak{H}_n(\underline{a}; \underline{w})}} \geq w_n + W_{n-1} \sqrt{\frac{\mathfrak{A}_{n-1}(\underline{a}; \underline{w})}{\mathfrak{H}_{n-1}(\underline{a}; \underline{w})}}, \qquad (1)$$
with equality only if $a_n = \sqrt{\mathfrak{A}_{n-1}(\underline{a}; \underline{w}) \mathfrak{H}_{n-1}(\underline{a}; \underline{w})}$.
(d)
$$1 + \mathfrak{G}_n(\underline{a}) \leq \mathfrak{G}_n(1 + \underline{a}^{-1}) \leq 1 + \frac{1}{\mathfrak{H}_n(\underline{a})},$$
with equality if and only if $\underline{a}$ is constant.

COMMENTS (i) (HG) is equivalent to (GA); and it follows easily from (GA) using the definition of the harmonic mean.

(ii) As (HG) is so closely related to (GA) it has many extensions similar to those discussed for that inequality.
In addition it is a special case of (r;s), and so again has many of the extensions that are discussed for that inequality.
(iii) Some interpretations of this inequality, in the equal weight case, are discussed in **Geometric Inequalities** (a).
(iv) The inequality

$$\mathfrak{H}_n(\underline{a};\underline{w}) \leq \mathfrak{A}_n(\underline{a};\underline{w}), \tag{HA}$$

follows by combining (HG) and (GA).
While (HA) is apparently weaker than (GA) it is actually equivalent to it; see **Walsh's Inequality**.
(v) Inequality (1) is a Rado-Popoviciu type extension of (HA).
(vi) Integral analogues of (HG) and (HA) are easily stated as particular cases of the integral analogues of (r;s).
(vii) See also **Binomial Function Inequalities** (j), **Kantorovič's Inequality**, **Knopp's Inequalities** COROLLARIES, **Sierpinski's Inequalities**.

REFERENCES   [MI, pp. 42–43, 98–99], [MPF, p. 73].

Harnack's Inequalities   Let h be a non-negative harmonic function in the domain $\Omega \subseteq \mathbb{R}^p$, $B = B(\underline{u}_0, r) = \{\underline{u}; |\underline{u} - \underline{u}_0| < r\}$, with $\overline{B} \subset \Omega$ then for all $\rho, 0 \leq \rho < r$.

$$\max_{\underline{u} \in B} h(\underline{u}) \leq \left(\frac{r+\rho}{r-\rho}\right)^n \min_{\underline{u} \in B} h(\underline{u}).$$

Further if $K \subset \Omega$ is compact there is a constant M depending only on Ω and K such that for all $\underline{u}, \underline{v} \in K$

$$\frac{1}{M} h(\underline{v}) \leq h(\underline{u}) \leq M h(\underline{v}).$$

COMMENTS   Harmonic functions are solutions of *Laplace's equation* $\nabla^2 h = 0$ and the important Harnack inequalities have been extended to solutions of more general elliptic partial differential equations; see [*Protter& Weinberger*]. Further extensions make these inequalities a subject of much research.

SPECIAL CASES   With the above notation for all $\underline{u}, \underline{v} \in B(\underline{u}_0, r/3)$,

$$\left(\frac{2}{3}\right)^n h(\underline{u}) \leq h(\underline{u}_0) \leq 2^n h(\underline{v}).$$

REFERENCES   [EM, vol. 4, pp. 387–388]; [Ahlfors,1966, pp. 235–236], [General Inequalities, vol. 3, pp. 333–339], [Lieb & Loss, p. 209], [Mitrović &Žubrinić, p.̃238], [Protter & Weinberger, pp. 106–121, 157, 194].

Hausdorff-Young Inequalities (a) If $1 < p \leq 2$, q the conjugate index, and $f \in \mathcal{L}^p([-\pi, \pi])$ and if

$$c_n = \frac{1}{2\pi} \int_{-\pi}^{\pi} f(t) e^{-int} \, dt, \quad n \in \mathbb{Z}, \tag{1}$$

then

$$||\underline{c}||_q \leq \left(\frac{1}{2\pi} \int_{-\pi}^{\pi} |f|^p \right)^{1/p}.$$

(b) If $1 < p \leq 2$, q the conjugate index, and $||\underline{c}||_p < \infty$, $\underline{c} = \{c_n, n \in \mathbb{Z}\}$ a complex sequence, then there is a function $f \in \mathcal{L}^q([-\pi, \pi])$ satisfying (1) and

$$\left(\frac{1}{2\pi} \int_{-\pi}^{\pi} |f|^q \right)^{1/q} \leq ||\underline{c}||_p.$$

EXTENSIONS [F & M RIESZ] (a) If $1 < p \leq 2$, q the conjugate index, $f \in \mathcal{L}^p([a, b])$ and if $\phi_n, n \in \mathbb{N}$ is a uniformly bounded orthonormal sequence of complex valued functions defined on an interval $[a, b]$, with $||\phi_n||_{\infty, [a, b]} \leq M, n \in \mathbb{N}$, then if $\underline{c}$ is the sequence of Fourier coefficients of f with respect to $\phi_n, n \in \mathbb{N}$,

$$||\underline{c}||_q \leq M^{(2/p)-1} ||f||_{p, [a, b]}.$$

(b) If $\phi_n, n \in \mathbb{N}$ is as in (a), $1 < p \leq 2$, q the conjugate index, and $||\underline{c}||_p < \infty$, $\underline{c} = \{c_n, n \in \mathbb{N}\}$ a complex sequence, then there is a function $f \in \mathcal{L}^q([a, b])$ having $\underline{c}$ as its sequence of Fourier coefficients with respect to $\phi_n, n \in \mathbb{N}$, and

$$||f||_{q, [a, b]} \leq M^{(2/p)-1} ||\underline{c}||_p.$$

COMMENTS (i) The two parts of all these results are equivalent in that either implies the other.
(ii) The case $p = 2$ of part (a) of the above results is Bessel's Inequality, see **Bessel's Inequality** (1).
(iii) For an integral analogue of the Hausdorff-Young result see **Fourier Transform Inequalities** (b).

REFERENCES [EM, vol. 4, pp. 394–395, vol. 8, pp. 154–155], [MPF, pp. 400–401]; [Rudin, 1966, pp. 247–249], [Zygmund, vol. II, pp. 101–105].

Heinig's Inequality If $\gamma, p \in \mathbb{R}$, with $p > 0$, and if $f : [0, \infty[\to [0, \infty[$ is measurable, then

$$\int_0^\infty x^\gamma \exp\left(\frac{p}{x^p} \int_0^x t^{p-1} \log f(t) \, dt \right) dx \leq e^{(\gamma+1)/p} \int_0^\infty x^\gamma f(x) \, dx.$$

COMMENTS (i) The constant is best possible and is due to Cochrane & Lee.
(ii) An extension has been given by Love.

DISCRETE ANALOGUES   Let p, γ be real numbers, $p \geq 1$, $\gamma \geq 0$. If $\underline{a}$ is a sequence with $0 \leq \underline{a} \leq 1$ and $\sum_{i=1}^{\infty} i^{\gamma} a_i < \infty$ then

$$\sum_{n=1}^{\infty} n^{\gamma} \Big(\prod_{i=1}^{n} a_i^{i^{p-1}} \Big)^{p/n^p} \leq e^{(\gamma+1)/p} \sum_{i=1}^{\infty} i^{\gamma} a_i.$$

COMMENTS   (iii) These inequalities generalize those of Carleman and Knopp; see **Carleman's Inequality** (1), INTEGRAL ANALOGUES.

REFERENCES   [*General Inequalities, vol. 5, pp. 87–93*]; [*Love, 1986*].

Heinz Inequality
If A, B are bounded positive linear operators on a Hilbert space X, if L is a bounded linear operator on X, and if $0 \leq \alpha \leq 1$, then:

$$\|AL + LB\| \geq \|A^{\alpha} L B^{1-\alpha} + A^{1-\alpha} L B^{\alpha}\|. \tag{1}$$

COMMENTS   (i) An operator A on the Hilbert space X is *positive* if, $\langle Ax, x \rangle \geq 0$ for all $x \in X$; we then write $A \geq 0$.
(ii) Inequality (1) implies the *Heinz-Kato Inequality*, **Heinz-Kato-Furuta Inequality** (1).
(iii) See also **Löwner-Heinz Inequality**.

REFERENCES   [*EM, Supp. pp. 288–289*].

Heinz-Kato-Furuta Inequality
Let X be a complex Hilbert space, A, B two positive bounded linear operators in X, L a bounded linear operator in X such that for all $x, y \in X$, $\|Lx\| \leq \|Ax\|$ and $\|L^* y\| \leq \|By\|$, then for all $x, y \in X$ and $0 \leq \alpha, \beta \leq 1, \alpha + \beta \geq 1$

$$|\langle L|L|^{\alpha+\beta} x, y \rangle| \leq \|A^{\alpha} x\| \|B^{\beta} y\|.$$

COMMENTS   (i) The definition of a positive operator is in **Heinz Inequality** COMMENTS (i).
(ii) L^* denotes the *adjoint* of L, defined by $\langle Lx, y \rangle = \langle x, L^* y \rangle$.
(iii) If $\alpha + \beta = 1$ we get the *Heinz-Kato inequality*,

$$|\langle Lx, y \rangle| \leq \|A^{\alpha} x\| \|B^{1-\alpha} y\|. \tag{1}$$

(iv) See also **Löwner-Heinz Inequality**, to which (1) is equivalent.

REFERENCES   [*EM, Supp. pp. 288–289*].

Heinz-Kato Inequality
See **Heinz-Kato-Furuta Inequality** (1).

Heisenberg-Weyl Inequality

If f is continuously differentiable on $[0, \infty[$, $p > 1$, q the conjugate index, and $r > -1$, then:

$$\int_0^\infty x^r |f(x)|^p \, dx \leq \frac{p}{r+1} \left(\int_0^\infty x^{q(r+1)} |f(x)|^p \, dx \right)^{1/q} \left(\int_0^\infty |f'|^p \right)^{1/p}.$$

There is equality if and only if $f(x) = A e^{-Bx^{q+r(q-1)}}$, for some positive B and non-negative A.

COMMENTS (i) In the case $p = 2, r = 0$, due to Weyl, this is the *Heisenberg Uncertainty Principle* of Quantum Mechanics.

(ii) This inequality can be deduced form **Benson's Inequalities**.

REFERENCES [AI, p. 128], [HLP, pp. 165–166]; [George, pp. 297–299], [Zwillinger, p. 209].

HELP Inequalities

These inequalities generalize the **Hardy-Littlewood-Pólya Inequalities** (1), (2) in asking whether, for some finite constant K,

$$\left(\int_a^b u|f'|^2 + v|f|^2 \right)^2 \leq K^2 \left(\int_a^b w|f|^2 \right) \left(\int_a^b w \left| \frac{-(uf')' + vf}{w} \right|^2 \right)$$

holds for any function making the right-hand side finite. As might be expected the detailed answer depends on properties of the differential equation

$$-(uf')' + vf = \lambda w f$$

The name refers to Hardy, Everitt-Evans, Littlewood and Pólya, because the above extension was introduced by Evans & Everitt.

REFERENCES [General Inequalities, vol. 4, pp. 15–23; vol. 5, pp. 337–346, vol. 6, pp. 269–305; vol. 7, pp. 179–192], [Zwillinger, p. 214].

Henrici's Inequality

If $\underline{a} \geq 1$,

$$\mathfrak{A}_n \left(\frac{1}{1+\underline{a}} \right) \geq \frac{1}{1 + \mathfrak{G}_n(\underline{a})},$$

with equality if and only if $\underline{a}$ is constant.

EXTENSIONS (a) If $1/(1+f)$ is strictly convex on $[c, d]$ and if $c \leq \underline{a} \leq d$ then:

$$\mathfrak{A}_n \left(\frac{1}{1 + f(\underline{a})}; \underline{w} \right) \geq \frac{1}{1 + f(\mathfrak{A}_n(\underline{a}; \underline{w}))},$$

with equality if and only if $\underline{a}$ is constant.
(b) If $0 \le r < s < \infty$ and if $\underline{a} \ge ((s-r)/(s+r))^{r/s}$,

$$\mathfrak{A}_n\left(\frac{1}{1+\underline{a}^s};\underline{w}\right) \ge \frac{1}{1+\left(\mathfrak{M}_n^{[r]}(\underline{a};\underline{w})\right)^s},$$

with equality if and only if $\underline{a}$ is constant.
(c) [RADO-TYPE] If $\mathfrak{G}_{n-1}(\underline{a}) \ge 1$ and $a_n^{(n+1)/n} \mathfrak{G}_{n-1}(\underline{a}) \ge 1$ then

$$n\left(\mathfrak{A}_n\left(\frac{1}{1+\underline{a}}\right) - \frac{1}{1+\mathfrak{G}_n(\underline{a})}\right)$$
$$\ge (n-1)\left(\mathfrak{A}_{n-1}\left(\frac{1}{1+\underline{a}}\right) - \frac{1}{1+\mathfrak{G}_{n-1}(\underline{a})}\right).$$

(d) [BORWEIN] If $0 < a_i \le \cdots \le a_n \le 1/\mathfrak{G}_{n-1}(\underline{a})$, then:

$$\frac{\mathfrak{A}_n(\underline{a})}{\mathfrak{A}_n(\underline{a}) + \mathfrak{G}_n(\underline{a})^n} \ge \mathfrak{A}_n\left(\frac{1}{1+\underline{a}}\right).$$

CONVERSE INEQUALITIES [KALAJDŽIĆ] If $1 - 1/n \le \alpha \le n$ then

$$\mathfrak{A}_n\left(\frac{1}{1+\underline{a}}\right) \ge \alpha \implies \frac{1}{1+\mathfrak{G}_n(\underline{a})} \ge \alpha.$$

REFERENCES   [AI, pp. 212–213], [MI, pp. 225–226, 240–241], [MPF, p. 73]; [General Inequalities, vol. 6, pp. 437–440].

Hermite-Hadamard Inequality   If f is convex on $[a,b]$, and if $a \le c < d \le b$, then

$$\frac{f(c)+f(d)}{2} \ge \frac{1}{d-c}\int_c^d f \ge f\left(\frac{c+d}{2}\right), \qquad (1)$$

and

$$\frac{1}{b-a}\int_a^b f \ge \frac{1}{n}\sum_{i=0}^{n-1} f\left(a + \frac{i}{n-1}(b-a)\right). \qquad (2)$$

COMMENTS   (i) The right inequality in (1) just says that the area under a convex curve is less than the area under the trapezoid formed by the lines:

$$x = c, x = d, y = 0, (y - f(c))(d - c) = (f(d) - f(c))(x - c).$$

Improvements can be obtained by applying this inequality to the two trapezoids obtained by using the lines:

$$x = c, x = c' = \frac{c+d}{2}, y = 0, (y - f(c))(c' - c) = (f(c') - f(c))(x - c),$$
$$x = c', x = d, y = 0, (y - f(c'))(d - c') = (f(d) - f(c'))(x - c').$$

(ii) The right-hand side of (2) actually tends to the left-hand side as $n \to \infty$.

EXTENSIONS (a) [LUPAŞ] If f is convex on $[a,b]$ then:

$$\mathfrak{A}_2\big(f(a),f(b);p,q\big) \geq \frac{1}{b-a}\int_a^b f \geq f\big(\mathfrak{A}_2(a,b;p,q)\big)$$

provided

$$\left|\frac{c+d}{2}\right| \leq \frac{b-a}{p+q}\min\{p,q\}.$$

(b) [FÉJER] If f is convex on $[a,b]$ and $g \geq 0$ is symmetric with respect to the mid-point $(a+b)/2$ then:

$$\frac{f(a)+f(b)}{2}\int_a^b g \geq \frac{1}{b-a}\int_a^b fg \geq f\left(\frac{a+b}{2}\right)\int_a^b g.$$

COMMENTS (iii) The Lupaş style extension can also be made to Féjer's result. Other extensions have been made to n-convex functions and to arithmetic means of order n.

REFERENCES [HLP, p. 98], [MI, p. 30], [MPF, p. 10], [PPT, pp. 137–151]; [Roberts & Varberg, p. 15]; [Alzer & Brenner,1990], [Neuman, 1988].

Higher Order Convex Function Inequalities
See n-Convex Function Inequalities.

Higher Order Convex Sequence Inequalities
See n-Convex Sequence Inequalities.

Hilbert's Inequalities
(a) If $\underline{a}$ is a non-negative n-tuple, then

$$\sum_{i=0}^n \sum_{j=0}^n \frac{a_i a_j}{i+j+1} \leq \pi \sum_{i=0}^n a_i^2. \tag{1}$$

(b) If $\underline{a}, \underline{b}$ are non-negative sequences, $p > 1$, q the conjugate index then

$$\sum_{i=0}^\infty \sum_{j=0}^\infty \frac{a_i b_j}{i+j+1} \leq \frac{\pi}{\sin(\pi/p)} \left(\sum_{i=0}^\infty a_i^p\right)^{1/p} \left(\sum_{i=0}^\infty b_i^q\right)^{1/q}, \tag{2}$$

with equality only if either $\underline{a}$ or $\underline{b}$ has all elements zero. The constant on the right-hand side is best possible.

COMMENTS (i) Inequalities (1) and (2) are both known as *Hilbert's inequality*.
(ii) The constant π on the right-hand side of (1) is not best possible, it can be replaced by $(n+1)\sin\pi/(n+1)$. An asymptotic form for the best possible constant has been given. The constant on the right-hand side of (2) is best possible.

EXTENSIONS (a) [WIDDER] If $\underline{a}$ is a non-negative n-tuple, then:

$$\sum_{i=0}^{m}\sum_{j=0}^{n}\frac{a_i a_j}{i+j+1} \le \pi \sum_{i=0}^{m}\sum_{j=0}^{n}\frac{(i+j)!}{i!j!}\frac{a_i a_j}{2^{i+j+1}}.$$

(b) [BENNETT] If $\underline{a}$ is a real sequence and if $p > 1$ then:

$$\frac{1}{(p-1)^{1/p}}!\underline{a}!_p \le \left(\sum_{i=1}^{\infty}\left(\sum_{j=1}^{\infty}\frac{|a_j|}{i+j-1}\right)^p\right)^{1/p} \le \frac{\pi}{\sin \pi/p}!\underline{a}!_p.$$

The constants are best possible and both inequalities are strict unless $\underline{a} = \underline{0}$.

(c) [KUBO[2]] If $\underline{a}$ is a real n-tuple then:

$$\sum_{i=0}^{m}\sum_{j=0}^{m}\frac{a_i a_j}{\sin((i+j)\pi/n)} \le (n \pm 1)\sum_{i=0}^{m}|a_i|^2, \ 1 \le m \le [n/2].$$

COMMENTS (iii) The notations in (b) are defined above **Hardy's Inequalities** EXTENSIONS (c).

INTEGRAL ANALOGUES If $f \in \mathcal{L}^p([0,\infty[), g \in \mathcal{L}^q([0,\infty[), f \ge 0, g \ge 0, p, q$ conjugate indices then

$$\int_0^{\infty}\int_0^{\infty}\frac{f(x)g(y)}{x+y}\,dxdy \le \frac{\pi}{\sin(\pi/p)}\|f\|_p\|g\|_q, \qquad (3)$$

with equality if and only if either $f = 0$ or $g = 0$.

COMMENTS (iv) There are extensions in which the function $K(x,y) = 1/(x+y)$ is replaced by a more general function; see [EM].
(v) See also **Bennett's Inequalities** (c), (d).

REFERENCES [AI, pp. 357–358], [EM, vol.4, pp. 417–418], [HLP, pp. 212–214, 227–259], [PPT, p. 234]; [Bennett, 1996, pp. 53–54]; [Kubo, F].

Hilbert's Transform Inequality

If $f \in \mathcal{L}^p(\mathbb{R})$ and if

$$\tilde{f}(x) = \frac{1}{\pi}\int_0^{\infty}\frac{f(x+t)-f(x-t)}{t}\,dt,$$

then $\tilde{f} \in \mathcal{L}^p(\mathbb{R})$ and

$$\|\tilde{f}\|_{p,\mathbb{R}} \le C\|f\|_{p,\mathbb{R}},$$

where C depends only on p.

COMMENTS $\tilde{f}$ is called the *Hilbert transform of f*.

REFERENCES [EM, vol. 4, pp. 433-434]; [Hirschman, pp. 168–169], [Titchmarsh, 1937 pp. 132–138], [Zygmund, vol. II, pp. 256–257].

---

[2] This is F Kubo.

Hinčin³-Littlewood Inequality If $r_n, n \in \mathbb{N}$, are the Rademacher functions, and $f(x) = \sum_{n \in \mathbb{N}} c_n r_n(x), 0 \leq x \leq 1$, with $\sum_{n \in \mathbb{N}} |c_n|^2 = C^2 < \infty$, then there are constants λ, Λ depending only on r, such that

$$\lambda C \leq \|f\|_{r,[0,1]} \leq \Lambda C.$$

COMMENTS (i) The Rademacher functions, $r_n, n \in \mathbb{N}$, are defined by:

$$r_n(x) = \text{sign} \circ \sin(2^{n+1}\pi x), \quad 0 \leq x \leq 1, \quad n \in \mathbb{N}.$$

(ii) If ν is the least even integer not less than r, then $\Lambda \leq \sqrt{2\nu}$.

REFERENCES [MPF, pp. 566–568]; [Zygmund, vol. I, pp. 213–214].

Hirsch's Inequalities If A is a complex $n \times n$ matrix then

$$|\lambda_s(A)| \leq n \max\{|a_{ij}|, 1 \leq i \leq n, 1 \leq j \leq n\};$$

$$|\Re(\lambda_s)| \leq n \max\left\{\left|\frac{a_{ij} + \overline{a}_{ji}}{2}\right|, 1 \leq i \leq n, 1 \leq j \leq n\right\};$$

$$|\Im(\lambda_s)| \leq n \max\left\{\left|\frac{a_{ij} - \overline{a}_{ji}}{2i}\right|, 1 \leq i \leq n, 1 \leq j \leq n\right\}.$$

$$\lambda_{[n]}\left(\frac{A + A^*}{2}\right) \leq \Re(\lambda_s(A)) \leq \lambda_{[1]}\left(\frac{A + A^*}{2}\right);$$

$$\lambda_{[n]}\left(\frac{A - A^*}{2i}\right) \leq \Im(\lambda_s(A)) \leq \lambda_{[1]}\left(\frac{A - A^*}{2i}\right);$$

COMMENTS The third inequality, in the real case, is due to Bendixson, who also gave the following result.

[BENDIXSON] If A is a real $n \times n$ matrix then

$$|\Im(\lambda_s)| \leq \max\left\{\left|\frac{a_{ij} - \overline{a}_{ji}}{2i}\right|, 1 \leq i \leq n, 1 \leq j \leq n\right\}\sqrt{\frac{n(n-1)}{2}}.$$

REFERENCES [Marcus & Minc, pp. 140–142].

Hlwaka's Inequality If $\underline{a}, \underline{b}, \underline{c}$ are real n-tuples,

$$|\underline{a} + \underline{b} + \underline{c}| + |\underline{a}| + |\underline{b}| + |\underline{c}| \geq |\underline{a} + \underline{b}| + |\underline{b} + \underline{c}| + |\underline{c} + \underline{a}|. \tag{1}$$

COMMENTS (i) This follows from (T) and the following, the *identity of Hlwaka*:

$$(A + B + C + D - E - F - G)(A + B + C + D) =$$
$$(A + B - G)(C + D - G) + (B + C - E)(D + A - E) \tag{2}$$
$$+ (C + A - F)(B + D - F).$$

(ii) Both the identity and the inequality hold in unitary spaces; for a definition see **Inner Product Inequalities**.

---

³ А Я Хинчин. Also transliterated as Khinchine.

EXTENSIONS [ADAMOVIĆ] If $\underline{a}_k, 1 \leq k \leq p$, are real n-tuples, $p \geq 2$, then

$$(p-2)\sum_{k=1}^{p}|\underline{a}_k| + \left|\sum_{k=1}^{p}\underline{a}_k\right| \geq \sum_{1\leq k<j\leq p}|\underline{a}_k + \underline{a}_j|.$$

COMMENTS (iii) This extension reduces to (1) if $p = 3$; its proof is based on a generalization of (2).
(iv) An inequality that can be considered a converse to (1) is given in **Beth & van der Corput Inequality**, COMMENTS (iv).
(V) See also **Hornich's Inequality**.

REFERENCES [AI, pp. 171–176], [MPF, pp. 521–534].

Hlwaka's Integral Inequality
If $f(0+) \leq 0$ and $f'^2(x)e^{-2x} \in \mathcal{L}([0,t])$, f not zero almost everywhere, then

$$\frac{1}{2}f^2(t)e^{-2t} + \int_0^t e^{-2x}|f(x)f'(x)|\,dx < \int_0^t e^{-2x}f'^2(x)\,dx.$$

COMMENTS This has been extended to higher dimensions by Redheffer.

REFERENCES [Inequalities II, pp. 273–276].

Hlwaka-Type Inequalities
[BURKILL[4]] If $\underline{a}, \underline{w}$ are positive triples with $W_3 = 1$ then

$$a_1^{w_1}a_2^{w_2}a_3^{w_3} + (w_1a_1 + w_2a_2 + w_3a_3)$$
$$\geq (w_1+w_2)a_1^{w_1/(w_1+w_2)}a_2^{w_2/(w_1+w_2)} + (w_2+w_3)a_2^{w_2/(w_2+w_3)}a_3^{w_3/(w_2+w_3)}$$
$$+ (w_1+w_2)a_3^{w_3/(w_3+w_1)}a_1^{w_1/(w_3+w_1)}.$$
(1)

COMMENTS (i) The methods of proof for this result are elementary.
(ii) The reason for the name given to this kind of inequality by Burkill is clear on comparing (1) with the **Hlwaka Inequality** (1).

EXTENSIONS [BURKILL[4]] If $f : [a,b] \to \mathbb{R}$ is convex, and if $a_i \in [a,b], 1 \leq i \leq 3$, $\underline{w}$ a positive triple, and defining $a_4 = a_1$, $w_4 = w_1$, then

$$f(\mathfrak{A}_3(\underline{a};\underline{w})) + \mathfrak{A}_3(f(\underline{a});\underline{w}) \geq \frac{1}{W_3}\left(\sum_{j=1}^{3}(w_j + w_{j+1})f\left(\frac{w_j a_j + w_{j+1}a_{j+1}}{w_j + w_{j+1}}\right)\right).$$

COMMENTS (iii) Many of the inequalities discussed have been concerned with the superadditivity of functions of index sets, see for instance **Rado's Geometric-Arithmetic Mean**

---
[4] This is J C Burkill.

Inequality Extension EXTENSIONS. Hlwaka-type inequalities are concerned with convexity of such functions,

$$\sigma(\mathcal{I} \cup \mathcal{J} \cup \mathcal{K}) + \sigma(\mathcal{I}) + \sigma(\mathcal{J}) + \sigma(\mathcal{K}) \geq \sigma(\mathcal{I} \cup \mathcal{J}) + \sigma(\mathcal{J} \cup \mathcal{K}) + \sigma(\mathcal{K} \cup \mathcal{K}).$$

For a further discussion and elaboration of this point see [MPF].
REFERENCES [MI, pp. 375–377], [MPF, pp. 527–534], [PPT, pp. 171–181].

Hölder Function Inequalities
See **Lipschitz Function Inequalities**.

Hölder's Inequality
If $\underline{a}, \underline{b}$ are positive n-tuples and if p, q are conjugate indices with $p > 0$ and $q > 0$, equivalently $p > 1$ or $q > 1$, then

$$\sum_{i=1}^{n} a_i b_i \leq \left(\sum_{i=1}^{n} a_i^p \right)^{1/p} \left(\sum_{i=1}^{n} b_i^q \right)^{1/q} ; \qquad (H)$$

or

$$||\underline{a}\,\underline{b}|| \leq ||\underline{a}||_p ||\underline{b}||_q.$$

If either $p < 0$, or $q < 0$ the inequality $(\sim H)$ holds. The inequality is strict unless $\underline{a}^p \sim \underline{b}^q$.

COMMENTS (i) This famous inequality has many proofs and perhaps the simplest is to rewrite (H) as

$$\sum_{i=1}^{n} \left(\frac{a_i^p}{\sum_{j=1}^{n} a_j^p} \right)^{1/p} \left(\frac{b_i^p}{\sum_{j=1}^{n} b_j^p} \right)^{1/q} \leq 1$$

and then apply **Geometric-Arithmetic Mean Inequality** (2) to the terms on the left-hand side. The case of $(\sim H)$ follows by an algebraic argument from (H).
(ii) The case $p = q = 2$ of (H) is just (C). It is of some interest that (C) is equivalent to (H).

EXTENSIONS (a) If $\underline{a}, \underline{b}$ are complex n-tuples, and p, q are as in (H),

$$\left| \sum_{i=1}^{n} a_i \bar{b}_i \right| \leq \sum_{i=1}^{n} |a_i||b_i| \leq \left(\sum_{i=1}^{n} |a_i|^p \right)^{1/p} \left(\sum_{i=1}^{n} |b_i|^q \right)^{1/q} ; \qquad (1)$$

There is equality if and only if $|\underline{a}|^p \sim |\underline{b}|^q$, and $\arg a_i b_i$ does not depend on i.
(b) If $\underline{a}, \underline{b}, \underline{w}$ are positive n-tuples, and p, q are as in (H) then

$$\sum_{i=1}^{n} w_i a_1 b_i \leq \left(\sum_{i=1}^{n} w_i a_i^p \right)^{1/p} \left(\sum_{i=1}^{n} w_i b_i^q \right)^{1/q}. \qquad (2)$$

(c) [GENERALIZED HÖLDER INEQUALITY; JENSEN] Suppose that $r_i > 0, 1 \leq i \leq m$, and $a_{ij} > 0, 1 \leq i \leq m, 1 \leq j \leq n$, and define

$$\frac{1}{\rho_m} = \sum_{i=1}^{m} \frac{1}{r_i},$$

then

$$\left(\sum_{j=1}^{n} \{\prod_{i=1}^{m} a_{ij}\}^{\rho_m}\right)^{1/\rho_m} \leq \prod_{i=1}^{m} \{\sum_{j=1}^{n} a_{ij}^{r_i}\}^{1/r_i}. \tag{3}$$

There is equality in (3) only if the n-tuples $(a_{i1}^{r_i}, \ldots, a_{in}^{r_i})$, $1 \leq i \leq m$, are linearly dependent. If $r_1 > 0, r_i < 0, 2 \leq i \leq m$, then ($\sim$2) holds.

(d) [POPOVICIU-TYPE] Write $H_m^{1/\rho_m}(\underline{a})$ for the ratio of the left-hand side of (3) to its right-hand side, then if $m \geq 2$,

$$\rho_m H_m(\underline{a}) \leq \rho_{m-1} H_{m-1}(\underline{a}) + \frac{1}{r_m}.$$

(e) [FUNCTIONS OF INDEX SETS] If $\mathcal{I}$ is an index set define

$$\chi(\mathcal{I}) = \left(\sum_{i \in \mathcal{I}} a_i^p\right)^{1/p} \left(\sum_{i \in \mathcal{I}} b_i^q\right)^{1/q} - \sum_{i \in \mathcal{I}} a_i b_i,$$

where $\underline{a}, \underline{b}, p, q$ are as in (H), then $\chi \geq 0$, and if $\mathcal{I} \cap \mathcal{J} = \emptyset$,

$$\chi(\mathcal{I}) + \chi(\mathcal{J}) \leq \chi(\mathcal{I} \cup \mathcal{J}), \tag{4}$$

with equality if and only if $\left(\sum_{i \in \mathcal{I}} a_i^p, \sum_{i \in \mathcal{J}} a_i^p,\right)$ is proportional to $\left(\sum_{i \in \mathcal{I}} b_i^q, \sum_{i \in \mathcal{J}} b_i^q,\right)$.

COMMENTS (iii) It is easily seen that (2) and (H) are identical.
(iv) In the notation of (d), (3) is just $H_m(\underline{a}) \leq 1$.
(v) Inequality (3) holds even if $\sum_{i=1}^{m} 1/r_i > 1/\rho_m$, and it is then strict. This follows from **Power Sums Inequalities** (1).
(vi) Inequality (4) just says that χ is superadditive, while (H) says it is non-negative.

SPECIAL CASES (a) If $\underline{a}$ is a positive m-tuple then

$$\prod_{i=1}^{m}(1+a_i) \geq (1+\mathfrak{G}_m(\underline{a}))^m.$$

(b) If $\underline{a}, \underline{b}$ are positive n-tuples and if

$$p > 0, \quad q > 0 \quad \text{and} \quad \frac{1}{p} + \frac{1}{q} = \frac{1}{r}$$

then
$$\left(\sum_{i=1}^{n} a_i^r b_i^r\right)^{1/r} \leq \left(\sum_{i=1}^{n} a_i^p\right)^{1/p} \left(\sum_{i=1}^{n} b_i^q\right)^{1/q}. \tag{5}$$

If $p < 0$, or $q < 0$ and $r > 0$, or if all three parameters are negative the ($\sim$5) holds; while if $p < 0$, or $q < 0$ and $r < 0$ then (5) holds.
(c) If $\underline{a}, \underline{b}, \underline{c}$ are positive n-tuples such that $\underline{a}\,\underline{b}\,\underline{c} = \underline{e}$ and if

$$\frac{1}{p} + \frac{1}{q} = \frac{1}{r},$$

then
$$\left(\sum_{i=1}^{n} a_i^p\right)^{1/p} \left(\sum_{i=1}^{n} b_i^q\right)^{1/q} \left(\sum_{i=1}^{n} c_i^r\right)^{1/r} \geq 1, \tag{6}$$

if all but one of the p, q, r are positive, while if all but one are negative ($\sim$ 6) holds.

COMMENTS (vii) (a) is (3) with $n = 2$, $a_{i1} = 1$, $a_{i2} = a_i$, $r_i = 1, 1 \leq i \leq m$. The result is related to **Weierstrass Inequalities**.
(viii) (b) is just the case $m = 2$ of (3), with a change of notation; and (c) is just a symmetric form for (b).

The generality of (H) has meant that many inequalities are special cases often in almost impenetrable disguises; see **Radon's Inequality, Liapunov's Inequality**.

[OTHER FORMS] (a) If $\underline{a}, \underline{b}$ are positive n-tuples and $0 < s < 1$ then

$$\sum_{i=1}^{n} a_i^s b_i^{1-s} \leq \left(\sum_{i=1}^{n} a_i\right)^s \left(\sum_{i=1}^{n} b_i\right)^{1-s}. \tag{7}$$

(b) If $\underline{a}, \underline{b}$ are positive n-tuples with $\sum_{i=1}^{n} a_i = \sum_{i=1}^{n} b_i = 1$ then (H) is equivalent to

$$\sum_{i=1}^{n} a_i^{1/p} b_i^{1/q} \leq 1$$

with equality if and only if $\underline{a} = \underline{b}$.
(c) If $f\underline{a}, \underline{b}$ are positive n-tuples, $p > 1$ and q is the conjugate index, and if $B > 0$ then a necessary and sufficient condition that $\left(\sum_{i=1}^{n} a^p\right)^{1/p} \leq A$ is that $\sum_{i=1}^{n} a_i b_i \leq AB$ for all $\underline{b}$ for which $\left(\sum_{i=1}^{n} b_i^q\right)^{1/q} < B$.

COMMENTS (ix) For converse Inequalities see **Barnes's Inequalities** (b).

INTEGRAL ANALOGUES If p, q are as in (H) and if $f \in \mathcal{L}^p([a,b])$, $g \in \mathcal{L}^q([a,b])$, $f, g \geq 0$ then $fg \in \mathcal{L}([a,b])$ and

$$\|fg\| \leq \|f\|_p \|g\|_q.$$

with equality if and only if for some A, B, not both zero $A|f|^p = B|g|^q$ almost everywhere.

COMMENTS (x) The integral analogue given above extends to general measure spaces; and other forms of (H) can also be given integral analogues; see [HLP].
(xi) For another integral analogue see **Young's Inequalities** (c); for a converse integral analogue see **Petschke's Inequality**.
(xii) There is another equivalent inequality that is sometimes called Hölder's Inequality see **Geometric Mean Inequalities** COMMENTS (i).
(xiii) See also **Beckenbach's Inequalities** COMMENTS (i), **Young's Convolution Inequality** COMMENTS (iii).

REFERENCES [AI, pp. 50–54], [EM, vol.4, pp. 438–439], [HLP, pp. 21–26, 61, 139–143], [MI, pp. 136–143, 146, 150–152], [MO, pp. 457–461], [MPF, pp. 99–107, 111–117, 135–209], [PPT, pp. 112–114, 126–128]; [Lieb & Loss, pp. 39–40], [Pólya & Szegö, 1972, pp. 68–69].

Holley's Inequality If X is a distributive lattice, let $\ell_i : X \to [0, \infty[, i = 1, 2$, satisfy

$$\ell_1(a)\ell_2(b) \leq \ell_1(a \vee b)\ell_2(a \wedge b), \quad \text{for all } a, b \in X. \tag{1}$$

If then $f : X \to \mathbb{R}$ is increasing

$$\sum_{x \in X} \ell_2(x) f(x) \leq \sum_{x \in X} \ell_1(x) f(x).$$

COMMENTS Condition (1) generalizes the log modularity condition of the **FKG Inequality**, and is in turn generalized by condition (1) of the **Ahlswede-Daykin Inequality**.
REFERENCES [EM, Supp., pp. 202, 292–293].

Holomorphic Function Inequalities
See **Analytic Function Inequalities**.

Hornich's Inequality If $\underline{a}_k, 0 \leq k \leq p$, are real n-tuples such that for some $\lambda \geq 1$

$$\lambda \underline{a}_0 + \sum_{k=1}^{p} \underline{a}_k = \underline{0},$$

then

$$\sum_{k=1}^{p} \left(|\underline{a}_k + \underline{a}_0| - |\underline{a}_k| \right) \leq (n-2)|\underline{a}_0|.$$

COMMENT (i) If $\lambda < 1$ the inequality need not hold.
(ii) This inequality can be deduced from **Hlawka's Inequality**.
REFERENCES [AI, pp. 172–173], [MPF, pp. 521–522].

Hua's Inequality If $\delta > 0, \alpha > 0$ and $\underline{a}$ is a real n-tuple then

$$\left(\delta - \sum_{i=1}^{n} a_i\right)^2 + \alpha \left(\sum_{i=1}^{n} a_i^2\right) \geq \frac{\alpha}{n+\alpha}\delta^2,$$

with equality if and only if $a_1 = \cdots = a_n = \delta/(n+\alpha)$.

COMMENTS (i) This has applications in the additive theory of primes.

EXTENSIONS [PEARCE & PEČARIĆ] If f is convex then, with the above notation,

$$f\left(\delta - \sum_{i=1}^{n} a_i\right) + \alpha \left(\sum_{i=1}^{n} f(a_i)\right) \geq \frac{n+\alpha}{\alpha} f\left(\frac{\alpha\delta}{n+\alpha}\right),$$

and if f is strictly convex there is equality if and only if $a_1 = \cdots = a_n = \delta/(n+\alpha)$.

COMMENTS (ii) This is deduced from (J), as is the following.

INTEGRAL ANALOGUES If $\delta, \alpha > 0$, $g \in \mathcal{L}([0,x])$, $g \geq 0$, and $f; [a,b] \to \mathbb{R}$, convex and such that $g[[0,x]] \subseteq [a/\alpha, b/\alpha]$, and $\left(\delta - \int_0^x g\right) \in [a,b]$ then

$$f\left(\delta - \int_0^x g\right) + \frac{1}{\alpha}\int_0^x f(\alpha g(t))\, dt \geq \frac{x+\alpha}{\alpha} f\left(\frac{\alpha\delta}{x+\alpha}\right).$$

If f is stricly convex there is equality if and only if $g(t) = \delta/(x+\alpha)$.

REFERENCES [Hua, p. 104]; [Pearce & Pečarić, 1994].

Hunter's Inequality
See **Geometric-Arithmetic Mean Inequality**, EXTENSIONS (d).

Hyperbolic Function Inequalities (a) If $z \in \mathbb{C}$ then

$$|\sinh \Im z| \leq |\sin z| \leq \cosh \Im z; \quad |\sinh \Im z| \leq |\cos z| \leq \cosh \Im z;$$
$$|\operatorname{cosec} z| \leq \operatorname{cosech}|\Im z|; \quad |\cos z| \leq \cosh |z|; \quad |\sin z| \leq \sinh |z|;$$

(b) [VAN DER CORPUT] If $a, b \geq 0$ then

$$|\cosh a - \cosh b| \geq |a-b|\sqrt{\sinh a \sinh b}.$$

(c) [LAZAREVIĆ, PÁLES] If $x \neq 0$ then

$$\frac{\sinh x}{x} < \cosh x < \left(\frac{\sinh x}{x}\right)^3$$

(d) If $-\infty < r < \infty$ and

$$r_1 = \begin{cases} \min\{(r+2)/3, (r\log 2)/(\log r + 1)\}, & \text{if } r > -1,\ r \neq 0, \\ \min\{2/3, \log 2\}, & \text{if } r = 0, \\ \min\{(r+2)/3, 0\} & \text{if } r \leq -1, \end{cases}$$

and r_2 is defined as r_1 but with min replaced by max, then for $t > 0$,

$$(\cosh r_1 t)^{1/r_1} \leq \left(\frac{\sinh(r+1)t}{r \sinh t}\right)^{1/r} \leq (\cosh r_2 t)^{1/r_2},$$

where the cases $r = 0, -1$ are taken to be the limits as $r \to 0$ of these values.

COMMENTS (i) The exponent 3 in (c) is best possible.
(ii) The inequality in (d) follows from **Pittenger's Inequalities** on putting $e^{2t} = b/a$.
(iii) See also **Function Inequalities** (a), **Lochs's Inequality**.

REFERENCES [AI, pp. 270, 323], [MI, p. 349]; [Páles, 1988[(2)]].

- end of H -

I

Increasing Function Inequalities (a) If f is an increasing function on $[a,b]$ then if $a \leq x < y \leq b$,
$$f(x) \leq f(y); \qquad (1)$$
if f is strictly increasing (1) is strict if $x \neq y$.

(b) [JENSEN] If $\underline{a}, \underline{w}$ are positive n-tuples, and if $f:]0, \infty[\to \mathbb{R}$ is increasing then
$$\mathfrak{A}_n(\underline{a}, \underline{w}) \leq f\left(\sum_{i=1}^n a_i\right). \qquad (2)$$

If f is strictly increasing and $n > 1$ then (2) is strict. If f is decreasing then ($\sim$2) holds.

(c) If μ is an increasing set function defined on the sets in the collection $\mathcal{A}$ then
$$A, B \in \mathcal{A}, \ A \subseteq B \implies \mu(A) \leq \mu(B). \qquad (3)$$
If μ is strictly increasing then (3) is strict if $A \subset B$.

COMMENTS (i) (a) is just the definition of *increasing* and *strictly increasing* functions. In addition, by definition a *decreasing function* is one for which ($\sim$1) holds. Similar comments hold for the set function in (c).

(ii) Inequality (2) characterizes increasing functions just as (J) characterizes convex functions.

EXTENSIONS [VASIĆ & PEČARIĆ] If $f : [0, a] \to \mathbb{R}$ is increasing and $\underline{a}, \underline{v}, \underline{w}$ are non-negative n-tuples satisfying
$$a_j \in [0, a]; \quad \sum_{i=1}^n v_i a_i \geq a_j, \quad 1 \leq j \leq n; \quad \sum_{i=1}^n v_i a_i \in [0, a];$$
then
$$\mathfrak{A}_n(f(\underline{a}); \underline{w}) \leq f\left(\sum_{i=1}^n v_i a_i\right).$$

COMMENTS (iii) This property also characterizes increasing functions. It has applications in **Starshaped Function Inequalities**.

(iv) See also **Integral Inequalities** (c), **Integral Test Inequality**.

REFERENCES [HLP, pp. 83–84], [PPT, pp. 151–152].

Inf and Sup Inequalities (a) If $\underline{a}, \underline{b}$ are real sequences then

$$\sup \underline{a} + \inf \underline{b} \le \sup(\underline{a} + \underline{b}) \le \sup \underline{a} + \sup \underline{b},$$

provided the terms in the inequalities are defined.
(b) If $\underline{a}, \underline{b}$ are non-negative sequences then

$$\sup \underline{a} \inf \underline{b} \le \sup \underline{a}\,\underline{b} \le \sup \underline{a} \sup \underline{b},$$

provided the terms in the inequalities are defined.

COMMENTS (i) The operations on these inequalities are understood to be in $\overline{\mathbb{R}}$. In addition we could assume the terms of the sequences to be taking values in $\overline{\mathbb{R}}$.
(ii) Similar results can be stated for functions with values in $\overline{\mathbb{R}}$.
(iii) For an application of these results see **Upper and Lower Limit Inequalities**.

REFERENCES [*Bourbaki, 1960, pp. 162–163*].

Inner Product Inequalities *If X is any inner product space and $\langle \cdot, \cdot \rangle : X \times X \to \mathbb{R}$ is the inner product on X then for any $x, y, z \in X, \lambda \in \mathbb{R}$*

$$\langle x, x \rangle > 0, \quad x \ne 0; \tag{1}$$

$$|\langle x, y \rangle| \le ||x||\,||y||; \tag{2}$$

$$2(||x||^2 + ||y||^2) = ||x + y||^2 + ||x - y||^2. \tag{3}$$

COMMENTS (i) (1), (2) and

$$\langle \lambda x, y \rangle = \lambda \langle x, y \rangle, \langle x + y, z \rangle = \langle x, z \rangle + \langle y, z \rangle, \langle x, y \rangle = \langle y, x \rangle; \tag{3}$$

just give the definition of an *inner product*; and then X together with $\langle \cdot, \cdot \rangle$ is called an *inner product space*, or *pre-Hilbert space*. If in the above we replace $\mathbb{R}$ by $\mathbb{C}$, and in (3) make the change $\langle x, y \rangle = \overline{\langle y, x \rangle}$, we get a *complex inner product*, and X is a *complex inner product space*, or a *unitary space*. A complete inner product or unitary space is called *Hilbert space*.
(ii) Inequality (2) is just (C) in this setting, and remains valid for unitary spaces.
(iii) Inequality (3) is the *parallelogram identity*; see **Parallelogram Inequality** COMMENTS (i), **Norm Inequalities** COMMENTS (vi), **von Neumann & Jordan Inequality** COMMENTS (ii).
(iii) $\mathbb{R}^n$ is an inner product space with $\langle \underline{a}, \underline{b} \rangle = \underline{a}.\underline{b}$. More generally ℓ_2 is an inner product space with $\langle \underline{a}, \underline{b} \rangle = \sum_{i=1}^{\infty} a_i b_i$; this series converges by (C).
The space $\mathcal{L}^2([a,b])$ can also be taken to be an inner product space, with $\langle f, g \rangle = \int_a^b fg$, provided we identify functions that are equal almost everywhere.
(iv) It is easy to see that in both the real and complex case $||x|| = \sqrt{\langle x, x \rangle}$ is a norm; see **Norm Inequalities**.
(vi) See also **Bessel's Inequality** COMMENTS (i), **Beth & van der Corput's Inequality** COMMENTS (v), **Clarkson's Inequalities** COMMENTS (v), **Cordes's Inequality, Dunk & Williams Inequality, Gram Determinant Inequalities** COMMENTS (i), **Grothendiek's Inequality, Heinz-Kato-Furuta Inequality, Hlwaka's Inequality** COMMENTS (ii), **Löwner-Heinz Inequality**.

REFERENCES [*EM, vol. 5, p. 89; vol. 9, p. 337*].

Integral Inequalities (a) If f is real, or complex, integrable function on $[a,b]$, with $|f|$ also integrable, then

$$\left|\int_a^b f\right| \le \int_a^b |f|,$$

with equality if and only if $\arg f$ is constant almost everywhere.

(b) [OSTROWSKI] If f, g are integrable on $[a,b]$ with f monotonic, $f(a)f(b) \ge 0$ and $|f(a)| \ge |f(b)|$ then

$$\left|\int_a^b fg\right| \le |f(a)| \max_{a \le x \le b} \left|\int_a^x g\right|.$$

(c) If f is an increasing function on the interval $[a,b]$ then

$$\int_a^b f' \le f(b) - f(a). \tag{1}$$

(d) If f is differentiable at all points of a measurable set D then

$$\lambda(f[D]) \le \int_D |f'|.$$

(e) [FREIMER & MUDHALKAR] Let $f : [0, \infty[\to \mathbb{R}$ be positive, decreasing and integrable, and let $X > 0$; then there is a $x, 0 < x < X$ such that

$$f(x)(X - x) \le \int_x^\infty f.$$

(f) [KARAMATA] Let $h, g, fg, fh \in \mathcal{L}_\mu([a,b])$, $f \ge 0$, decreasing and $\int_a^b fh\,d\mu > 0$; write $G(x) = \int_a^x g\,d\mu$, $H(x) = \int_a^x h\,d\mu$, $a \le x \le b$ and assume that $H > 0$; then

$$\inf_{a \le x \le b} \frac{G(x)}{H(x)} \le \frac{\int_a^b fg\,d\mu}{\int_a^b fh\,d\mu} \le \sup_{a \le x \le b} \frac{G(x)}{H(x)}.$$

COMMENTS (i) (a) is an integral analogue of (T). A convers can be found in **Wilf's Inequality** INTEGRAL ANALOGUES.

(ii) It is known that (1) can be strict even if f is continuous; if f is absolutely continuous there is always equality.

DISCRETE ANALOGUES [KARAMATA] If $\underline{a}, \underline{u}, \underline{v}$ are positive n-tuples with $\underline{a}$ decreasing then

$$\min_{1 \le i \le n} \frac{U_i}{V_i} \le \frac{\sum_{i=1}^n a_i u_i}{\sum_{i=1}^n a_i v_i} \le \max_{1 \le i \le n} \frac{U_i}{V_i}.$$

EXTENSIONS (a) If on $[a,b]$, f is continuous, g of bounded variation, or f, g are both of bounded variation and g is continuous, then

$$\left|\int_a^b f\,dg\right| \le \int_a^b |f|\,|dg|.$$

(b) [OSTROWSKI] *let f be a monotone integrable function on $[a,b]$, and g a real or complex function integrable on $[a,b]$ then*

$$\left|\int_a^b fg\right| \leq |f(a)| \max_{a\leq x\leq b}\left|\int_a^x g\right| + |f(b)| \max_{a\leq x\leq b}\left|\int_x^b g\right|.$$

COMMENTS (iii) The integrals in (a) are Riemann-Stieltjes integrals.
(iv) See also **Abel's Inequalities** INTEGRAL ANALOGUE, **Bounded Variation Function Inequalities** (c), (d) and EXTENSIONS, **Complex Function Inequalities, Erdös & Grünwald's Inequality, Gauss's Inequality, Integral Mean Value Theorems, Steffensen's Inequalities**, as well as the integral analogues of many other inequalities.

REFERENCES [AI, pp. 301–302], [MPF, p. 116, 337, 476–481]; [Hewitt & Stromberg, p. 284], [Titchmarsh, 1939, pp. 361–362, 373], [Widder, pp. 8–10].

Integral Mean Value Theorems

(a) If $m \leq f(x) \leq M, a \leq x \leq b$, then

$$m \leq \frac{1}{b-a}\int_a^b f \leq M. \tag{1}$$

(b) [BONNET] *If $f \geq 0$, continuous and increasing, and g is of bounded variation then*

$$f(b)\inf_{a\leq x\leq b}\int_x^b dg \leq \int_a^b f\,dg \leq f(b)\sup_{a\leq x\leq b}\int_x^b dg,$$

and if f is decreasing,

$$f(a)\inf_{a\leq x\leq b}\int_a^x dg \leq \int_a^b f\,dg \leq f(a)\sup_{a\leq x\leq b}\int_a^x dg.$$

COMMENTS (i) inequality (1) is an integral analogue of **Arithmetic Mean Inequalities** (1).
(ii) The integrals in (b) are Riemann-Stieltjes integrals.

EXTENSIONS (a) Let f be μ-measurable on E, $m = \inf_{x\in E} f(x)$, $M = \sup_{x\in E} f(x)$, $g, fg \in \mathcal{L}_\mu(E)$ then

$$m\int_E |g|\,d\mu \leq \int_E f|g|\,d\mu \leq M\int_E |g|\,d\mu.$$

(b) [FEJÉR] If T_n is a trigonometric polynomial of degree n, $m = \min_{0\leq x\leq 2\pi} T_n(x)$, $M = \max_{0\leq x\leq 2\pi} T_n(x)$ then

$$m + \frac{M-m}{n+1} \leq \frac{1}{2\pi}\int_0^{2\pi} T_n \leq M - \frac{M-m}{n+1}.$$

(c) [LUKÁCS] If p_n is a polynomial of degree n, $m = \min_{a \leq x \leq b} p_n(x)$, and $M = \max_{a \leq x \leq b} p_n(x)$ then

$$m + \frac{M-m}{\tau_n} \leq \frac{1}{2\pi} \int_0^{2\pi} p_n \leq M - \frac{M-m}{\tau_n},$$

where

$$\tau_n = \begin{cases} (m+1)^2, & \text{if } n = 2m, \\ (m+1)(m+2), & \text{if } n = 2m+1. \end{cases}$$

The constant τ_n is best possible.

COMMENTS (iii) An extension of Bonnet's results, analogous to **Abel's Inequalities** (4), has been given by Bromwich.
(iv) A definition of trigonometric polynomial of degree n is given in **Trigonometric Polynomial Inequalities**.
(v) See also **Integral Inequalities** (b), and EXTENSIONS.

REFERENCES [Apostol, vol.1, p. 358], [Bromwich, pp. 473–475], [Saks, p. 26], [Szegö, pp. 178–181], [Widder, pp. 16–20].

Integral Test Inequality
If f is a non-negative strictly decreasing function defined on the interval $[m-1, n]$, $m, n \in \mathbb{N}$ then

$$\sum_{m}^{n} f(i) < \int_{m-1}^{n} f < \sum_{m-1}^{n-1} f(i). \tag{1}$$

COMMENTS (i) This follows by considering the areas the terms in the inequalities represent.
(ii) The name is given since this inequality is the basis of the convergence test of the same name; the series $\sum_{i=m}^{\infty} f(i)$ and the improper integral $\int_m^{\infty} f$ either both converge or both diverge.

REFERENCES [Knopp, pp. 294–295].

Internal Function Inequalities
If $f : [a, b] \to \mathbb{R}$ is an internal function then for all x, y, $a \leq x, y \leq b$

$$\min\{f(x), f(y)\} \leq f\left(\frac{x+y}{2}\right) \leq \max\{f(x), f(y)\}.$$

COMMENTS (i) This is just the definition of an *internal function* on $[a, b]$.
(ii) See **Fischer's Inequalities** COMMENTS.

REFERENCES [MPF, p. 641].

Irrationality Measure Inequalities
If p, q are positive integers, $p \geq 2$ and q large enough, then

$$\left|\pi - \frac{p}{q}\right| > q^{-23.72}.$$

REFERENCES [Borwein & Borwein, pp. 362–386].

Isodiametric Inequality [CLASSICAL RESULT] If A is the area of a plane region with inner and outer radii R, r respectively, then

$$R^2 \geq \frac{A}{\pi} \geq r^2,$$

with equality when the region is circular.

COMMENTS   (i) This is an example of **Symmetrization Inequalities**.
(ii) The *inner radius of a set* is the radius of the largest disk covered by the set; the *outer radius of a set* is the radius of the smallest disk that covers the set.

EXTENSIONS   For all sets $A \subseteq \mathbb{R}^n$ if $d(A)$ the diameter of A then

$$|A|_n \leq v_n \left(\frac{d(A)}{2}\right)^n.$$

COMMENTS   (iii) This inequality is of interest since A need not be contained in a ball of diameter $d(A)$.
(iv) See also **Isoperimetric Inequalities, n-Simplex Inequality**.

REFERENCES   [Evans & Gariepy, p. 69], [Pólya & Szegö, 1951, p. 8].

Isoperimetric Inequalities (a) [THE CLASSICAL ISOPERIMETRIC INEQUALITIES]
(a) If A is the area, and ℓ the perimeter of a plane region then

$$A \leq \frac{\ell^2}{4\pi},$$

with equality only when the region is a circular.
(b) If V is the volume, S the surface area of a domain in $\mathbb{R}^3$ then

$$V^2 \leq \frac{S^3}{36\pi},$$

with equality only if the domain is spherical.

EXTENSIONS   If V is the n-dimensional volume of a domain in $\mathbb{R}^n, n \geq 2$ with A the $n-1$-dimensional area of its bounding surface then

$$\frac{V^{n-1}}{A^n} \leq \frac{v_n^{n-1}}{a_n^n}, \quad \text{or} \quad A_n \geq n v_n^{1/n} V_n^{1/m}, \tag{1}$$

where m is the conjugate of n..

COMMENTS   (i) Inequality (1) is, surprisingly equivalent to **Sobolev's Inequalities** (b).
(ii) See also **Blaschke-Santaló Inequality, Bonnensen's Inequality, Gale's Inequality, Geometric Inequalities, Isodiametric Inequality, Mahler's Inequalities, Mixed-Volume Inequalities, Symmetrization Inequalities, Young's Inequalities** (e).

REFERENCES   [EM, vol. 5, pp. 203–208]; [Bobkov & Houdré, p. 1], [Pólya & Szegö, 1951, p. 8].

- end of I -

J

Jackson's Inequality Let $f : \mathbb{R} \to \mathbb{R}$ be continuous and of period 2π then

$$\inf_{T_n \in \mathcal{T}_n} \|f - T_n\|_\infty \leq C\omega(f; n^{-1}), \tag{1}$$

where $\mathcal{T}_n$ is the set of all trigonometric polynomials of degree n, ω is the modulus of continuity, and C is an absolute constant.

COMMENTS   (i) The definition of ω is given in **Modulus of Continuity Inequalities**, COMMENTS (i).

EXTENSIONS   If $f : \mathbb{R} \to \mathbb{R}$ has continuous derivative of order k, $k \geq 1$, and of period 2π then, with the above notation,

$$\inf_{T_n \in \mathcal{T}_n} \|f - T_n\|_\infty \leq \frac{C_k}{n^k} \omega(f^{(k)}; n^{-1}).$$

COMMENTS   (ii) A similar result can be stated with trigonometric polynomials replaced by polynomials; see [EM].

REFERENCES   [EM, vol. 5, p. 219]; [General Inequalities, vol. 1, pp. 85–114].

Jensen's Inequality   (a) If I is an interval in $\mathbb{R}$ and $f : I \to \mathbb{R}$ is convex, $\underline{a} \in I^n$, $n \geq 2$ and $\underline{w}$ a positive n-tuple, then

$$f\left(\frac{1}{W_n} \sum_{i=1}^n w_i a_i\right) \leq \frac{1}{W_n} \sum_{i=1}^n w_i f(a_i); \tag{J}$$

or equivalently,

$$f(\mathfrak{A}_n(\underline{a}; \underline{w})) \leq \mathfrak{A}_n(f(\underline{a}); \underline{w}). \tag{1}$$

If f is strictly convex then (J) is strict unless $\underline{a}$ is constant.
(b) [VASIĆ & PEČARIĆ] If $\underline{w}$ is a real n-tuple with $W_n > 0$, $w_1 > 0$ and $w_i \leq 0$, $2 \leq i \leq n$, and $\mathfrak{A}_n(\underline{a}; \underline{w}) \in I$, then $(\sim J)$ holds.

COMMENTS   (i) The case $n = 2$ of (J) is just the definition of convexity, see **Convex Function Inequalities** (1). The general case follows by induction. See also **Order Inequalities** COMMENTS (iii).
(ii) (J) can be interpreted as saying:

the centroid of the points $P_i = (a_i, f(a_i))$ $1 \leq i \leq n$, lies above the graph of f.

(iii) (b) can be proved by applying (J) to the n-tuple $(\mathfrak{A}_n(\underline{a};\underline{w}), a_2, \ldots, a_n)$ with the weights $(W_n, -w_2, \ldots, -w_n)$.

EXTENSIONS (a) [FUNCTIONS OF INDEX SETS; VASIC&MIJALKOVIĆ] Let $I, f, \underline{a}, \underline{w}$ be as in (J) and define F on the index sets by

$$F(\mathcal{I}) = W_{\mathcal{I}}\left\{\mathfrak{A}_{\mathcal{I}}(f(\underline{a});\underline{w}) - f(\mathfrak{A}_{\mathcal{I}}(\underline{a};\underline{w}))\right\}. \qquad (2)$$

Then $F \geq 0$, and if $\mathcal{I} \cap \mathcal{J} = \emptyset$,

$$F(\mathcal{I}) + F(\mathcal{J}) \leq F(\mathcal{I} \cup \mathcal{J}). \qquad (3)$$

Further, if f is strictly convex (3) is strict unless $A_{\mathcal{I}}(\underline{a};\underline{w}) = A_{\mathcal{J}}(\underline{a};\underline{w})$.
(b) [DRAGOMIR & CRSTICI] Let $I, f, \underline{a}, \underline{w}$ be as in (J) and if $\underline{v}$ is a k-tuple, with $1 \leq k \leq n$ then

$$f(\mathfrak{A}_n(\underline{a};\underline{w}))$$
$$\leq \frac{1}{W_n^n} \sum_{i_1,\ldots,i_k=1}^{n} \left(\prod_{j=1}^{k} w_{i_j}\right) f\left(\frac{1}{V_k}\sum_{j=1}^{k} v_j a_{i_j}\right)$$
$$\leq \mathfrak{A}_n(f(\underline{a});\underline{w}).$$

COMMENTS (iv) (3) is a simple deduction from (J). While (J) says that the function F of (2) is non-negative, the refinement (3) says that F is super-additive.

COROLLARIES [RADO TYPE] If $\mathcal{I}_i = \{1, 2, \ldots, i\}$, $1 \leq i \leq n$ then, with $F_{\mathcal{I}}$ as in (2),

$$F_{\mathcal{I}_n} \geq \cdots \geq F_{\mathcal{I}_2},$$

and in particular

$$F_{\mathcal{I}_n} \geq \sup_{1 \leq i,j \leq n, i \neq j} \left\{w_i f(a_i) + w_j f(a_j) - (w_i + w_j)f\left(\frac{w_i a_i + w_j a_j}{w_i + w_j}\right)\right\}.$$

COMMENTS (v) The first part of the Corollary is an immediate consequence of (3) and the second is just $F_{\mathcal{I}_n} \geq F_{\mathcal{I}_2}$, noting that the terms in the sum in $F_{\mathcal{I}_n}$ can be rearranged.
(vi) A very interesting extension given in **Jensen-Steffensen Inequality** allows negative weights in both (J) and ($\sim$J).

CONVERSE INEQUALITIES (a) [LAH & RIBARIĆ] Let $I, f, \underline{a}, \underline{w}$ be as in (J), and let

$$a = \min \underline{a}; \qquad A = \max \underline{a}; \qquad \overline{a} = \mathfrak{A}_n(\underline{a};\underline{w}), \qquad (4)$$

then

$$\mathfrak{A}_n(f(\underline{a});\underline{w}) \leq \frac{A - \overline{a}}{A - a}f(a) + \frac{\overline{a} - a}{A - a}f(A),$$

with equality if and only if for each i, either $a_i = A$ or $a_i = a$.

(b) [MITRINOVIĆ & VASIĆ] If I is an interval in $\mathbb{R}$ and $f: I \to \mathbb{R}$ is positive, strictly convex and twice differentiable, $\underline{a} \in I^n$, $n \geq 2$, then

$$\frac{1}{W_n}\sum_{i=1}^n w_i f(a_i) \leq \lambda f\left(\frac{1}{W_n}\sum_{i=1}^n w_i a_i\right),$$

where, if $\phi = (f')^{-1}$ and, using the notation of (4), with

$$\mu = \frac{f(A) - f(a)}{A - a}, \qquad \nu = \frac{Af(a) - af(A)}{A - a},$$

λ is the unique solution of

$$f \circ \phi\left(\frac{\mu}{\lambda}\right) = \frac{\mu}{\lambda}\phi\left(\frac{\mu}{\lambda}\right) + \frac{\nu}{\lambda}.$$

COMMENTS (vii) The result of Lah & Ribarić is a simple consequence of **Convex Function Inequalities**, (2). It extends COMMENTS (ii) to:

the centroid of the points $(a_i, f(a_i))$, $1 \leq i \leq n$ lies below the chord joining $(a, f(a))$ to $(A, f(A))$; a, A as in (4) above.

(viii) The proof of the Mitrinović & Vasić result (b) is based on the geometrical remarks in COMMENTS (ii) and (vii); they called this the *centroid method* of proving inequalities.

(ix) For another form of (J) see **Quasi-Arithmetic Mean Inequalities** (1).

INTEGRAL ANALOGUES If $w\phi \in \mathcal{L}(a,b)$, $w \geq 0$, $\int_a^b w > 0$, f convex on the interval I, I containing the range of ϕ, and if $wf \circ \phi \in \mathcal{L}(a,b)$ then

$$f\left(\frac{\int_a^b w\phi}{\int_a^b w}\right) \leq \frac{\int_a^b wf \circ \phi}{\int_a^b w}.$$

If f is strictly convex then this inequality is strict unless f is constant.

COMMENTS (x) This can be extended to general measure spaces; see, for instance, [Hewitt & Stromberg] or [Lieb & Loss].

(xi) See also **Slater's Inequality**; and for another inequality also called Jensen's inequality see **Power Sum Inequalities** COMMENTS (i).

(xii) This important inequality has been the subject of much research in recent years see in particular [MPF, PPT]. Another inequality of the same name is **Power Sums Inequalities** (1).

REFERENCES [EM, vol. 5, pp. 234–235], [MI, pp. 23–30], [MO, pp. 454–456], [MPF, pp. 1–19, 681–695], [PPT, pp. 43–57, 83–87, 105, 133–134]; [Conway, vol. II, p. 225], [Hájos et al, pp. 73–77], [Hewitt & Stromberg, p. 202], [Lieb & Loss, pp. 38–39], [Pólya & Szegö, 1972, pp. 65–67], [Roberts & Varberg, pp. 89, 189–190]; [Dragomir & Crstici].

Jensen-Steffensen Inequality   (a) If I is some interval in $\mathbb{R}$, $f: I \to \mathbb{R}$ convex, $\underline{a}$ a monotone n-tuple with $a_i \in I, 1 \leq i \leq n$, and $\underline{w}$ is a real n-tuple with

$$W_n \neq 0, \qquad 0 \leq \frac{W_i}{W_n} \leq 1, \ 1 \leq i \leq n, \tag{1}$$

then (J) holds, and holds strictly unless $\underline{a}$ is constant.
(b) [PEČARIĆ] Let $I, f, \underline{a}, \underline{w}$ be as in (J) and $\underline{w}$ a real n-tuple with

$$W_n > 0, \quad \mathfrak{A}_n(\underline{a}; \underline{w}) \in I;$$

and for some $m, 1 \leq m \leq n$,

$$W_k \leq 0, \ 1 \leq k < m, \qquad W_n - W_{k-1} \leq 0, \ m < k \leq n;$$

then ($\sim$J) holds.

COMMENTS   (i) A proof of (a) based on the **Steffensen Inequality** was given by Steffensen. A simple inductive proof has been given; see [Bullen, 1998].
(ii) A discussion of condition (1) can be found in **Geometric-Arithmetic Mean Inequality** COMMENTS (ix); in particular inequality (5) in that reference shows that (1) implies that $\mathfrak{A}_n(\underline{a}; \underline{w}) \in I$.
(iii) Many of the inequalities can be derived from (J), and so usually they will have generalizations that allow weights satisfying (1); see for instance **Geometric-Arithmetic Mean Inequality**, EXTENSIONS (m).

EXTENSION [CIESIELSKI]   Let $\underline{w}$ be a real n-tuple with

$$W_k \geq 0, 1 \leq k \leq n, \qquad \text{and} \qquad \sum_{i=1}^{n} |w_i| > 0.$$

If f, f' are both convex functions on $[0, \ell]$ with $f(0) \leq 0$ then for any decreasing n-tuple, $\underline{a}$ with terms in $[0, \ell]$

$$f\left(\frac{\sum_{i=1}^{n} w_i a_i}{\sum_{i=1}^{n} |w_i|}\right) \leq \frac{\sum_{i=1}^{n} w_i f(a_i)}{\sum_{i=1}^{n} |w_i|}.$$

INTEGRAL ANALOGUES   [STEFFENSEN] If f is convex on an interval containing the range of ϕ, an increasing function, and if

$$0 \leq \int_a^x w \leq \int_a^b w, \ a \leq x \leq b; \quad \text{and} \quad \int_a^b w > 0,$$

then

$$f\left(\frac{\int_a^b w\phi}{\int_a^b w}\right) \leq \frac{\int_a^b w f \circ \phi}{\int_a^b w}.$$

REFERENCES   [AI, pp. 109–110, 115–116], [MI, pp. 25–26], [MPF, pp. 6–7], [PPT, pp. 57–63, 83–105, 161]; [General Inequalities, vol. 4, pp. 87–92]; [Bullen, 1998].

Jessen's Inequality Let $\underline{a}^{(j)} = (a_{1j}, \ldots, a_{mj}), 1 \leq j \leq n$, and $\underline{u}$ be positive m-tuples, $\underline{a}_{(i)} = (a_{i1}, \ldots, a_{1n}), 1 \leq i \leq m$, and $\underline{v}$ positive n-tuples. If $-\infty \leq r < s \leq \infty$ then:

$$\mathfrak{M}_n^{[s]}\left(\mathfrak{M}_m^{[r]}(\underline{a}^{(j)}; \underline{u}); \underline{v}\right) \leq \mathfrak{M}_m^{[r]}\left(\mathfrak{M}_n^{[s]}(\underline{a}_{(i)}; \underline{v}); \underline{u}\right).$$

There is equality if and only if $a_{ij} = b_i c_j, 1 \leq i \leq m, 1 \leq j \leq n$.

COMMENTS (i) This is a form of (M); see **Minkowski's Inequality** (2).
(ii) An integral analogue can be found in **Minkowski's Inequality** INTEGRAL ANALOGUES (b); see also **Hölder's Inequality** EXTENSIONS (c).

REFERENCES [HLP, pp. 31–32], [MPF, p. 109].

Jordan's Inequality If $0 < |x| \leq \pi/2$ then

$$\frac{2}{\pi} \leq \frac{\sin x}{x} < 1.$$

COMMENTS (i) This is an immediate consequence of the strict concavity of the sine function on the interval $[0, \pi/2]$.

EXTENSIONS (a) [REDHEFFER] For all $x \neq 0$

$$\frac{\sin x}{x} \geq \frac{\pi^2 - x^2}{\pi^2 + x^2}.$$

(b) [EVERITT] If $0 < x \leq \pi$ then

$$\frac{\sin x}{x} < \min\left\{\frac{2(1 - \cos x)}{x^2}, \frac{2 + \cos x}{3}\right\};$$

and

$$\frac{\sin x}{x} > \max\left\{\frac{4 - 4\cos x - x^2}{x^2}, \frac{3\cos x + 12 - x^3}{15}\right\}.$$

(c) [BECKER & STARK] If $0 < \alpha < 1$ and $0 \leq x \leq 2\alpha\pi$ then

$$1 - \cos x \geq \left(\frac{\sin \alpha\pi}{\alpha\pi}\right)^2 \frac{x^2}{2}.$$

COMMENTS (ii) See also **Trigonometric Function Inequalities** (a).
REFERENCES [AI, pp. 33, 354]; [Becker & Stark].

- end of J -

K

Kaczmarz & Steinhaus's Inequalities   If $p > 2$ then for some constant α depending only on p,

$$|1+x|^p \leq 1 + px + \sum_{i=2}^{[p]} \binom{p}{i} x^i + \alpha |x|^p, \quad x \in \mathbb{R}.$$

COROLLARIES   If $p > 2$ and $f, g \in \mathcal{L}^p([a,b])$ then there are constants α, β depending only on p such that

$$\int_a^b |f+g|^p \leq \int_a^b |f|^p + p\int_a^b |f|^{p-2} fg + \alpha \int_a^b |g|^p + \beta \sum_{i=2}^{[p]} \int_a^b |f|^{p-1} |g|^i.$$

COMMENTS   This inequality is important in the theory of orthogonal series.

REFERENCES   [MPF, p. 66]; [Kacmarz & Steinhaus, p. 247].

Kakeya's Maximal Function Inequality   If R is a rectangle of sides $a, b, a < b$, put $e(R) = \alpha = a/b, 0 < \alpha \leq 1$, and for $f : \mathbb{R}^2 \to \mathbb{R}$ define

$$K_\alpha f(\underline{x}) = \sup_{\underline{x} \in R, e(R) = \alpha} \frac{1}{|R|} \int_R |f|, \underline{x} \in \mathbb{R}^2.$$

Then

$$K_\alpha f \leq \frac{C}{\alpha} M_f,$$

and if $f \in \mathcal{L}^2(\mathbb{R}^2), \beta > 0$,

$$\|K_\alpha f\|_2 \leq C_1 \left(\log \frac{2}{\alpha}\right)^2 \|f\|_2;$$

$$|\{\underline{x}; K_\alpha f(\underline{x}) > \beta\}| \leq C_2 \left(\log \frac{2}{\alpha}\right)^3 \frac{\|f\|_2^2}{\beta^2}.$$

COMMENTS   (i) M_f is the Hardy-Littlewood maximal function, defined in **Hardy-Littlewood Maximal Inequalities** COMMENTS (v); the extension of that definition to higher dimensions is immediate.
(ii) The quantity $e(R)$ is called the *eccentricity* of R; and $K_\alpha f$ is the *Kakeya maximal function*.
(iii) These results have partial extensions to higher dimensions.

REFERENCES   [Igari], [Müller].

Kalajdžić Inequality Let $b > 1$, $\underline{a}, \underline{w}$ positive n-tuples with $W_n = 1$ then

$$\sum! b^{a_1^{w_{i1}} \ldots a_n^{w_{in}}} \leq (n-1)! \sum_{i=1}^{n} b^{a_i}, \qquad (1)$$

with equality if and only if $\underline{a}$ is constant.

COMMENTS (i) This follows from (GA) and the method of proof shows that (a) below holds.

EXTENSIONS (a) Under the above hypotheses but allowing $\underline{w}$ to have zero values

$$\sum! b^{a_1^{w_{i1}} \ldots a_n^{w_{in}}} \leq (n-1)! \sum_{i=1}^{n} b^{a_i^{W_n}};$$

equality can occur if all but at most one of the w_i are zero, or if $\underline{a}$ is constant.
(b) If $\underline{a}, \underline{w}$ are positive n-tuples, $W_n = 1$, then

$$\sum! a_1^{w_{i1}} \cdots a_n^{w_{in}} \leq (n-1)! \sum_{i=1}^{n} a_i.$$

COMMENTS (ii) In the case of equal weights (b) is an extension of (GA).
REFERENCES [MI, pp. 92–93].

Kallman-Rota Inequality If A is the infinitesimal generator of a strongly continuous semigroup of contractions on a Banach space X, and if x is in the domain of A^2 then

$$||Ax||^2 \leq 4||x||\,||A^2x||.$$

COMMENTS (i) This inequality can be used to prove **Hardy-Littlewood-Pólya Inequality** (1).
(ii) Definitions of the terms used can be found in the first reference.
REFERENCES [EM, vol.8, p. 254]; [Biler & Witkowski, p. 123].

Kaluza-Szegö Inequality If $\underline{a}$ is a positive sequence,

$$\sum_{i=1}^{n} \mathfrak{G}_i(\underline{a}) \leq \sum_{i=1}^{n} \left(1 + \frac{1}{i}\right)^i a_i.$$

COMMENTS This is an extension of the finite form of **Carleman's Inequality**.
REFERENCES [Redheffer, p. 684].

Kantorovič's[1] Inequality

If $0 < m \leq \underline{a} \leq M$ then

$$\mathfrak{A}_n(\underline{a};\underline{w})\mathfrak{A}_n(\underline{a}^{-1};\underline{w}) = \frac{\mathfrak{A}_n(\underline{a};\underline{w})}{\mathfrak{H}_n(\underline{a};\underline{w})} \leq \frac{(M+m)^2}{4Mm} = \left(\frac{M+m}{2}\right)\left(\frac{M^{-1}+m^{-1}}{2}\right),$$

with equality if and only if there is a $\mathcal{I} \subseteq \{1,\ldots,n\}$ such that $W_\mathcal{I} = 1/2$, $a_i = M, i \in \mathcal{I}$ and $a_i = m, i \notin \mathcal{I}$.

COMMENTS (i) This inequality, the equal weight case of which is due to Schweitzer, is equivalent to the **Pólya & Szegö's Inequality**. It is also a special case of a result of Specht, see **Power Mean Inequalities** CONVERSE INEQUALITIES (b).

(ii) In the equal weight case, and when n is odd equality is impossible; a better constant has been obtained in this case by Clausing; see [General Inequalities].

INTEGRAL ANALOGUES [SCHWEITZER] If $f, 1/f \in \mathcal{L}([a,b])$. with $0 < m \leq f \leq M$ then

$$\int_a^b f \int_a^b \frac{1}{f} \leq \frac{(M+m)^2}{4Mm}(b-a)^2.$$

COMMENTS (iii) For a converse inequality see **Walsh's Inequality**; and for a matrix analogue see **Matrix Inequalities** COMMENTS (i). For another generalization see **Symmetric Mean Inequalities** COMMENTS (iv).

REFERENCES [AI, pp. 59–66], [BB, pp. 44-45], [MI, pp. 198–204], [MO, p. 71], [MPF, pp. 684–685]; [General Inequalities, vol. 3, p. 61].

Karamata's Inequalities

If f, g are integrable on $[0,1]$ with $0 < \alpha \leq f \leq A$, and $0 < \beta \leq g \leq B$ then

$$\left(\frac{\sqrt{\alpha\beta}+\sqrt{AB}}{\sqrt{\alpha B}+\sqrt{A\beta}}\right)^{-2} \leq \frac{\mathfrak{A}_{[0,1]}(fg)}{\mathfrak{A}_{[0,1]}(f)\mathfrak{A}_{[0,1]}(g)} \leq \left(\frac{\sqrt{\alpha\beta}+\sqrt{AB}}{\sqrt{\alpha B}+\sqrt{A\beta}}\right)^{2}.$$

COMMENTS (i) The constant on the right-hand side is greater than or equal to 1.

(ii) This result is a converse inequality for (Č); for another see **Grüsses' Inequalities** (a).

REFERENCES [MPF, p. 298].

Khinchine-Littlewood Inequality
See **Hinčin-Littlewood Inequality**.

Klamkin's Inequality
See **Beth & van der Corput's Inequality** EXTENSIONS.

---

[1] Л В Канторович. Also transliterated as Kantorovich.

Klamkin & Newman Inequalities Let $\underline{a}$ be a non-negative n-tuple, $a_0 = 0$ and $0 \leq \tilde{\Delta} a_i \leq 1, 0 \leq i \leq n-1$; if $r \geq 1, s+1 \geq 2(r+1)$ then

$$\left((s+1)\sum_{i=1}^{n} a_i^s\right)^{1/(s+1)} \leq \left((r+1)\sum_{i=1}^{n} a_i^r\right)^{1/(r+1)}.$$

EXTENSIONS (a) [MEIR] Let $\underline{a}$ be a non-negative n-tuple, $a_0 = 0, 0 \leq w_0 \leq \cdots \leq w_n$, and $0 \leq \tilde{\Delta} a_i \leq (w_i + w_{i-1})/2, 0 \leq i \leq n-1$; if $r \geq 1, s+1 \geq 2(r+1)$ then

$$\left((s+1)\sum_{i=1}^{n} w_i a_i^s\right)^{1/(s+1)} \leq \left((r+1)\sum_{i=1}^{n} w_i a_i^r\right)^{1/(r+1)}.$$

(b) [MILOVANOVIĆ] Let $f, g : [0, \infty[\to \mathbb{R}$ satisfy (i) $f(0) = f'(0) = g(0) = g'(0) = 0$, (ii) f', g' are convex. If $\underline{a}, \underline{w}$ are as in (a) and $h = g \circ f$ then

$$h^{-1}\left(\sum_{i=1}^{n} w_i h'(a_i)\right) \leq f^{-1}\left(\sum_{i=1}^{n} w_i f'(a_i)\right).$$

COMMENTS (i) (a) is the case $h(x) = x^{s+1}, f(x) = x^{r+1}$ of (b).
(ii) Another pair of functions satisfying the hypotheses of (b) are $f(x) = x^2, g(x) = x^3 e^x$.
(iii) Compare these to the **Gauss-Winkler Inequality**.
REFERENCES [PPT, pp. 166–169].

Kneser's Inequality If p is a polynomial of degree n, $p = qr$, where q, r are polynomials of degrees $m, n - m$ respectively, then

$$\|q\|_{\infty,[-1,1]}\|r\|_{\infty,[-1,1]} \leq \frac{1}{2} C_{n,m} C_{n,n-m} \|p\|_{\infty,[-1,1]},$$

where

$$C_{n,m} = 2^m \prod_{k=1}^{m}\left(1 + \cos\frac{2k-1}{2n}\pi\right).$$

In particular

$$\|q\|_{\infty,[-1,1]}\|r\|_{\infty,[-1,1]} \leq 2^{n-1} \prod_{k=1}^{[n/2]}\left(1 + \cos\frac{2k-1}{2n}\pi\right)\|p\|_{\infty,[-1,1]}.$$

COMMENTS (i) The constant on the right-hand side of the last inequality is approximately $(3.20991\cdots)^n$.
(ii) This result has been extensively discussed by P. B. Borwein.
REFERENCES [Borwein].

Knopp's Inequalities (a) If $\underline{a}$ is a positive ℓ_p sequence, $p > 0$, and if $\alpha > 0$ then

$$\sum_{n \in \mathbb{N}} \binom{n+\alpha}{n}^p \left(\sum_{i=0}^{n} \frac{\binom{n+\alpha-1-i}{n-i}}{a_i} \right)^{-p} \leq \frac{(\alpha + p^{-1})!}{\alpha!(p^{-1})!} \sum_{n \in \mathbb{N}} a_n^p.$$

(b) If $\underline{a}$ is a positive sequence and $\alpha > 0$ then

$$\sum_{n \in \mathbb{N}} \left(\prod_{i=0}^{n} a_i^{\binom{n-\alpha-1-i}{n-i}} \right)^{1/\binom{n+\alpha}{n}} \leq e^{\psi(\alpha)-\psi(0)} \sum_{n \in \mathbb{N}} a_n.$$

COMMENTS (i) For a definition of ψ see **Digamma Function Inequalities**.
(ii) These inequalities are due to Knopp; but the exact constants on the right-hand sides were given by Bennett.

COROLLARIES If $\underline{a}$ is a positive ℓ_p sequence, $p > 0$, then

$$\sum_{n=1}^{\infty} (\mathfrak{H}_n(\underline{a}))^p \leq \left(\frac{p+1}{p} \right)^p \sum_{n=1}^{\infty} a_n^p \quad \text{or} \quad \|\mathfrak{H}(\underline{a})\|_p \leq \frac{p+1}{p} \|\underline{a}\|_p.$$

COMMENTS (iii) This is obtained from (a) by taking $\alpha = 1$.
(iv) See also **Copson's Inequalities** EXTENSIONS (b).
REFERENCES [Bennett, 1996, pp. 37–39].

Kober & Diananda's Inequalities Assume that $\underline{a}$ is a non-constant positive n-tuple, that $\underline{w}$ is a positive n-tuple with $W_n = 1$ and that

$$S_n(\underline{a}) = \frac{1}{2} \sum_{i,j=1}^{n} (\sqrt{a_i} - \sqrt{a_j})^2, \quad S_n(\underline{a}; \underline{w}) = \frac{1}{2} \sum_{i,j=1}^{n} w_i w_j (\sqrt{a_i} - \sqrt{a_j})^2;$$

then

$$\frac{\min \underline{w}}{n-1} < \frac{\mathfrak{A}_n(\underline{a}; \underline{w}) - \mathfrak{G}_n(\underline{a}; \underline{w})}{S_n(\underline{a})} < \max \underline{w},$$

and

$$\frac{1}{1 - \min \underline{w}} < \frac{\mathfrak{A}_n(\underline{a}; \underline{w}) - \mathfrak{G}_n(\underline{a}; \underline{w})}{S_n(\underline{a}; \underline{w})} < \frac{1}{\min \underline{w}}.$$

COMMENTS (i) If $\underline{a}$ is allowed to be non-negative the bounds in these inequalities can be attained.
(ii) If $\underline{a} \to \alpha \underline{e} \neq \underline{0}$ then the central term in the second set of inequalities tends to 2.
(iii) These results can be used to improve both (H) and (M); see [MI].
REFERENCES [AI, pp. 81–83], [MI, pp. 109–113, 152–154], [MFP, pp. 113–114].

Kolmogorov's[2] Inequalities
If $f \in \mathcal{L}^r(\mathbb{R})$, $f^{(n)} \in \mathcal{L}^p(\mathbb{R})$ then

$$\|f^{(k)}\|_q \le C_{p,r,k,n} \|f\|_r^\nu \|f^{(n)}\|_p^{1-\nu}$$

where

$$\nu = \frac{n - k - p^{-1} + q^{-1}}{n - p^{-1} + r^{-1}}$$

COMMENTS (i) In the case $k = 1, n = p = q = r = 2$ this is **Hardy-Littlewood-Pólya Inequalities** (2), while in the case $p = q = r = \infty$ it is **Hardy-Littlewood-Landau Derivative Inequalities** (1), and the case $k = 2, n = 4, p = q = r = 2$ is due to Kurepa. See also **Sobolev's Inequalities**.

(ii) For other inequalities due to Kolmogorov see **Kolmogorov's Probability Inequality**.

REFERENCES [EM, vol. 5, p. 295]; [Biler & Witkowski, p. 123].

Kolmogorov's[2] Probability Inequality
If X_i, $1 \le i \le n$, are independent random variables of zero mean and finite variances, and if $S_j = \sum_{i=1}^{j} X_i$, $1 \le j \le n$, then

$$P\big(\max_{1 \le j \le n} |S_j| > r\big) \le \frac{\sigma^2 S_n}{r^2}.$$

COMMENTS (i) This extends **Čebišev's Probability Inequality**.

(ii) A converse inequality can be found in the last reference, and an extension is in **Lévy's Inequalities**.

(ii) See also **Martingale Inequalities** (a), COMMENTS (i).

REFERENCES [EM, vol. 2, p. 120]; [Feller, vol. II, p. 156], [Loève, p. 235].

König's Inequality
If $\underline{a}, \underline{b}$ are decreasing non-negative n-tuples and if $p > 0$ then

$$\prod_{i=1}^{k} b_i \le \prod_{i=1}^{k} a_i, \ 1 \le k \le n, \quad \Longrightarrow \quad \sum_{i=1}^{k} b_i^p \le \sum_{i=1}^{k} a_i^p, \ 1 \le k \le n.$$

COMMENTS This was used by König to give a proof of **Weyl's Inequalities** (c).

REFERENCES [König, p. 24].

Korn's Inequality
If $\Omega \subset \mathbb{R}^p$ is a bounded domain, $f_i : \Omega \to \mathbb{R}$, $1 \le i \le p$, differentiable, then for some constant C

$$\int_\Omega \left\{ \sum_{i,j=1}^{p} \left(\frac{\partial f_i}{\partial x_j} \right)^2 + \sum_{i=1}^{p} f_i^2 \right\} d\underline{x} \le C \int_\Omega \left\{ \sum_{i,j=1}^{p} \left(\frac{\partial f_i}{\partial x_j} + \frac{\partial f_j}{\partial x_i} \right)^2 + \sum_{i=1}^{p} f_i^2 \right\} d\underline{x}.$$

COMMENTS If the second term on the right-hand side is omitted this is *Korn's first inequality*, and then the present result is *Korn's second inequality*.

REFERENCES [EM, vol. 5, p. 299].

---

[2] А Н Колмогоров. Also transliterated as Kolmogoroff.

Korovkin's[3] Inequality If $n \geq 2$ then

$$\frac{a_1}{a_2} + \frac{a_2}{a_3} + \cdots + \frac{a_{n-1}}{a_n} + \frac{a_n}{a_1} \geq n, \tag{1}$$

with equality if and only if $\underline{a}$ is constant.

COMMENTS (i) The proof is by induction; the $n = 2$ case being the well known elementary inequality,

$$x > 0 \quad \text{and} \quad x \neq 1 \quad \Longrightarrow \quad x + \frac{1}{x} > 2.$$

EXTENSIONS (a) [JANIĆ & VASIĆ] If $1 \leq k < n$ then

$$\frac{a_1 + \cdots + a_k}{a_{k+1} + \cdots + a_n} + \frac{a_2 + \cdots + a_{k+1}}{a_{k+2} + \cdots + a_1} + \cdots + \frac{a_n + \cdots + a_{k-1}}{a_k + \cdots + a_{n-1}} \geq \frac{nk}{n-k}.$$

(b) [CHONG] If $a \geq 1$ and $\alpha \geq 1$ then

$$\alpha a + \frac{1}{a} \geq \alpha + 1.$$

COMMENTS (ii) A generalization of (1) is in **Cyclic Inequalities** EXTENSIONS, COMMENTS (ii).

REFERENCES [MI, pp. 45, 73–75, 87].

Ky Fan's Inequality If $\underline{w}$ is a positive n-tuple and if $0 \leq a_i \leq 1/2$, $1 \leq i \leq n$, then

$$\frac{\mathfrak{G}_n(\underline{a}, \underline{w})}{\mathfrak{G}_n(1 - \underline{a}, \underline{w})} \leq \frac{\mathfrak{A}_n(\underline{a}, \underline{w})}{\mathfrak{A}_n(1 - \underline{a}, \underline{w})},$$

with equality if and only if $\underline{a}$ is constant.

COMMENTS (i) The equal weight case was due to Ky Fan. This inequality has been much studied and there are many generalizations.

EXTENSIONS (a) [WANG, CHEN & LI] If $\underline{w}$ is a positive n-tuple and if $0 \leq a_i \leq 1/2$, $1 \leq i \leq n$, and if $r < s$ then

$$\frac{\mathfrak{M}_n^{[r]}(\underline{a}, \underline{w})}{\mathfrak{M}_n^{[r]}(1 - \underline{a}, \underline{w})} \leq \frac{\mathfrak{M}_n^{[s]}(\underline{a}, \underline{w})}{\mathfrak{M}_n^{[s]}(1 - \underline{a}, \underline{w})},$$

if and only if $|r + s| \leq 3$, and $2^r/r \geq 2^s/s$ when $r > 0$, and $r2^r < s2^s$ when $s < 0$.

(b) [WANG & WANG] If $0 \leq a_i \leq 1/2$, $1 \leq i \leq n$, and if $1 \leq r < s \leq n$ then

$$\frac{\mathfrak{P}_n^{[s]}(\underline{a})}{\mathfrak{P}_n^{[s]}(1 - \underline{a})} \leq \frac{\mathfrak{P}_n^{[r]}(\underline{a})}{\mathfrak{P}_n^{[r]}(1 - \underline{a})}.$$

---

[3] П П Коровкин.

COMMENTS   (ii) Some restrictions on r, s are necessary in (a), as Chan, Goldberg & Gonek showed.

(iii) A further of extension is in **Levinson's Inequality**; and a converse inequality has been given by Alzer. The same author has written a survey article on this inequality.

(iv) See also **Logarithmic Mean Inequalities** EXTENSIONS (b), **Weierstrass's Inequalities** RELATED INEQUALITIES (a). For another inequality due to Ky Fan see **Determinant Inequalities** EXTENSIONS (a).

REFERENCES   [AI, p. 363], [BB. pp. 5], [MI, pp. 281–282], [MPF, pp. 25–32]; [Alzer, 1995].

Ky Fan-Taussky-Todd Inequalities   If $\underline{a}$ is a real n-tuple and if a_0, a_{n+1} are defined to be zero then

$$4\sin^2\left(\frac{\pi}{2(n+1)}\right)\sum_{i=1}^{n} a_i^2 \leq \sum_{i=0}^{n}(\Delta a_i)^2,$$

with equality if and only if $a_i = c\sin i\pi/(n+1)$, $1 \leq i \leq n$; also

$$4\sin^2\left(\frac{\pi}{2(2n+1)}\right)\sum_{i=1}^{n} a_i^2 \leq \sum_{i=0}^{n-1}(\Delta a_i)^2,$$

with equality if and only if $a_i = c\sin i\pi/(2n+1)$, $1 \leq i \leq n$.

COMMENTS   (i) This is a discrete analogue of **Wirtinger's Inequality**.

EXTENSIONS   (a) With $\underline{a}$ as above

$$16\sin^4\left(\frac{\pi}{2(n+1)}\right)\sum_{i=1}^{n} a_i^2 \leq \sum_{i=0}^{n}(\Delta^2 a_i)^2,$$

with equality if and only if $a_i = c\sin i\pi/(n+1)$, $1 \leq i \leq n$.

(b) [REDHEFFER] If $\underline{a}$ is a real n-tuple and if for some θ, $0 \leq \theta < \frac{\pi}{n}$,

$$\mu \leq 4\sin^2\theta/2, \qquad \lambda \geq 1 - \frac{\sin(n+1)\theta}{\sin n\theta},$$

then

$$\mu\sum_{i=1}^{n} a_i^2 \leq a_1^2 + \sum_{i=1}^{n-1}\Delta a_i^2 + \lambda a_n^2.$$

(c) [ALZER] If $\underline{a}$ is a complex n-tuple with $A_n = 0$ then

$$\max_{1\leq k \leq n} |a_k|^2 \leq \frac{n^2-1}{2n}\sum_{k=1}^{n}(\Delta a_k)^2,$$

where $a_{n+1} = a_n$.

COMMENTS (ii) (a) has been extended to higher order differences, and converse inequalities have been given; see [*Milovanović & Milovanović*].
(iii) Redheffer's result is a recurrent inequality; see **Recurrent Inequalities**.

REFERENCES [*AI, pp. 131, 150*]; [*Alzer, 1992$^{(2)}$*], [*Milovanović & Milovanović*].

Ky Fan & Todd Inequality
See **Ostrowski's Inequalities**, EXTENSIONS (a).

- end of K -

L

Labelle's Inequality If $p(x) = \sum_{i=0}^{n} a_i x^i$ then

$$|a_n| \leq \frac{(2n)!}{2^n (n!)^2} \sqrt{\frac{2n+1}{2}} \, \|p\|_{\infty,[-1,1]}.$$

EXTENSIONS [TARIQ] If $p_n(x) = \sum_{i=0}^{n} a_i x^i$ and $p(1) = 0$, then

$$|a_n| \leq \frac{n}{n+1} \frac{(2n)!}{2^n (n!)^2} \sqrt{\frac{2n+1}{2}} \, \|p_n\|_{\infty,[-1,1]}.$$

REFERENCES [Tariq].

Laguerre Function Inequalities

The entire functions

$$L_\nu^\mu(z) = \binom{\mu+\nu}{\nu} \sum_{n=0}^{\infty} \frac{\nu(\nu+1)\cdots(\nu-n+1)}{(\mu+1)\cdots(\mu+n)} \frac{(-z)^n}{n!}$$

are called *Laguerre functions*, or in the case that $\nu \in \mathbb{N}$, *Laguerre polynomials*.

If $\mu \geq 0, \nu \in \mathbb{N}, x \geq 0$ then

$$|L_\nu^\mu(x)| \leq \binom{\mu+\nu}{\nu} e^{x/2}.$$

EXTENSIONS [LOVE] If $\Re\mu \geq 0, \nu \in \mathbb{N}, x > 0$ then

$$|L_\nu^\mu(x)| \leq \left|\frac{\mu}{\Re\mu} \binom{\mu+\nu}{\nu}\right| e^{x/2}.$$

COMMENTS Other extensions can be found in [Love, 1997]; see also **Enveloping Series Inequalities** COMMENTS (iii).

REFERENCES [EM, vol. 5, p. 341]; [Abramowitz & Stegun, p. 775], [Szegö], [Widder, pp. 168–171]; [Love, 1997].

Laguerre Polynomial Inequalities

See **Laguerre Function Inequalities**.

Landau's Constant Let L_f denote the inner radius of the image of D by a function f analytic in D, and put $L = \inf L_f$ where the inf is taken over all functions analytic in D with $f(0) = f'(0) = 0$, then

$$B \leq L, \quad \text{and} \quad \frac{1}{2} \leq L \leq \frac{(-2/3)!(-1/6)!}{(-5/6)!} \approx 0.5432588\ldots;$$

where B is Bloch's constant.

COMMENTS   (i) L is called *Landau's constant*.
(ii) A definition of inner radius can be found in **Isodiametric Inequality** COMMENTS (ii).
(iii) See also **Bloch's Constant**.

REFERENCES   [EM, vol. 5, pp. 346–347]; [Conway, vol. I, pp. 297–298].

Landau's Inequality
See **Hardy-Littlewood-Landau Derivative Inequalities**.

Lattice Inequalities
See **Ahlswede-Daykin Inequality, FKG Inequality, Holley's Inequality, Subadditive Function Inequalities** (b).

Lebedev-Milin Inequalities   If $f(z) = \sum_{n=1}^{\infty} a_n z^n$ is analytic in some neighbourhood of the origin, and if $g(z) = e^{f(z)} = \sum_{n=0}^{\infty} b_n z^n$ then :

$$\sum_{n=0}^{\infty} |b_n|^2 \leq \exp\left(\sum_{n=1}^{\infty} n|a_n|^2\right); \tag{1}$$

$$\sum_{n=0}^{N} |b_n|^2 \leq (N+1) \exp\left(\frac{1}{N+1} \sum_{m=1}^{N} \sum_{n=1}^{m} \left(n|a_n|^2 - \frac{1}{n}\right)\right); \tag{2}$$

$$|b_N|^2 \leq \exp\left(\sum_{n=1}^{N} \left(n|a_n|^2 - \frac{1}{n}\right)\right). \tag{3}$$

(a) If the right-hand side of (1) is finite there is equality if and only if for some $\gamma \in D$, and all $n \geq 1$, $a_n = \gamma^n/n$;
(b) there is equality in (2), or (3), for a given N, if and only if for some $\gamma = e^{i\theta}, \theta \in \mathbb{R}$, and all $n, 1 \leq n \leq N$, $a_n = \gamma^n/n$.

COMMENTS   These inequalities are important in the solution of the **Bieberbach Conjecture**.
REFERENCES   [Conway, vol. II, pp. 149–155].

Legendre Polynomial Inequalities   (a) If P_n is a Legendre polynomial then

$$\sqrt{\sin \theta} |P_n(\cos \theta)| < \sqrt{\frac{2}{\pi n}}; \quad 0 \leq \theta \leq \pi.$$

The constant is best possible.

(b) [TURÁN] If $-1 \leq x \leq 1$ then

$$P_{n+1}^2(x) \geq P_n(x)P_{n+2}(x).$$

COMMENTS (i) For generalizations of (a) see **Martin's Inequalities, Ultraspherical Polynomial Inequalities** (a).

(ii) (b) is an example of one of many similar inequalities satisfied by various special; functions; see **Bessel Function Inequalities** (c), **Ultraspherical Polynomials Inequalities** (b).

REFERENCES [*General Inequalities*, vol. 1, pp. 35–38], [*Szegö*, p. 165].

Lehmer Mean Inequalities
See **Gini-Dresher Mean Inequalities**.

Leindler's Inequality If $z \in \mathbb{C}$ and $p \geq 2$ then

$$1 + p\Re z + a_p|z|^2 + b_p|z|^p \leq |1+z|^p \leq 1 + p\Re z + A_p|z|^2 + B_p|z|^p,$$

where a_p, b_p, A_p, B_p are any positive real numbers that satisfy either

$$0 < a_p < \frac{p}{2},\ 0 < b_p \leq \mu_1(p),\ 1 < B_p < \infty,\ M_1(p) \leq A_p < \infty,$$

where

$$\mu_1(p) = \inf_{t \geq 2} \frac{(t-1)^p + pt - 1 - a_p t^2}{t^p};\ M_1(p) = \sup_{t > 0} \frac{(t+1)^p - 1 - pt - B_p t^p}{t^2};$$

or

$$0 < b_p < \mu_2(p),\ 0 < a_p \leq \mu_3(p),\ \frac{p(p-1)}{2} < A_p < \infty,\ M_2(p) \leq B_p < \infty,$$

where

$$\mu_2(p) = \inf_{t \geq 2} \frac{(t-1)^p + pt - 1}{t^p};\ \mu_3(p) = \inf_{t \geq 2} \frac{(t-1)^p + pt - 1 - b_p t^p}{t^2};$$

$$M_2(p) = \sup_{t > 0} \frac{(t+1)^p - 1 - pt - A_p t^2}{t^p}.$$

The ranges $]a_p, b_p[,\ ]A_p, B_p[$ are best possible.

COMMENTS The case $p = 4$, with $A_4 = 3,\ B_4 = 3,\ a_4 = 1,\ b_4 = 1/5$ is due to Shapiro.

REFERENCES [*MPF*, pp. 66–68].

Levinson's Inequality Let f be 3-convex on $[a,b]$ and let $\underline{a}, \underline{b}$ be n-tuples with elements in $[a,b]$ such that

$$a_1 + b_1 = \cdots = a_n + b_n; \qquad \max \underline{a} \leq \min \underline{b}. \qquad (1)$$

If $\underline{w}$ is a positive n-tuple then

$$\mathfrak{A}_n\bigl(f(\underline{a});\underline{w}\bigr) - f\bigl(\mathfrak{A}_n(\underline{a};\underline{w})\bigr) \leq \mathfrak{A}_n\bigl(f(\underline{b});\underline{w}\bigr) - f\bigl(\mathfrak{A}_n(\underline{b};\underline{w})\bigr). \qquad (2)$$

Conversely if (1) holds for all n and all n-tuples $\underline{a}, \underline{b}$ satisfying (1) then f is 3-convex.

COMMENTS Inequality (2), which generalizes the **Ky Fan Inequalities**, was proved by Levinson under slightly more restrictive conditions, see [AI]; and a stronger version has been proved by Pečarić.

SPECIAL CASES Let $\underline{a}, \underline{b}$ satisfy (1) and $\underline{w}$ be a positive n-tuple then:
(i) if $s > 0, s > t$ or $t > 2s$; or $s = 0, t > 0$; or $s < 0, s > t > 2s$ and if $t \neq 0$

$$\left(\left(\mathfrak{M}_n^{[t]}(\underline{a};\underline{w})\right)^t - \left(\mathfrak{M}_n^{[s]}(\underline{a};\underline{w})\right)^t\right)^{1/t} \leq \left(\left(\mathfrak{M}_n^{[t]}(\underline{b};\underline{w})\right)^t - \left(\mathfrak{M}_n^{[s]}(\underline{b};\underline{w})\right)^t\right)^{1/t};$$

while if $t = 0$

$$\frac{\mathfrak{G}_n(\underline{a};\underline{w})}{\mathfrak{M}_n^{[s]}(\underline{a};\underline{w})} \leq \frac{\mathfrak{G}_n(\underline{b};\underline{w})}{\mathfrak{M}_n^{[s]}(\underline{b};\underline{w})};$$

(ii) these inequalities are reversed if $s > 0, s < t < 2s$ or $s = 0, t < 0$ or $s < 0, t \neq 0$. These inequalities are strict unless $\underline{a}$ is constant.

REFERENCES [AI. p. 363], [MI, pp. 279–282], [MPF, pp. 32–36]; [PPT, pp. 71–75].

Lévy's Inequalities If $X_i, 1 \leq i \leq n$, are independent random variables and if $S_k = \sum_{i=1}^{k} X_i, 1 \leq k \leq n$, then

$$P\left\{\max_{1 \leq k \leq n}\{S_k - m(S - k - S_n)\} \geq r\right\} \leq 2P\{S_n \geq r\};$$

$$P\left\{\max_{1 \leq k \leq n}\{|S_k - m(S - k - S_n)|\} \geq r\right\} \leq 2P\{|S_n| \geq r\}.$$

In particular if the random variables are symmetrically distributed

$$P\{\max_{1 \leq k \leq n} S_k \geq x\} \leq 2P\{S_n \geq r\};$$

$$P\{\max_{1 \leq k \leq n} |S_k| \geq x\} \leq 2P\{|S_n| \geq r\}.$$

COMMENTS These are extensions of **Kolmogorov's Probability Inequality**.
REFERENCES [EM, vol. 5, pp. 404–405]; [Loève, pp. 247–248].

L'Hôpital's[1] **Rule** *Hypotheses: $a, b \in \overline{\mathbb{R}}$; $f, g :]a, b[\to \mathbb{R}$ are differentiable; g is never zero; g' is never zero; $c = a$ or b; either (a) or (b) holds;*

$$(a) \quad \lim_{x \to c} f(x) = \lim_{x \to c} g(x) = 0,$$

$$(b) \quad \lim_{x \to c} |g(x)| = \infty.$$

Conclusion:

$$\liminf_{x \to c} \frac{f'(x)}{g'(x)} \leq \liminf_{x \to c} \frac{f(x)}{g(x)} \leq \limsup_{x \to c} \frac{f(x)}{g(x)} \leq \limsup_{x \to c} \frac{f'(x)}{g'(x)}.$$

In particular, if $\lim_{x \to c} \frac{f'(x)}{g'(x)} = \ell$, $\ell \in \overline{\mathbb{R}}$, *then* $\lim_{x \to c} \frac{f(x)}{g(x)} = \ell$.

COMMENTS A discrete analogue has been given by Agarwal.

REFERENCES [General Inequalities, vol. 4, pp. 95–98]; [Taylor].

Liapounoff's Inequality
See **Liapunov's Inequality**.

Liapunov's[2] Inequality *If $0 < t < s < r$ then*

$$\Big(\sum_{i=1}^{n} w_i a_i^s\Big)^{r-t} \leq \Big(\sum_{i=1}^{n} w_i a_i^t\Big)^{r-s} \Big(\sum_{i=1}^{n} w_i a_i^r\Big)^{s-t}.$$

COMMENTS (i) This is **Hölder's Inequality** (2) with $p = (r-t)/(r-s)$, $q = (r-t)/(s-t)$ and $\underline{a}^p$, $\underline{b}^q$ replaced by $\underline{a}^t$, $\underline{a}^r$ respectively.
(ii) Liapounov's inequality can be written in terms of the Gini means

$$\mathfrak{B}_n^{s,t}(\underline{a}; \underline{w}) \leq \mathfrak{B}_n^{r,t}(\underline{a}; \underline{w});$$

and so is a particular case of **Gini-Dresher Mean Inequalities** (2).
(iii) Liapunov's inequality is just the statement that a certain function is log-convex; see [MO].

INTEGRAL ANALOGUES *If $0 < t < s < r$ and $f \in \mathcal{L}^r([a,b])$ then*

$$\Big(\int_a^b |f|^s\Big)^{r-t} \leq \Big(\int_a^b |f|^t\Big)^{r-s} \Big(\int_a^b |f|^r\Big)^{s-t},$$

with equality if and only if either f is constant on some measurable subset and zero on its complement, or $r = s$, or $s = t$, or $t(2r - s) = rs$.

COMMENTS (iv) This last can be extended to more general measure spaces.

REFERENCES [AI, p. 54], [MI, pp. 139–140], [MO, pp. 74, 459, 462], [MPF, p. 101].

---

[1] Also spelt L'Hospital.
[2] А М Ляпунов. Also transliterated as Liapounoff, Lyapunov.

Lipschitz Function Inequalities *If f is a real-valued Lipschitz function of order $\alpha, 0 < \alpha \le 1$, defined on a set E in $\mathbb{R}^n$ then for all $\underline{u}, \underline{v} \in E$ and some constant C*

$$|f(\underline{u}) - f(\underline{v})| \le C|\underline{u} - \underline{v}|^\alpha.$$

COMMENTS   (i) This is just the definition of a *Lipschitz function of order α*. The lower bound of the constants C is called the *Lipschitz constant of order α* of the function.
(ii) In many references a Lipschitz function means a Lipschitz function of order 1, and the other classes are called *Hölder functions of order $\alpha, 0 < \alpha < 1$*.
REFERENCES   [EM, vol. 5, p. 532].

Littlewood's Conjecture *If $\underline{a}$ is a non-negative sequence and $p, q \ge 1$ then*

$$\sum_{n=1}^{\infty} a_n^p \left(A_n \sum_{m \ge n} a_m^{1+p/q} \right)^q \le \left(\frac{q(2p-1)}{p} \right)^q \sum_{n=1}^{\infty} (a_n^p A_n^q)^2.$$

COMMENTS   The cases $p = 2, q = 1$ and $p = 1, q = 2$ were conjectured in a set of problems by Littlewood. The conjecture was solved, in this general form, by Bennett.
REFERENCES   [Inequalities I, pp. 151–162]; [Bennett, 1987].

Littlewood-Paley Inequalities *Let $1 < p < \infty$, $F(z) = \sum_{n \in \mathbb{N}} c_n z^n$ be analytic D and let $\sum_{n \in \mathbb{N}} c_n e^{in\theta}$ be the Fourier series of $f(\theta) = F(e^{i\theta}) = \lim_{r \to 1} F(re^{i\theta})$, and suppose that $f \in \mathcal{L}^p([-\pi, \pi])$. If $\lambda > 1$ and $n_0 = 0, n_1 = 1 < n_2 < n_3 < \cdots$ is such that $n_{k+1}/n_k > \lambda$, $k = 1, 2, \ldots$ put $\Delta_0 = c_0, \Delta_k(\theta) = \sum_{m=n_{k-1}+1}^{n_k} c_m e^{im\theta}$, $k = 1, 2, \ldots, -\pi \le \theta \le \pi$, and define*

$$\gamma(\theta) = \left(\sum_{k \in \mathbb{N}} |\Delta_k(\theta)|^2 \right)^{1/2}, \quad -\pi \le \theta \le \pi.$$

Then

$$A_{p,\lambda} \|f\|_p \le \|\gamma\|_p \le B_{p,\lambda} \|f\|_p,$$

where $A_{p,\lambda}, B_{p,\lambda}$ are constants that depend only on p and λ.

COMMENTS   This very important result in Fourier analysis has been the subject of much research. Extensions can be found in the reference and in more recent literature.
REFERENCES   [Zygmund, vol. II, pp. 222–241].

Littlewood's Problem *If $n_k \in \mathbb{Z}$, $1 \le k \le n$, then*

$$\frac{1}{2\pi} \int_{-\pi}^{\pi} \left| \sum_{k=1}^{n} e^{in_k t} \right| dt \ge \frac{4 \log n}{\pi^3}.$$

COMMENTS (i) Littlewood conjectured the size of the right-hand side in 1948, and the conjecture was settled by McGhee, Pigno & Smith, and also by Konyagin[4], in 1981. Of course there are several problems and conjectures by Littlewood; this is sometimes called the *Littlewood problem for integrals*.

(ii) The value of the constant on the right-hand side was obtained by Stegeman; the best value is not known but it is conjectured that the π^3 can be replaced by π^2.

REFERENCES [EM, vol. 5, p. 534]; [General Inequalities, vol. 3, pp. 141–148].

Lochs's Inequality If $x > 0$ then

$$\tanh x > \sin x \cos x.$$

REFERENCES [AI, p. 270].

Logarithmic Capacity Inequalities

If $E \subseteq \mathbb{C}$ put $v(E) = \inf \left\{ t;\ t = \int \int \log \frac{1}{|z-w|}\, d\mu(z)\, d\mu(w) \right\}$, where the infimum is over all measures μ of total mass 1, and having as support a compact subset of E; we assume that E is such that this collection of measures is not empty. Then, the *logarithmic capacity* of E is $C(E) = e^{-v(E)}$.

If E is a Borel set then

$$C(E) \geq \sqrt{\frac{|E|}{e\pi}}.$$

COMMENTS See also **Capacity Inequalities**.

REFERENCES [Conway, vol. II, pp. 331–336].

Logarithmic Function Inequalities (a) If $n \geq 1$,

$$\frac{1}{n+1} < \log\left(1 + \frac{1}{n}\right) < \frac{1}{n}.$$

(b) If $n \geq 2$,

$$1 + \frac{1}{12n + 1/4} - \frac{1}{12(n+1) + 1/4} < \left(n + \frac{1}{2}\right) \log\left(1 + \frac{1}{n}\right) < 1 + \frac{1}{12n} - \frac{1}{12(n+1)}.$$

(c) If $x > 0, x \neq 1$,

$$\log x < x - 1; \qquad \frac{\log x}{x-1} \leq \frac{1}{\sqrt{x}}; \qquad \frac{\log x}{x-1} \leq \frac{1+\sqrt[3]{x}}{x+\sqrt[3]{x}};$$

more generally if $n \geq 1$

$$\log x < n\left(x^{1/n} - 1\right).$$

---

[4] С В Конягин.

(d) If $x > -1$,
$$\frac{2|x|}{2+x} < |\log(1+x)| < \frac{|x|}{\sqrt{1+x}}.$$

(e) If $x > 0$ then
$$xy \le x \log x + e^{y-1},$$
with equality if and only if $y = 1 + \log x$.

(f) If $a, x > 0$,
$$\log x \le \frac{a}{e} \sqrt[a]{x}; \qquad -\log x \le \frac{1}{eax^a}.$$

(g) [LEHMER] If $a, b, x \ge 0$ then
$$\log\left(1 + \frac{x}{a}\right) \log\left(1 + \frac{b}{x}\right) < \frac{a}{b}.$$

(h) If $z \in \mathbb{C}$ with $|z| < 1$ with
$$|\log(1+z)| \le -\log(1-|z|),$$
and
$$\frac{|z|}{1+|z|} \le |\log(1+z)| \le |z| \frac{1+|z|}{|1+z|}.$$

(j) If $a > 1$ then
$$\frac{\log_a(n^{1/e^n})}{\log_a e} < \sum_{i=1}^{n} \frac{1}{i} < \frac{\log_a(ne)}{\log_a e};$$
in particular,
$$\frac{1}{n} + \log n < \sum_{i=1}^{n} \frac{1}{i} < 1 + \log n.$$

COMMENTS (i) The inequalities in (b) are of importance in the deduction of **Stirling's Formula**.

(ii) For (c) just note that $1 - x + \log x$ has a unique maximum at $x = 1$; or use the strict concavity of the logarithmic function at $x = 1$. For another use of the strict concavity of the logarithmic function see **Exponential Function Inequalities** COMMENTS (iii).

(iii) (d) follows by considering the differences between the centre term and the two outside terms, or from special cases of the **Logarithmic Mean Inequalities** (1), namely:
$$\mathcal{L}_{-2}(a,b) < \mathcal{L}_{-1}(a,b) < \mathcal{L}_0(a,b) < \mathcal{L}_1(a,b).$$

(iv) (e) can be deduced from **Young's Inequalities** (b); alternatively let $p \to 0$ in **Geometric-Arithmetic Mean Inequality** EXTENSIONS (ℓ); see also **Conjugate Convex Function Inequalities** COMMENTS (i).

(v) Inequalities (f) are of importance for large values of a.

(vi) See also **Euler's Constant Inequalities** (a), **Logarithmic Mean Inequalities**, **Schlömilch-Lemonnier Inequality**, **Series Inequalities** (c).

REFERENCES [AI, pp. 49, 181–182, 266, 272–273, 326–328], [HLP, pp. 61, 107], [MI, pp. 6–8, 130], [MPF, p. 79]; [Abramowicz & Stegun, p. 68].

Logarithmic Mean Inequalities (a) If $0 < a \le b$ and if $-\infty < r < s < \infty$ then

$$a \le \mathfrak{L}^{[r]}(a,b) \le \mathfrak{L}^{[s]}(a,b) \le b, \qquad (1)$$

with equality if and only if $a = b$.

(b) [SÁNDOR]
$$\frac{3}{\mathfrak{L}(a,b)} < \frac{1}{\mathfrak{G}_2(a,b)} + \frac{2}{\mathfrak{H}_2(a,b)}.$$

(c) [SÁNDOR]
$$\mathfrak{I}(a,b) > \mathfrak{G}_2^2(a,b) \exp\left(\mathfrak{G}_2(\log a, \log b)\right).$$

(d) [LIN]
$$\mathfrak{G}_2(a,b) \le \mathfrak{L}(a,b) \le \mathfrak{M}_2^{[1/3]}(a,b). \qquad (2)$$

COMMENTS (i) Inequality (1) follows from the integral analogue of (r;s).
(ii) Lin's result is a particular case of **Pittenger's Inequalities**.

COROLLARIES (a) If $0 < a \le b$, and if $-2 < r < -1/2$ then

$$\mathfrak{G}_2(a,b) \le \mathfrak{L}^{[r]}(a,b) \le \mathfrak{M}_2^{[1/2]}(a,b),$$

with equality only if $a = b$.
(b) If $0 < a \le b$, and if $-2 < r < 1$ then

$$\mathfrak{G}_2(a,b) \le \mathfrak{L}^{[r]}(a,b) \le \mathfrak{A}_2(a,b),$$

with equality only if $a = b$.
(c)
$$\mathfrak{H}_2(a,b) \le \frac{1}{\mathfrak{I}(a,b)} \le \frac{1}{\mathfrak{L}(a,b)} \le \mathfrak{G}_2(a,b) \le \mathfrak{L}(a,b) \le \mathfrak{I}(a,b) \le \mathfrak{A}_2(a,b).$$

COMMENTS (iii) Inequalities (a), (b) result from (1) and the simple identities between the various means.
(iv) The last three inequalities in (c) are particular cases of (1), and the whole set follows on letting $n \to \infty$ in **Nanjundiah Inequalities** (1).

EXTENSIONS [ALZER] (a) If $0 < a < b$ then

$$\sqrt{\mathfrak{A}_2(a,b)\mathfrak{G}_2(a,b)} < \sqrt{\mathfrak{L}(a,b)\mathfrak{I}(a,b)} < \mathfrak{M}_2^{[1/2]}(a,b);$$
$$\mathfrak{L}(a,b) + \mathfrak{I}(a,b) < \mathfrak{A}_2(a,b) + \mathfrak{G}_2(a,b)$$
$$\sqrt{\mathfrak{G}_2(a,b)\mathfrak{I}(a,b)} < \mathfrak{L}(a,b) < \frac{1}{2}[\mathfrak{G}_2(a,b) + \mathfrak{I}(a,b)];$$
$$\mathfrak{G}_2(a,b) < \sqrt{\mathfrak{L}^{[-r]}(a,b)\mathfrak{L}^{[r]}(a,b)} < \mathfrak{I}(a,b), \quad r \ne 0.$$

(b) If $0 < a, b \le 1/2$, $a \ne b$ then

$$\frac{\mathfrak{G}_2(a,b)}{\mathfrak{G}_2(1-a,1-b)} \le \frac{\mathfrak{L}(a,b)}{\mathfrak{L}(1-a,1-b)} \le \frac{\mathfrak{M}_2^{[1/3]}(a,b)}{\mathfrak{M}_2^{[1/3]}(1-a,1-b)}.$$

COMMENTS (v) Alzer's result (b) is a Ky Fan extension of (2).

These means have been generalized. For instance suppose that $a \ne b$, $r \ne s$, $rs \ne 0$ and either $ab \ne 0$ or if $ab = 0$ then r, s are positive, then define

$$\mathfrak{E}_{r,s}(a,b) = \left(\frac{r(a^s - b^s)}{s(a^r - b^r)}\right)^{1/s-r}.$$

For other values the definition is obtained by taking limits. Clearly $\mathfrak{L}^{[p]}(a,b) = \mathfrak{E}_{p+1,1}(a,b)$. These are the so called *extended means of Leach & Sholander*.

EXTENDED MEAN INEQUALITIES (a) [LEACH & SHOLANDER] If $r \ne s, p \ne q, r+s \le p+q$ then

$$\mathfrak{E}_{r,s}(a,b) \le \mathfrak{E}_{p,q}(a,b).$$

(b) [ALZER] If $0 < a < b, r \ne 0$ then

$$\mathfrak{G}_2(a,b) < \sqrt{\mathfrak{E}_{-r,-r+1}(a,b)\mathfrak{E}_{r,r+1}(a,b)} < \mathfrak{L}(a,b)$$
$$< \frac{\mathfrak{E}_{-r,-r+1}(a,b) + \mathfrak{E}_{r,r+1}(a,b)}{2} < \mathfrak{A}_2(a,b).$$

COMMENTS (vi) Neuman has defined the logarithmic mean of general n-tuples.
(vii) See also **Arithmetic-geometric Compound Mean Inequalities, Muirhead Symmetric Function and Mean Inequalities** (d), **Rado's Inequality** EXTENSIONS.

REFERENCES [AI, p. 273], [MI, pp. 130, 345–351], [MO, pp. 98–99], [MPF, pp. 40–46]; [Neuman, 1994], [Páles, 1988$^{(2)}$], [Sándor].

Logarithmic Sobolev Inequalities

(a) If $f \in \mathcal{C}([-\pi, \pi])$ with $f' \in \mathcal{L}^2([-\pi, \pi])$, $\int_{-\pi}^{\pi} f'^2 \ge 0$, then

$$\frac{1}{2\pi}\int_{-\pi}^{\pi} f^2 \log f^2 \le \left(\frac{1}{2\pi}\int_{-\pi}^{\pi} f^2\right) \log \left(\frac{1}{2\pi}\int_{-\pi}^{\pi} f^2\right) + \frac{1}{\pi}\int_{-\pi}^{\pi} f'^2.$$

(b) [GROSS] If $f : \mathbb{R} \to \mathbb{C}$ and $\int_{-\infty}^{\infty} |f'(t)|^2 \frac{e^{-t^2/2}}{\sqrt{2\pi}} dt < \infty$, then

$$\int_{-\infty}^{\infty} |f(t)|^2 \log |f(t)|^2 \frac{e^{-t^2/2}}{\sqrt{2\pi}} dt$$
$$\le \frac{1}{2}\left(\int_{-\infty}^{\infty} |f(t)|^2 \frac{e^{-t^2/2}}{\sqrt{2\pi}} dt\right) \log \left(\int_{-\infty}^{\infty} |f(t)|^2 \frac{e^{-t^2/2}}{\sqrt{2\pi}} dt\right)$$
$$+ \int_{-\infty}^{\infty} |f'(t)|^2 \frac{e^{-t^2/2}}{\sqrt{2\pi}} dt.$$

COMMENTS   (i) There are many extensions of these results; in particular Gross has extended his result to higher dimensions.
(ii) See also **Sobolev's Inequalities**.

REFERENCES   [EM, Supp., p. 307]; [Emery & Yukich], [Gross], [Pearson].

Log-convex Function Inequalities   (a) If f is log-convex and if $0 \leq \lambda \leq 1$ then

$$f\big((1-\lambda)x + \lambda y\big) \leq f(x)^{1-\lambda} f(y)^{\lambda}. \tag{1}$$

(b) The necessary and sufficient condition that

$$f\big(\mathfrak{A}_n(\underline{a};\underline{w})\big) \leq \mathfrak{G}_n\big(f(\underline{a});\underline{w}\big), \tag{2}$$

is that f be log-convex.

COMMENTS   (i) The inequality (1) is just the definition; a function f is *log-convex* precisely when $\log \circ f$ is convex. Such functions are sometimes called *multiplicatively convex*, because of the form of inequality (1).
(ii) A log-convex function is convex. An interesting example of a log-convex function is $x!, x > -1$.
(iii) Inequality (2) is a particular case of **Quasi-Arithmetic Mean Inequalities** (c). In turn a particular case of (2) is **Shannon's Inequality**.
(iv) Many inequalities say that a certain function is log-convex; see for instance **Hadamard's Three Circles Theorem, Hardy's Analytic Function Inequality, Liapunov's Inequality** COMMENTS (iii), **Phragmen-Lindelöf's Inequality, Riesz-Thorin Theorem** COMMENTS (i), **Series' Inequalities** COMMENTS (ii), **Trigonometric Function Inequalities** COMMENTS (ix).

REFERENCES   [MI, p. 30], [MPF, p. 3], [PPT, pp. 294–295]; [Roberts & Varberg, pp. 18–19].

Log-convex Sequence Inequalities   If $\alpha \geq 0$ and if the real sequence $\underline{a}$ is α-log-convex then

$$a_n^2 \leq \frac{(n+1)(\alpha+n-1)}{n(\alpha+n)} a_{n+1} a_{n-1}, \quad n \geq 2. \tag{2}$$

COMMENTS   (i) This is just the definition of an α-log-convex sequence. When $\alpha = 1$ we say *log-convex* instead of 1-log-convex. Letting $\alpha \to \infty$ in (2) gives definition of the case $\alpha = \infty$, such a sequence is said to be *weakly log-convex*.
(ii) If instead we have the inequality ($\sim$2) we get the definitions of α-log-concave, log-concave, and in the case $\alpha = \infty$, *strongly log-concave*.
(iii) The sequence $(-1)^n \binom{-\alpha}{n}$, $n \geq 1$ is α'-log-convex,(-concave), for all $\alpha' \geq \alpha$, ($\leq \alpha$). The sequence $1/n$, $n \geq 1$ is α-log-convex if $\alpha \geq 0$. The sequence $1/n!$, $n \geq 1$ is α-log-concave if $\alpha \leq \infty$.
(iv) See also **Elementary Symmetric Function Inequalities** EXTENSIONS (f), **Complete Symmetric Function Inequalities** (c).

REFERENCES   [MI, pp. 13–15], [PPT, pp. 288–292].

Love-Young[5] Inequalities (a) If $\underline{a}, \underline{b}$ are complex n-tuples and if $p, q > 0$ then for some $k, 1 \leq k \leq n$,
$$|a_k b_k| \leq \mathfrak{M}_n^{[p]}(\underline{a})\mathfrak{M}_n^{[q]}(\underline{b}).$$

(b) Hypotheses: $\underline{x} = (x_1, \ldots, x_m)$ is defined from the n-tuple $\underline{a} = (a_1, \ldots, a_n)$, $m \leq n$, by replacing certain of the consecutive commas, ',' , by plus signs, '+'; and the m-tuple $\underline{y}$ is obtained in the same way from the n-tuple $\underline{b}$;

$$S_{p,q}(\underline{a}, \underline{b}) = \max\left\{ (\sum_{i=1}^m |x_k|^p)^{1/p}, (\sum_{i=1}^m |y_k|^q)^{1/q} \right\},$$

where the maximum is over all possible choices of $\underline{x}, \underline{y}$; $p, q > 0$; $1/p + 1/q > 1$,
Conclusions:
$$\left| \sum_{1 \leq r \leq s \leq n} a_r b_s \right| \leq \left\{ 1 + \zeta\left(\frac{1}{p} + \frac{1}{q}\right) \right\} S_{p,q}(\underline{a}, \underline{b}).$$

COMMENTS (i) (a) can be used to prove (H).
(ii) (b) is of importance in certain work with Stieltjes integral; ζ denotes the zeta function.
REFERENCES [*Young*].

Lower and Upper Limit Inequalities
See **Upper and Lower Limit Inequalities**.

Lower Semi-continuous Function Inequalities
See **Semi-continuous Function Inequalities**.

Löwner-Heinz Inequality If A, B are positive bounded linear operators on the Hilbert space X and if $0 \leq \alpha \leq 1$ then
$$A \geq B \implies A^\alpha \geq B^\alpha.$$

COMMENTS (i) The definition of a positive operator is given in **Heinz Inequality** COMMENTS (i).
(ii) This result is based on the fact that $x^\alpha, 0 \leq \alpha \leq 1$ is an operator monotone function; see **Monotone Matrix Function Inequalities** COMMENTS (ii).
(iii) This inequality is equivalent to the Heinz-Kato Inequality, see **Heinz-Kato-Furata Inequality** (1); see also **Cordes's Inequality**.
REFERENCES [*EM, Supp. p. 359*]; [*General Inequalities, vol. 7, pp. 65–76*].

Lupaş & Mitrović Inequality
See **Geometric-Arithmetic Mean Inequality**, EXTENSIONS (b).

- end of L -

---

[5] This is L C Young.

M

Mahajan's Inequality If J_ν is the Bessel function of the first kind then

$$(x+1)^{\nu+1} J_\nu\left(\frac{\pi}{x+1}\right) - x^{\nu+1} J_\nu\left(\frac{\pi}{x}\right) > \left(\frac{\pi}{2}\right)^\nu \frac{1}{\nu!}.$$

[Lorch & Muldoon].

Mahler's Inequalities If $K \subset \mathbb{R}^n$ is convex and compact, with $\mathring{K} \neq \emptyset$ and centroid the origin, then

$$|K||K^*| \geq \frac{(n+1)^{n+1}}{(n!)^2},$$

with equality if and only if K is the simplex with centroid the origin. If K is symmetric with respect to the origin, then

$$|K||K^*| \geq \frac{4^n}{n!},$$

with equality if and only if K is the n-cube $[-1,1]^n$.

COMMENTS (i) For the notation K^* see **Blaschke-Santaló Inequality** COMMENTS (i).
(ii) These inequalities are only known in the case $n = 2$, they are conjectured for the other cases. The special case when K is a zonoid is *Reisner's Inequality*.
(iii) See also **Minkowski-Mahler Inequality**.

REFERENCES [EM, Supp., pp. 129–130].

Marchaud's Inequality

If $f : \mathbb{R} \to \mathbb{R}$ is continuous and of period 2π the r-th modulus of continuity is

$$\omega_r(\delta; f) = \sup_{|u| < \delta} \left\{ \left\| \sum_{k=0}^{r} \binom{r}{k} (-1)^k f(x+ku) \right\|_{\infty,[-\pi,\pi]} \right\}.$$

(a) If $r \leq s$ then

$$\omega_s(\delta; f) \leq 2^{s-r} \omega_r(\delta; f). \tag{1}$$

(b) [MARCHAUD] *Under the same hypotheses as in (a) there is a constant depending only on r, s such that*

$$\omega_r(\delta; f) \le C_{r,s}\delta^r \int_\delta^1 \omega_s(y; f) y^{-(r+1)}\, dy.$$

COMMENTS (i) Inequality (1) is an easy consequence of the definition.
(ii) See also **Modulus of Continuity Inequalities**.

REFERENCES [*General Inequalities, vol. 4, pp. 221–238*].

Marcus & Lopes's Inequality
(a) *If r, s are integers, $1 \le r \le s \le n$, then*

$$\left(\frac{e_n^{[s]}(\underline{a}+\underline{b})}{e_n^{[s-r]}(\underline{a}+\underline{b})}\right)^{1/r} \ge \left(\frac{e_n^{[s]}(\underline{a})}{e_n^{[s-r]}(\underline{a})}\right)^{1/r} + \left(\frac{e_n^{[s]}(\underline{b})}{e_n^{[s-r]}(\underline{b})}\right)^{1/r},$$

with equality if and only if either $\underline{a} \sim \underline{b}$, or $r = s - 1$.
(b) *If r is an integer and $1 \le r \le n$ then*

$$\mathfrak{P}_n^{[r]}(\underline{a}+\underline{b}) \ge \mathfrak{P}_n^{[r]}(\underline{a}) + \mathfrak{P}_n^{[r]}(\underline{b}),$$

with equality if and only if either $\underline{a} \sim \underline{b}$, or $r = 1$.

COMMENTS (i) (b) is just the case $s = r$ of (a).
(ii) For extensions see **Muirhead Symmetric Function and Mean Inequalities** EXTENSIONS (b), **Complete Symmetric Function Inequalities** (b), COMMENTS (ii), **Complete Symmetric Mean Inequalities** (c), **Whiteley Mean Inequalities** (d).

REFERENCES [*AI, pp. 102–104*], [*BB, pp. 33–35*], [*MI, pp. 306–310*], [*MPF, pp. 163–164*], [*MO, pp. 79–81*], [*PPT, p. 303*].

Markov's[1] Inequality
If p_n is a polynomial of degree at most n then

$$\|p_n'\|_{\infty, [a,b]} \le \frac{2n^2}{b-a} \|p_n\|_{\infty, [a,b]}.$$

COMMENTS (i) This inequality is best possible as is shown by T_n, the Čebišev polynomial of degree n; for a definition see **Čebišev Polynomial Inequalities**.
(ii) The inequality implied for higher derivatives is not best possible; a best possible result is given in (b) below.

---

[1] А А Марков.

EXTENSIONS   (a) [DUFFIN & SCHAEFFER] *If p_n is a polynomial of degree at most n on $[-1,1]$. and if $|p_n(\cos(k\pi/n)| \leq 1, 0 \leq k \leq n$, then*

$$\|p'_n\|_{\infty,[-1,1]} \leq n^2.$$

(b) [V.A. MARKOV[2]] *If p_n is a polynomial of degree at most n on $[-1,1]$. and if $|p_n(x)| \leq 1$ then*

$$\|p_n^{(k)}\|_{\infty,[-1,1]} \leq T_n^{(k)}(1) = \frac{n^2(n^2-1)(n^2-2^2)\cdots(n^2-\overline{k-1}^2)}{(2k-1)!!}, \quad k \geq 1.$$

COMMENTS   (iv) The notation is defined in **Factorial Function Inequalities** (f).
(v) See also **Bernšteĭn's Inequality** COMMENTS (iv), **Erdós's Inequality**.

REFERENCES   [EM, vol. 6, p. 100; Supp., pp. 365–366]; [General Inequalities, vol. 6, pp. 161–174].

Markov's[1] Probability Inequality   *If X is a random variable and $k > 0$ then*

$$P(X \geq r) \leq \frac{E|X|^k}{r^k}. \tag{1}$$

COMMENTS   (i) (1) is a simple deduction from **Probability Inequalities**; just take $f(x) = |x|^k$. The same result also gives a converse inequality for (1);

$$P(X \geq r) \geq \frac{E|X|^k - r^k}{\|X^k\|_{\infty,P}}.$$

(ii) The case $k = 2$ of (1) is **Čebišev's Probability Inequality**.

REFERENCES   [EM, vol. 2, pp. 119–120]; [Loève, p. 158].

Martingale Inequalities   (a) [DOOB'S INEQUALITY] *If $\mathcal{X} = (X_n, \mathcal{F}_n, n \in \mathbb{N})$ is a non-negative sub-martingale in $(\Omega, \mathcal{F}, P)$, $X_n^* = \max_{1 \leq i \leq n} X_i, n \in \mathbb{N}$, then, writing $\|\cdot\|_p = (E|\cdot|^p)^{1/p}$,*

$$P\{X_n^* \geq r\} \leq \frac{EX_n}{r}; \tag{1}$$

and

$$\|X_n\|_p \leq \begin{cases} \|X_n^*\|_p \leq \dfrac{p}{p-1}\|X_n\|_p, & \text{if } p > 1, \\ \dfrac{e}{e-1}\left[1 + \|X_n \log^+ X_n\|\right], & \text{if } p = 1. \end{cases}$$

---

[2] В А Марков.
[1] А А Марков.

(b) [BURKHOLDER'S INEQUALITIES] If $\mathcal{X} = (X_n, \mathcal{F}_n, n \in \mathbb{N})$ is a martingale and if $p > 1$ then

$$\frac{p-1}{18p^{3/2}}||\sqrt{\{X\}_n}||_p \leq ||X_n||_p \leq \frac{18p^{3/2}}{\sqrt{p-1}}||\sqrt{\{X\}_n}||_p,$$

where

$$\{X\}_n = \sum_{i=1}^n (\Delta X_i)^2, \text{ and } X_0 = 0.$$

COMMENTS (i) Inequality (1) is just **Kolmogorov's Probability Inequality** extended to this situation.

(ii) Using the above results we can deduce that if $p > 1$,

$$\frac{p-1}{18p^{3/2}}||\sqrt{\{X\}_n}||_p \leq ||X_n^*||_p \leq \frac{18p^{3/2}}{\sqrt{(p-1)^3}}||\sqrt{\{X\}_n}||_p.$$

(iii) The inequality in (ii) has been extended to the cases $p = 1, \infty$ by Davis, and Gundy; as a result the inequality is known as the *Burkholder-Davis-Gundy Inequality*.

(iv) See also **Doob's Upcrossing Inequality**.

REFERENCES [EM, vol. 6, pp. 109–110; Supp., p. 163]; [Feller, vol. II, pp. 241–242], [Loève, pp. 530–535], [Tong ed., pp. 78–83].

Martin's Inequalities If P_n is a Legendre polynomial then if $0 \leq \theta \leq \pi$,

$$\sqrt{\sin\theta}\, P_n(\cos\theta) \leq \frac{\sqrt{2}}{\sqrt{\pi(n+1/2)}};$$

$$P_n(\cos\theta) \leq \frac{1}{\sqrt[4]{1 + n(n+1)\sin^2\theta}}.$$

COMMENTS These have been generalized by Common; see **Ultraspherical Polynomial Inequalities** COMMENTS (ii).

REFERENCES [Common].

Mathieu's Inequality If $c \neq 0$ then

$$\frac{1}{c^2 + 1/2} < \sum_{i=1}^{\infty} \frac{2n}{(n^2+c^2)^2} < \frac{1}{c^2}.$$

COMMENTS The inequality on the right was conjectured by Mathieu, and proved 60 years later by Berg.

EXTENSIONS [EMERSLEBEN, WANG & WANG] If $c \neq 0$ then

$$\frac{1}{c^2} - \frac{5}{16c^4} < \sum_{i=1}^{\infty} \frac{2n}{(n^2+c^2)^2} = \frac{1}{c^2} - \frac{1}{16c^4} + O(\frac{1}{c^6}).$$

REFERENCES [AI, pp. 360–361], [MPF, pp. 629–634]; [Russell D C]

Matrix Inequalities [MARSHALL & OLKIN] *(a) If A is a $m \times n$ complex matrix, B a $p \times q$ complex matrix of rank p then*

$$AA^* \geq AB^* (BB^*)^{-1} BA^*,$$

where by $C \geq D$ we mean that $C - D$ is positive definite.
(b) If the eigenvalues of the $n \times n$ complex matrix A are all positive and lie in the interval $[m, M]$ and if B is a $p \times n$ complex matrix with $BB^* = I$ then, with the same notation as in (i),

$$BA^{-1}B^* \leq \frac{M+m}{Mm}I - \frac{BAB^*}{Mm} \leq \frac{(M+m)^2}{4Mm}(BAB^*)^{-1}.$$

(c) [SAGAE & TANABE] If $\underline{C}$ is an n-tuple of $m \times m$ positive definite matrices and $\underline{w}$ is a positive n-tuple with $W_n = 1$ then

$$\mathfrak{H}_n(\underline{C};\underline{w}) \leq \mathfrak{G}_n(\underline{C};\underline{w}) \leq \mathfrak{A}_n(\underline{C};\underline{w}),$$

with equality if and only if $C_1 = \cdots = C_n$.
(d) If A, B are two $n \times n$ Hermitian positive definite matrices and if $0 \leq \lambda \leq 1$, then

$$(\lambda A + (1-\lambda)B)^{-1} \leq \lambda A^{-1} + (1-\lambda)B^{-1}.$$

COMMENTS (i) The first inequality of Marshall & Olkin is a matrix analogue of (C), the second a matrix analogue of **Kantorovič's Inequality**.
(ii) The definition of the arithmetic mean, the harmonic mean, in fact every power mean except the geometric mean, is obvious; care must be taken in the case of the geometric mean as a product of positive definite matrices need not be positive definite: so we define $\mathfrak{G}_n(\underline{C};\underline{w})$ as

$$C_n^{1/2}\left(C_n^{-1/2}C_{n-1}^{1/2}\cdots\right.$$
$$\cdots\left(C_3^{-1/2}C_2^{1/2}\left(C_2^{-1/2}C_1C_2\right)^{v_1}C_2^{1/2}C_3^{-1/2}\right)^{v_2}\cdots$$
$$\left.\cdots C_{n-1}^{1/2}C_n^{-1/2}\right)^{v_{n-1}}C_n^{-1/2},$$

where $v_k = 1 - \frac{w_{k+1}}{W_{k+1}}, 1 \leq k \leq n-1$. Of course (c) is just a matrix analogue of (GA).
(iii) See also **Circulant Matrix Inequalities, Convex Matrix Function Inequalities, Determinant Inequalities, Eigenvalue Inequalities, Gram Determinant Inequalities, Grunsky's Inequalities**

COMMENTS (iii), **Hadamard's Determinant Inequality, Hirsch's Inequalities, Matrix Norm Inequalities, Monotone Matrix Function Inequalities, Permanent Inequalities, Rank Inequalities, Rayleigh-Ritz Ratio, Trace Inequalities, Weyl's Inequalities.**

REFERENCES [MPF, p. 96, 225]; [General Inequalities, vol. 7, pp. 77–91]; [Furuta & Yanagida].

Matrix Norm Inequalities

If we write C_n for the complex vector space of $n \times n$ complex matrices then a function $||\cdot|| : C_n \to \mathbb{R}$ is called a *matrix norm* if it is a norm in the sense of **Norm Inequalities** COMMENTS (i), and if in addition if $A, B \in C_n$ then $||AB|| \leq ||A||\,||B||$; see **Banach Algebra Inequalities**.

If $A = (a_{ij})_{\substack{1 \leq i \leq m \\ 1 \leq j \leq n}}$ is an $m \times n$ complex matrix and if $1 \leq p < \infty$, we write

$$|A|_p = \left(\sum_{i=1}^m \sum_{j=1}^n |a_{ij}|^p \right)^{1/p}.$$

[OSTROWSKI] *If A, B are two complex matrices such that AB exists, and if $1 \leq p \leq 2$, then*

$$|AB|_p \leq |A|_p |B|_p.$$

In particular $|\cdot|_p$, $1 \leq p \leq 2$, is a matrix norm on C_n.

COMMENTS (i) This result is in general false if $p \geq 2$; consider $p = \infty$ and $A = B \in C_2$, matrices all of whose entries are equal to 1.

(ii) The case $p = 2$ is called the *Euclidean, Frobenius, Schur or Hilbert-Schmidt norm*; it is equal to $\sqrt{tr(AA^*)}$.

EXTENSIONS [GOLDBERG & STRAUS] *If A is a $m \times k$ complex matrix, and B a $k \times n$ complex matrix, and $p \geq 2$, then*

$$|AB|_p \leq k^{(p-2)/p} |A|_p |B|_p.$$

COMMENTS (iii) These results have been further extended by Goldberg to allow different values of p for A and B.

REFERENCES [HLP. p. 36]; [General Inequalities, vol. 4, pp. 185–189], [Horn & Johnson, pp. 290–320], [Marcus & Minc, p. 18].

Max and Min Inequalities
See **Inf and Sup Inequalities.**

Maximal Function Inequalities
See **Hardy-Littlewood Maximal Inequalities**.

Maximin Theorem
See **Minimax Theorem**.

Maximum-Modulus Principle
If f is analytic, not constant, in a closed domain Ω that has a simple closed boundary then

$$|f(z)| \leq M,\ z \in \partial\Omega \implies |f(z)| < M,\ z \in \overset{\circ}{\Omega}.$$

COMMENTS   (i) This result can be stated as: if f is analytic and not constant in a region Ω then $|f(z)|$ has no maximum in Ω. A similar result holds for harmonic and subharmonic functions, and is then called the *Maximum Principle*. The result has been extended to solutions of certain types of differential equations; see **Harnack's Inequalities** COMMENTS, [Protter & Weinberger].

(ii) This result is exceedingly important in the study of analytic functions and many basic results such as those in **Hadamard's Three Circle Theorem, Harnack's Inequalities, Phragmen-Lindelöf Inequality** follow from it.

REFERENCES   [EM, vol. 6, pp. 174–175]; [Conway, vol. I, pp. 124–125, 255–257], [Mitović & Žubrinić, pp. 233–238], [Pólya & Szegö, 1972, pp. 131, 157–160, 165–166], [Protter & Weinberger, pp. 1–9, 61–67, 72–75], [Titchmarsh, 1939 pp. 165–187].

Maximum Principle for Continuous Functions
If K is a compact set and f is continuous real-valued on K then for some $c, d \in K$

$$f(d) \leq f(x) \leq f(c), \qquad x \in K.$$

COMMENTS   This is a basic property of continuous functions.

REFERENCES   [Apostol, vol.I, p. 377], [Hewitt & Stromberg. p. 74], [Rudin, 1964, p. 77].

Mean Inequalities
See **Basic Mean Inequalities**.

Mean Monotonic Sequence Inequalities
Let us say that *the sequence $\underline{a}$ is mean monotonic, or monotonic in the mean*, if the sequence of arithmetic means, $\{\mathfrak{A}_1(\underline{a}; \underline{w}), \mathfrak{A}_2(\underline{a}; \underline{w}), \ldots\}$, is monotonic; in particular we can define *increasing, decreasing, in mean*.

[BURKILL[3] & MIRSKY] If the k non-negative sequences $\underline{a}_i, 1 \leq i \leq k$, are mean monotonic in the same sense then

$$\prod_{i=1}^k \mathfrak{A}_n(\underline{a}_i, \underline{w}) \leq \mathfrak{A}\left(\prod_{i=1}^k \underline{a}_i; \underline{w}\right).$$

COMMENTS In the case $k = 2$ the sequences can be taken to be real. This should be compared to (Č) and **Čebišev's Inequality** (2).

REFERENCES [MI, pp. 232–233], [MPF, pp. 265, 271], [PPT, p. 199].

Mean Value Theorem of Differential Calculus
If the continuous function $f: [a,b] \to \mathbb{R}^n$ is differentiable on $]a,b[$ then for some $c, a < c < b$

$$|f(b) - f(a)| \leq (b-a)|f'(c)|;$$

in particular

$$\frac{|f(b) - f(a)|}{|b-a|} \leq \sup_{a<c<b} |f'(c)|.$$

COMMENTS For the case $n = 1$ the inequalities above can be replaced by

$$f(b) - f(a) = (b-a)f'(c);$$

such a c is called a *mean value point* of f on $[a,b]$.

Further if $a \leq x < y \leq b$,

$$\inf_{a<c<b} f'(c) \leq \frac{f(y) - f(x)}{y-x} \leq \sup_{a<c<b} f'(c).$$

In fact the bounds of $f'(c)$ and $(f(y) - f(x))/(y-x)$ are the same.

This result can be extended to unilateral derivatives, and to more general situations.

EXTENSIONS [GARSIA, FICHERA, SNEIDER] If $f' \in \mathcal{L}^p([0,1]), p \geq 1$, then

$$\int_0^1 \int_0^1 \left|\frac{f(x) - f(y)}{y - x}\right|^p dx\, dy \leq C_p \int_0^1 |f'|^p,$$

where $C_p \leq C_1 = \log 4$.

REFERENCES [Bourbaki, 1949, pp. 18–31], [General Inequalities, vol. 1, p. 319], [Mitrović & Žubrinić, p. 352], [Rudin, 1964, pp. 92–93, 99].

---

[3] This is H Burkill.

Measure Inequalities (a) If μ is a measure on X and if $A \subseteq \bigcup_{n \in \mathbb{N}} B_n \subseteq X$ then

$$\mu A \leq \sum_{n \in \mathbb{N}} \mu B_n. \tag{1}$$

(b) [KAUFMAN & RICKERT] (i) If μ is a measure with values in $\mathbb{R}^n$ of total variation 1, then there is a measurable set E such that

$$|\mu(E)| \geq \frac{((n-2)/2)!}{2\sqrt{n}((n-1)/2)!}$$

(ii) If μ is a complex measure of total variation 1 there is a measurable set E such that $|\mu(E)| \geq \pi^{-1}$.

COMMENTS A measure on X, or what is often called a *Carathéodory outer measure* on X, is a function μ on the subsets of X, taking values in $[0, \infty]$, with $\mu(\emptyset) = 0$, and satisfying (1); that is, it is *increasing* and *subadditive*.

REFERENCES [EM, vol. 6, pp. 183], [MPF, pp. 509]; [Saks, p. 43]

Mediant Inequalities (a) If $a, b, c, d \in \mathbb{R}$ and if $b, d > 0$ then

$$\frac{a}{b} < \frac{c}{d} \implies \frac{a}{b} < \frac{a+c}{b+d} < \frac{c}{d}.$$

(b) If C is a regular continued fraction with n-th convergent $C_n = P_n/Q_n$ then the mediant of C_n and C_{n+1} lies between C and C_n; hence

$$|C - C_n| \geq \frac{1}{Q_n(Q_n + Q_{n+1})}. \tag{1}$$

COMMENTS (i) Given two fractions $a/b, c/d$ the quantity $(a+c)/(b+d)$ is called the *mediant* of these fractions.

(ii) For the notation in (b) see **Continued Fraction Inequalities**. Inequality (1) is a converse of **Continued Fraction Inequalities** (1).

(iii) If all the reduced fractions in $]0, 1[$ with denominators at most n, $n > 0$, are written in increasing order then each term, except the first and last, is the mediant of its neighbours. Such a sequence of fractions is called a *Farey sequence*.

REFERENCES [EM, vol. 6, p. 190].

Metric Inequalities If X is any space and $\rho : X \times X \to \mathbb{R}$ a metric on X then for all $x, y, z \in X$

$$\rho(x, z) \leq \rho(x, y) + \rho(y, z). \tag{1}$$

COMMENTS (i) A function on $X \times X$ with values in $[0, \infty[$ is a metric if it is symmetric, positive except on $\Delta = \{(x, x), x \in X\}$, where it is zero, and satisfies (1); then X is called a *metric space*.

(ii) It follows from (T) that $|\cdot|$ is a metric on $\mathbb{R}^n$, the *Euclidean metric*; and from **Norm Inequalities** (2) it follows that if $||\cdot||$ is a norm on X then $\rho(x, y) = ||x - y||$ gives a metric on X.

(iii) If $z, w \in \overline{\mathbb{C}}$ then

$$\rho(z, w) = \frac{2|z - w|}{\sqrt{(1 + |z|^2)(1 + |w|^2)}}$$

is a bounded metric on $\overline{\mathbb{C}}$. It is the distance between the stereographic images of z, w on the Riemann sphere in $\mathbb{R}^3$, $|\underline{x}| = 1$.

(iv) Another metric associated with the complex plane is the *hyperbolic metric*, ρ_h, defined for $z, w \in D$ by:

$$\rho_h(z, w) = \log \frac{1 + \delta(z, w)}{1 - \delta(z, w)}, \text{ where } \delta(z, w) = \left| \frac{z - w}{1 - \overline{z}w} \right|;$$

for an application see **Schwarz's Lemma** COMMENTS (ii), EXTENSIONS (d).

(v) If the requirement that ρ be zero only on Δ is dropped we get a *pseudo-metric*; the $\mathcal{L}^p([a, b])$ spaces, are examples of pseudo-metric spaces. Equivalently we say they are metric spaces if we identify functions that are equal almost everywhere.

EXTENSIONS (a) If $x_i \in X, 1 \leq i \leq n$, and if ρ is a metric on X then

$$\rho(x_1, x_n) \leq \sum_{i=1}^{n-1} \rho(x_i, x_{i+1}).$$

(b) [QUADRILATERAL INEQUALITY] If $x_i \in X, 1 \leq i \leq 4$, and if ρ is a metric on X then

$$\left| \rho(x_4, x_1) - \rho(x_4, x_3) \right| \leq \rho(x_2, x_1) + \rho(x_2, x_3).$$

COMMENTS (vi) (a) follows from (1) by a simple induction and is a generalization of **Triangle Inequality** EXTENSIONS (a) (ii); (b) is a corollary of (a), and is an extension of the **Quadrilateral Inequality**.

REFERENCES [EM, vol. 6, p. 206], [MPF, pp. 481–483]; [Ahlfors, 1966, pp. 18–20,51; 1973, pp. 1–3].

Mills' Ratio Inequalities
See Error Function Inequalities.

Min and Max Inequalities
See Inf and Sup Inequalities.

Mingarelli's Inequality
If $\alpha > 0$ and $f \in \mathcal{C}^1(\mathbb{R})$ has compact support then

$$\int_{\mathbb{R}} |f(x)|^2 e^{\alpha x^2}\, \mathrm{d}x \leq \frac{1}{2\alpha} \int_{\mathbb{R}} |f'(x)|^2 e^{\alpha x^2}\, \mathrm{d}x.$$

REFERENCES [Mingarelli].

Minimax Theorems
(a) If $f : [a,b] \times [c,d] \to \mathbb{R}$ is continuous then

$$\max_{c \leq y \leq d} \min_{a \leq x \leq b} f(x,y) \leq \min_{a \leq x \leq b} \max_{c \leq y \leq d} f(x,y). \tag{1}$$

(b) [KY FAN] Let $f : X \times X \to \mathbb{R}$, where X is a compact set in $\mathbb{R}^n$, be such that: for all $x \in X$, $f(x, \cdot)$ is lower semicontinuous, and for all $y \in X$, $f(\cdot, y)$ is quasi-concave. Then

$$\min_{y \in X} \sup_{x \in X} f(x,y) \leq \sup_{x \in X} f(x,x).$$

COMMENTS (i) For the definition of quasi-concave see **Quasi-convex Function Inequalities**.
(ii) If there is equality in (1) the *minimax principle* is said to hold.

REFERENCES [EM, vol. 6, p. 239]; [Pólya & Szegö, 1972, pp. 101–102], [Inequalities III, pp. 103–114].

Minkowski's Inequality
If $\underline{a}, \underline{b}$ are positive n-tuples, and $p > 1$ or $p < 0$ then

$$\left(\sum_{i=1}^n (a_i + b_i)^p\right)^{1/p} \leq \left(\sum_{i=1}^n a_i^p\right)^{1/p} + \left(\sum_{i=1}^n b_i^p\right)^{1/p}. \tag{M}$$

If $0 < p < 1$ then $(\sim M)$ holds. There is equality if and only if $\underline{a} \sim \underline{b}$.

COMMENTS (i) There are several proofs of this famous inequality; the easiest approach is to deduce it from (H).
(ii) One of the most important uses of (M) is to show that $\|\underline{a}\|_p$, $1 \leq p < \infty$, is a norm, (1) below; see **Norm Inequalities** (1).

EXTENSIONS (a) If $\underline{a}, \underline{b}$ are complex n-tuples adn p is as in (M),

$$\left(\sum_{i=1}^{n} |a_i + b_i|^p\right)^{1/p} \leq \left(\sum_{i=1}^{n} |a_i|^p\right)^{1/p} + \left(\sum_{i=1}^{n} |b_i|^p\right)^{1/p},$$

or

$$\|\underline{a} + \underline{b}\|_p \leq \|\underline{a}\|_p + \|\underline{b}\|_p. \tag{1}$$

with equality if and only if $\underline{a} \sim^+ \underline{b}$.

(b) [GENERALIZED MINKOWSKI'S INEQUALITY] If $a_{ij} > 0$, $1 \leq i \leq m, 1 \leq j \leq n$, and if $p > 1$ then

$$\left\{\sum_{j=1}^{n}\left(\sum_{i=1}^{m} a_{ij}\right)^p\right\}^{1/p} \leq \sum_{i=1}^{m}\left(\sum_{j=1}^{n} a_{ij}^p\right)^{1/p}. \tag{2}$$

There is equality in (2) if and only if the n-tuples $(a_{i1}, \ldots, a_{in})$, $1 \leq i \leq m$, are linearly dependent.

(c) [FUNCTIONS OF THE INDEX SET] If $\mathcal{I}$ is an index set let

$$\mu(\mathcal{I}) = \left(\left(\sum_{i \in \mathcal{I}} a_i^p\right)^{1/p} + \left(\sum_{i \in \mathcal{I}} b_i^p\right)^{1/p}\right)^p - \sum_{i \in \mathcal{I}} (a_i + b_i)^p,$$

where $\underline{a}, \underline{b}, p$ are as in (M) then $\mu \geq 0$, and if $\mathcal{I} \cap \mathcal{J} = \emptyset$

$$\mu(\mathcal{I}) + \mu(\mathcal{J}) \leq \mu(\mathcal{I} \cup \mathcal{J});$$

this inequality is strict unless $\left(\sum_{i \in \mathcal{I}} a_i^p, \sum_{i \in \mathcal{I}} b_i^p\right)$ and $\left(\sum_{i \in \mathcal{J}} a_i^p, \sum_{i \in \mathcal{J}} b_i^p\right)$ are proportional.

(d) [MIKOLÁS] If $a_{ij} > 0, 1 \leq i \leq m, 1 \leq j \leq n$, then if $0 < r \leq 1$, $0 < rp \leq 1$ then

$$\sum_{i=1}^{m}\left(\sum_{j=1}^{n} a_{ij}^r\right)^p \leq n^{1-pr} \sum_{j=1}^{n}\left(\sum_{i=1}^{m} a_{ij}\right)^{pr}; \tag{3}$$

if $r \geq 1$ and $pr \geq 1$ then ($\sim$3) holds.

COMMENTS (iii) If we take $p = 1/r$ then (3) reduces to (M).

(iv) Further extensions can be found in [AI, MPF].

CONVERSE INEQUALITIES [MOND & SHISHA] If $\underline{a}, \underline{b}$ are positive n-tuples such that

$$0 < m \leq \frac{a_i}{a_i + b_i}, \frac{b_i}{a_i + b_i} \leq M, M \neq m;$$

and if $p > 1$ then

$$\left(\sum_{i=1}^n a_i^p\right)^{1/p} + \left(\sum_{i=1}^n b_i^p\right)^{1/p} \leq \Lambda \left(\sum_{i=1}^n (a_i + b_i)^p\right)^{1/p}, \tag{4}$$

where

$$\Lambda = \frac{M^p - m^p}{(p(M-m))^{1/p} |q(Mm^p - mM^p)|^{1/q}},$$

q being the conjugate index. If $p < 1, p \neq 0$ then ($\sim$4) holds.

INTEGRAL ANALOGUES (a) If p is as in (M) and if $f_i \in \mathcal{L}^p([a,b]), 1 \leq i \leq n$, then $\sum_{i=1}^n f_i \in \mathcal{L}_p([a,b])$ and

$$\left\|\sum_{i=1}^n f_i\right\|_{p,[a,b]} \leq \sum_{i=1}^n \|f_i\|_{p,[a,b]}.$$

There is equality if $A_i f_i = B_i g$ almost everywhere, where not both A_i, B_i are zero, $1 \leq i \leq n$.

(b) If p is as in (M) then

$$\left(\int_a^b \left(\int_c^d |f(x,y)|\,dy\right)^p dx\right)^{1/p} \leq \int_c^d \left(\int_a^b |f(x,y)|^p\,dx\right)^{1/p} dy,$$

with equality if and only if $f(x,y) = g(x)h(y)$ almost everywhere.

COMMENTS (v) These can be extended to general measure spaces.

(vi) See also **Jessen's Inequality, Kober & Dianada's Inequality** COMMENTS (ii), **Power Mean Inequalities** (3), **Quasi-arithmetic Mean Inequalities** COMMENTS (iv), **Rahmail's Inequality**. For other inequalities of the same name see **Determinant Inequalities** (1), **Minkowski-Mahler Inequality, Mixed-volume Inequalities** (a), (b).

REFERENCES [AI, pp. 55–57], [BB, pp. 19–29], [EM, vol. 6. pp. 247–248], [HLP, pp. 30–32, 85–88, 146–150], [MI, pp. 147–149, 151–159], [MO, pp. 459–460], [MPF, pp. 107–117, 135–209], [PPT, pp. 114–118, 126–128, 166–167]; [Lieb & Loss, pp. 41–42], [Pólya & Szegö, 1972, p. 71].

Minkowski-Mahler Inequality

Let F be a generalized norm on $\mathbb{R}^n$, define

$$G(\underline{b}) = \max_{\underline{a} \in \mathbb{R}^n} \left(\frac{\underline{a}.\underline{b}}{F(\underline{a})}\right), \quad \underline{b} \in \mathbb{R}^n.$$

Then

$$F(\underline{a}) = \max_{\underline{b} \in \mathbb{R}^n} \left(\frac{\underline{a}.\underline{b}}{G(\underline{b})}\right), \quad \underline{a} \in \mathbb{R}^n,$$

and for all $\underline{a}, \underline{b} \in \mathbb{R}^n$,

$$\underline{a}.\underline{b} \leq F(\underline{a})G(\underline{b}).$$

COMMENTS   (i) For the definition of generalized norm see **Norm Inequalities** COMMENTS (i). (ii) The function G is called the *polar function* of F, when, from the above, F is the polar function of G.

REFERENCES   [BB, pp. 28–29], [EM, vol. 6, p. 248].

Minkowski's Mixed Volume Inequality
See **Mixed-Volume Inequalities**.

Mitrinović & Djoković's Inequality
If $\underline{a}$ is a positive n-tuple with $A_n = 1$ and if $p \geq 0$, then

$$\mathfrak{M}_n^{[p]}\left(\underline{a} + \frac{1}{\underline{a}}\right) \geq n + \frac{1}{n},$$

COMMENTS   (i) This inequality can be proved using (J) or **Schur Convex Function Inequalities** (b). The case $p = 0$ is due to Li & Chen; of course by (r;s) this is the critical case.

EXTENSIONS [WANG]   Let $p, r \geq 0, q > 0$, $\underline{a}, \underline{w}$ positive n-tuples with $\max \underline{a} \leq \sqrt{r/q}$ then writing

$$f(x) = \left(qx + \frac{r}{x}\right)^p,$$

we have

$$\mathfrak{A}_n(f(\underline{a}), \underline{w}) \geq f\left(\mathfrak{G}_n(\underline{a}, \underline{w})\right) \geq f\left(\mathfrak{A}_n(\underline{a}, \underline{w})\right).$$

COMMENTS   (ii) This follows by an application of (J) and (GA).

REFERENCES   [AI, pp. 282–283], [MO, pp. 72–73], [MPF, pp. 36–37].

Mitrinović & Pečarić's Inequality
Let $p, f_i, g_i, 1 \leq i \leq n$, be continuous functions defined on $[a, b]$ and denote by $(\Pi\Phi\Gamma)_n$ the $n \times n$ matrix with i, j-th entry $\int_a^b p f_i g_j$; let $\phi(\underline{x}), \gamma(\underline{x})$ denote the $n \times n$ matrices with i, j-th entries $f_i(x_j), g_i(x_j)$ respectively, where $a \leq x_j \leq b, 1 \leq j \leq n$: then,

$$\det \phi(\underline{x}) \geq 0, \ \det \gamma(\underline{x}) \geq 0 \ \text{ for all } \ \underline{x} \implies \det(\Pi\Phi\Gamma)_n \geq 0.$$

COMMENTS   (i) This inequality is deduced from an identity of Andréief,

$$\det(\Pi\Phi\Gamma)_n = \frac{1}{n!}\int_a^b \cdots \int_a^b \det \phi(\underline{x}) \det \gamma(\underline{x}) \, d\underline{x}.$$

(ii) Special cases of this result are (C), (Č), **Gram Determinant Inequalities** (1).

REFERENCES   [BB, p. 61], [MPF, pp. 244, 600–601]; [Mitrinović & Pečarić, 1991$^{(2)}$, p. 51].

Mixed Mean Inequalities

If $\underline{a}$ is a positive n-tuple, $1 \leq k \leq n$, put $\kappa = \binom{n}{k}$, and denote by $\underline{a}_1^{(k)}, \ldots, \underline{a}_\kappa^{(k)}$ the κ k-tuples formed from the elements of $\underline{a}$. If then $r, s \in \mathbb{R}$ the mixed mean of order r and s of $\underline{a}$ taken k at a time is

$$\mathfrak{M}_n(r, s; k; \underline{a}) = \mathfrak{M}_\kappa^{[r]}\big(\mathfrak{M}_k^{[s]}(\underline{a}_i^{(k)}); 1 \leq i \leq \kappa\big)$$

The following relations with other means are immediate:

$$\mathfrak{M}_n(r, s; 1; \underline{a}) = \mathfrak{M}_n^{[r]}(\underline{a}); \quad \mathfrak{M}_n(r, s; n; \underline{a}) = \mathfrak{M}_n^{[s]}(\underline{a});$$
$$\mathfrak{M}_n(r, r; k; \underline{a}) = \mathfrak{M}_n^{[r]}(\underline{a}); \quad \mathfrak{M}_n(k; 0; k; \underline{a}) = \mathfrak{P}_n^{[r]}(\underline{a}).$$

[CARLSON[4]] (a) $\mathfrak{M}_n(r, s; k; \underline{a})$ *is increasing both as a function of* r *and* s.
(b) *If* $-\infty \leq r < s \leq \infty$ *then*

$$\mathfrak{M}_n(r, s; k - 1; \underline{a}) \leq \mathfrak{M}_n(r, s; k; \underline{a}). \tag{1}$$

If $r > s$ *then* (~ 1) *holds. In both cases there is equality if and only if* $\underline{a}$ *is constant.*
(c) *If* $-\infty \leq r < s \leq \infty$ *and* $k + \ell > n$ *then*

$$\mathfrak{M}_n(s, r; k - 1; \underline{a}) \leq \mathfrak{M}_n(r, s; \ell; \underline{a}), \tag{2}$$

with equality if and only if $\underline{a}$ *is constant.*

COMMENTS (i) Both results follow by applications of (r;s), and contain (r;s) as a special case.
(ii) Let $r < s$ and consider the following $2 \times n$ matrix in which $\underline{a}$ is not constant;

$$\mathbf{M} = \begin{pmatrix} \mathfrak{M}_n(r, s; 1; \underline{a}) & \mathfrak{M}_n(r, s; 2; \underline{a}) & \cdots & \mathfrak{M}_n(r, s; n-1; \underline{a}) & \mathfrak{M}_n(r, s; n; \underline{a}) \\ \mathfrak{M}_n(r, s; n; \underline{a}) & \mathfrak{M}_n(r, s; n-1; \underline{a}) & \cdots & \mathfrak{M}_n(r, s; 2; \underline{a}) & \mathfrak{M}_n(r, s; 1; \underline{a}) \end{pmatrix}.$$

The inequalities (1), (2) can be summarized as saying: (i) the rows of $\mathbf{M}$ are strictly increasing to the right; (ii) the columns, except the first and last, strictly increase downwards; of course the entries in the first column both are equal to $\mathfrak{M}_n^{[r]}(\underline{a})$, while both in the last column are equal to $\mathfrak{M}_n^{[s]}(\underline{a})$. Another matrix of the same type is given in **Symmetric Mean Inequalities** COMMENTS (v).

---

[4] This is B C Carlson.

SPECIAL CASES (a) If a, b, c are positive and not all equal then

$$\frac{\sqrt{ab} + \sqrt{bc} + \sqrt{ca}}{3} < \sqrt[3]{\frac{(a+b)(b+c)(c+a)}{8}}. \quad (3)$$

(b) If $\underline{a}$ is a positive n-tuple then

$$\mathfrak{A}_n(\mathfrak{G}_{n-1}(\underline{a}'_k), 1 \leq k \leq n) \leq \mathfrak{G}_n(\mathfrak{A}_{n-1}(\underline{a}'_k), 1 \leq k \leq n).$$

COMMENTS (iii) Both are obtained from (2), and (3) should be compared with **Symmetric Mean Inequalities** (1)

(iv) See also **Nanjundiah's Mixed Mean Inequalities**.

REFERENCES [AI, p. 379], [MI, pp. 191–194].

Mixed-volume Inequalities

If K_i is a convex body in $\mathbb{R}^p$, $\lambda_i \geq 0, 1 \leq i \leq n$, then the linear combination $B = \sum_{i=1}^n \lambda_i K_i$ has a volume, $V(B)$, that is a homogeneous polynomial of degree n in the $\lambda_i, 1 \leq i \leq n$. The coefficients of this polynomial are called the *mixed-volumes of the bodies* $K_i, 1 \leq i \leq n$. In particular $V(K_{i_1}, \ldots K_{i_n})$ denotes the coefficient of $\lambda_{i_1} \ldots \lambda_{i_n}$. Inequalities between these quantities are the subject of this entry.

With the above notation

$$V^n(K, L, \ldots, L) \geq V(K)V^{n-1}(L);$$

and

$$V^2(K, L, \ldots, L) \geq V(L)V(K, K, L, \ldots, L).$$

COMMENTS (i) The first inequality is called *Minkowski's mixed-volume inequality*, while the second is called *Minkowski's quadratic mixed-volume inequality*.

(ii) Many isoperimetric inequalities are special cases of these results.

EXTENSIONS [ALEKSANDROV[5]-FENCHEL]

$$V^m(K_1, \ldots, K_m, L_1, \ldots, L_{n-m}) \geq \prod_{j=1}^m V(K_j, \ldots, K_j, L_1, \ldots, L_{n-m});$$

in particular

$$V^n(K_1, \ldots, K_n) \geq V(K_1) \ldots V(K_n).$$

COMMENTS (iii) See also **Brunn-Minkowski Inequalities**.

REFERENCES [EM, vol. 6, pp. 262–264].

---

[5] П С Александров. Also transliterated Alexandroff.

Modulus of Continuity Inequalities [HAYMAN] If $f \in \mathcal{C}([0,1])$ then

$$\omega(\delta; f^*) \leq \omega(\delta; f);$$
$$H(f^*; \alpha) \leq H(f; \alpha).$$

COMMENTS (i) $\omega(\delta; f)$ is the modulus of continuity of f;

$$\omega(\delta; f) = \sup_{|x-y|\leq \delta} |f(x) - f(y)|;$$

and if $0 < \alpha \leq 1$ then $H(f; \alpha)$ is the coefficient of Hölder continuity of f:

$$H(f; \alpha) = \sup_{0 \leq x, y \leq y, x \neq y} \frac{|f(y) - f(x)|}{|y - x|^\alpha}.$$

(ii) See also **Marchaud's Inequality**.

REFERENCES [Yangihara].

Moment Inequalities [PETSCHKE] Let $f : [0,1] \to \mathbb{R}$ be decreasing, $p, q \geq 0$, $p < q$, then if $0 < x < 1$

$$x^{q-p} \int_x^1 t^p f(t)\,dt \leq \left(\frac{q-p}{q+1}\right)^{(q-p)/(p+1)} \int_0^1 t^q f(t)\,dt.$$

COMMENTS (i) This is a variant of **Steffensen's Inequality**.

(ii) Other similar results are given in the reference; see also **Convex FunctionIntegral Inequalities, Gauss's Inequality, Ting's Inequalities**.

REFERENCES [MPF, pp. 329–330].

Moments of Inertia Inequality If A is the area of a plane domain and I the polar moment of inertia about the centre of gravity then

$$I \geq \frac{A^2}{2\pi}.$$

Equality occurs only when the domain is a circle.

COMMENTS (i) The *polar moment of inertia about the centre of gravity* of a plane domain A is

$$I = \int_A (x - \overline{x})^2 + (y - \overline{y})^2\,dx\,dy,$$

where

$$\overline{x} = \frac{1}{|A|} \int_A x\,dx\,dy,$$

and $\bar{y}$ is defined analogously; $(\bar{x},\bar{y})$ is the *centre of gravity of A*.
(ii) This is an example of **Symmetrization Inequalities**.

REFERENCES   [*Pólya & Szegö, 1951, pp. 2, 8, 153, 195–196*].

Mond & Shisha Inequalities
See **Geometric-Arithmetic Mean Inequality**, CONVERSE INEQUALITIES, **Minkowski's Inequality**, CONVERSE INEQUALITIES.

Monotone Matrix Function Inequalities
(a) if A, B are $n \times n$ Hermitian matrices and if f is a monotone matrix function of order n then

$$A \geq B \quad \Longrightarrow \quad f(A) \geq f(B). \tag{1}$$

(b) If f is a non-constant monotone matrix function of order $n, n \geq 2$, on $[0, \infty[$ and if A is a positive semi-definite $n \times n$ Hermitian matrix then

$$f^{-1}(A)_{i_1,\ldots,i_m} \geq f^{-1}(A_{i_1,\ldots,i_m}) \quad \Longrightarrow \quad f(A)_{i_1,\ldots,i_m} \geq f(A_{i_1,\ldots,i_m}).$$

(c) If A is a rank one positive semi-definite Hermitian matrix, and if $\alpha \geq 1$ then

$$(A_{i_1,\ldots,i_m})^{1/\alpha} \geq A^{1/\alpha}_{i_1,\ldots,i_m}.$$

COMMENTS   (i) If I is a interval in $\mathbb{R}$ then a function $f : I \to \mathbb{R}$ is said to be a *monotone matrix function of order n* if (1) holds for all A, B with eigenvalues in I; here $A \geq B$ means that $A - B$ is positive semi-definite.
$x^\alpha, 0 < \alpha \leq 1, \log x, -x^{-1}$ are all monotone matrix function of all orders on $]0, \infty[$.
(ii) A matrix function that is monotone of all orders on I is called an *operator monotone function on I*.
(iii) See also **Convex Matrix Function Inequalities**.

REFERENCES   [*Roberts & Varberg, pp. 259–261*]; [*Chollet*].

Monotonic Function Inequalities
See **Increasing Function Inequalities, Monotone Matrix Function Inequalities**.

Morrey's Inequality
If $n < p < \infty$, $B = B_{\underline{x},r} = \{\underline{u}; |\underline{u} - \underline{x}| < r\} \subset \mathbb{R}^n$, and if $f \in \mathcal{W}^{1,p}(\mathbb{R}^n)$ then for almost all y, z

$$|f(y) - f(z)| \leq rC \left(\frac{1}{|B|} \int_B |Df| \right)^{1/p}$$

where C depends only on n and p.

In particular $\lim_{r\to 0} \frac{1}{|B|} \int_B f$ is Lipschitz of order $(p-n)/p$.

COMMENTS   For the definition of $\mathcal{W}^{1,p}(\mathbb{R}^n)$ see **Sobolev's Inequalities**; and for the definition of Lipschitz of order α see **Lipschitz Function Inequalities**.

REFERENCES   [Evans & Gariepy, pp. 143–144].

Muirhead's Inequality

See **Muirhead Symmetric Function and Mean Inequalities** (1).

Muirhead Symmetric Function and Mean Inequalities

If $\underline{a}$ a positive n-tuple and $\underline{\alpha}$ is a non-negative n-tuple, put $|\underline{\alpha}| = \alpha_1 + \cdots + \alpha_n$ and define

$$e_n(\underline{a}; \underline{\alpha}) = e_n(\underline{a}; \alpha_1, \ldots, \alpha_n) = \frac{1}{n!} \sum! \prod_{j=1}^n a_{i_j}^{\alpha_j};$$

$$\mathfrak{A}_{n,\underline{\alpha}}(\underline{a}) = \mathfrak{A}_{n,\{\alpha_1,\ldots,\alpha_n\}}(\underline{a}) = (e_n(\underline{a}; \underline{\alpha}))^{1/|\underline{\alpha}|}.$$

These are the *Muirhead symmetric functions and means* respectively. These means are called symmetric means in [HLP].

There is no loss in generality in assuming that $\underline{\alpha}$ is decreasing.

It is easily seen that some earlier means are special cases of the Muirhead means:

$$\mathfrak{A}_{n,\{p,0,\ldots,0\}}(\underline{a}) = \mathfrak{M}_n^{[p]}(\underline{a}); \qquad \mathfrak{A}_{n,\{1,1,\ldots,1\}}(\underline{a}) = \mathfrak{G}_n(\underline{a});$$

and if $1 \leq r < n$,

$$\mathfrak{A}_{n,\underbrace{\{1,1,\ldots 1,0,\ldots,0\}}_{r\text{ terms}}}(\underline{a}) = \mathfrak{P}_n^{[r]}(\underline{a}).$$

(a) If $\underline{\alpha}, \underline{\beta}$ are non-identical decreasing non-negative n-tuples then

$$e_n(\underline{a}; \underline{\alpha}) \leq e_n(\underline{a}; \underline{\beta}) \tag{1}$$

if and only if $\underline{\alpha} \prec \underline{\beta}$; further there is equality if and only if $\underline{a}$ is constant.

(b) If $\underline{a}$ is a positive n-tuple and $\underline{\alpha}$ is a non-negative n-tuple then

$$\min \underline{a} \leq \mathfrak{A}_{n,\underline{\alpha}}(\underline{a}) \leq \max \underline{a}.$$

(c) If $\underline{a}$ is a positive n-tuple and if $\underline{\alpha}_1, \underline{\alpha}_2$ are non-identical decreasing non-negative n-tuples with $\underline{\alpha}_1 \prec \underline{\alpha}_2$ then

$$\mathfrak{A}_{n,\underline{\alpha}_1}(\underline{a}) \leq \mathfrak{A}_{n,\underline{\alpha}_2}(\underline{a}), \tag{2}$$

with equality if and only if $\underline{a}$ is constant.

(d) [CARLSON[4], PITTENGER] If $0 < a < b$ and if $\underline{\alpha} = \left((1+\sqrt{\delta})/2, (1-\sqrt{\delta})/2\right)$ where $0 \le \delta \le 1/3$ then

$$\mathfrak{A}_{2,\underline{\alpha}}(a,b) \le \mathfrak{L}(a,b).$$

COMMENTS (i) Inequality (1) is called *Muirhead's Inequality*; (2) follows from (1).
(ii) Inequality (2) implies (GA), and (r;s). In addition if $\underline{\alpha}$ has at least two non-zero entries and $|\underline{\alpha}| = 1$,

$$\mathfrak{G}_n(\underline{a}) \le \mathfrak{A}_{n,\underline{\alpha}}(\underline{a}) \le \mathfrak{A}_n(\underline{a})$$

(iii) Inequality (2) identifies a class of order preserving functions from the pre-ordered positive n-tuples to $\mathbb{R}$; see **Schur Convex Function Inequalities** COMMENTS (ii).
(iv) The inequality in (d) can fail if $1/3 < \delta < 1$.

EXTENSIONS (a) If $f: \mathbb{R}^n \to \mathbb{R}$ is convex and symmetric, and $\underline{\alpha}, \underline{\beta}, \underline{a}$ are as in (a) above with $\underline{\alpha} \prec \underline{\beta}$, then

$$\sum! f(\alpha_1 a_1, \ldots, \alpha_n a_n) \le \sum! f(\beta_1 a_1, \ldots, \beta_n a_n).$$

(b) [MARCUS & LOPEZ TYPE] If $\underline{a}, \underline{b}, \underline{\alpha}$ are positive n-tuples, and if $0 < |\underline{\alpha}| < 1$ then

$$\mathfrak{A}_{n,\underline{\alpha}}(\underline{a}) + \mathfrak{A}_{n,\underline{\alpha}}(\underline{b}) \le \mathfrak{A}_{n,\underline{\alpha}}(\underline{a}+\underline{b}).$$

INTEGRAL ANALOGUES [RYFF] Let α, β be two bounded measurable functions on $[0,1]$, $f \in \mathcal{L}^p([0,1])$ for all $p \in \mathbb{R}$, $f > 0$; then

$$\int_0^1 \log\left(\int_0^1 f(t)^{\alpha(s)} \, dt\right) ds \le \int_0^1 \log\left(\int_0^1 f(t)^{\beta(s)} \, dt\right) ds,$$

if and only if $\alpha \prec \beta$.

COMMENTS (v) A geometric mean analogue of $\mathfrak{A}_{n,\underline{\alpha}}(\underline{a})$ has also been defined, and appropriate inequalities obtained; see [MI] for this and further extensions of these means.

REFERENCES [AI, p. 167], [HLP, pp. 44–51], [MI, pp. 333–342, 350], [MO, pp. 87, 110–111], [PPT, pp. 361–364].

---

[4] This is B C Carlson.

Mulholland's Inequality If μ is a probability measure on $\mathbb{C}$, and if $p > 0$ then

$$\mathfrak{G}_{\mathbb{C}\times\mathbb{C}}(|z-w|;\mu\otimes\mu) \leq \left(\frac{2}{\sqrt{e}}\right)^{1/p}\mathfrak{M}_{\mathbb{C}}^{[p]}(|z|;\mu).$$

There is equality if μ is defined by

$$\mu(B) = K\int_B 1_{\{z;|z|<a\}}|z|^{k-2}\,d\lambda_2,$$

where B a Borel set in $\mathbb{C}$.

COMMENTS   (i) The best value for the constant has been conjectured as $\left(\dfrac{2}{\sqrt{e}}\dfrac{(p/2)!}{p!}\right)^{1/p}$.
(ii) For a discrete analogue of the left-hand side see the **Difference Means of Gini**.

REFERENCES   [EM, vol. 4, p. 277]; [Mulholland].

Multigamma Function Inequalities
See **Digamma Function Inequalities**.

Multilinear Form Inequalities   [HARDY-LITTLEWOOD-PÓLYA]   Hypotheses: $\underline{a}_i = \{a_{ij}, j \in \mathbb{Z}\}$ are non-negative sequences with $\|\underline{a}_i\|_{p_i} \leq A_i$, $p_i \geq 1$, $1 \leq i \leq n$; $\underline{c} = \{c_{j_1,j_2,\ldots,j_n}, j_i \in \mathbb{Z}, 1 \leq i \leq n\}$ are non-negative with $\sum_i{}' \underline{c}^{r_i} \leq C_i, r_i > 0, 1 \leq i \leq n$;

$P = \dfrac{1}{n-1}\left(\sum_{i=1}^n \dfrac{1}{p_i} - 1\right)$, p_i, $1 \leq i \leq n$, satisfying $0 \leq P \leq \min_{1\leq i\leq n}\{p_i^{-1}\}$; $\overline{p}_i$ is defined by $\dfrac{1}{\overline{p}_i} = \dfrac{1}{p_i} - P$, $1 \leq i \leq n$, then r_i, $1 \leq i \leq n$, satisfies $\sum_{i=1}^n \dfrac{r_i}{\overline{p}_i} = 1$.

Conclusion:

$$\sum_{j_i\in\mathbb{Z},1\leq i\leq n} c_{j_1,j_2,\ldots,j_n} a_{1j_1}\cdots a_{nj_n} \leq \prod_{i=1}^n C_i^{1/\overline{p}_i} \prod_{i=1}^n A_i. \quad (1)$$

COMMENTS   (i) The notation $\sum_i{}' \underline{c}$ means that in the multiple summation there is no summing over the i-th suffix.

SPECIAL CASES   [$n = 2$] Let $p, q \geq 1$, $\dfrac{1}{p} + \dfrac{1}{q} \geq 1$, $r, s > 0$, $\dfrac{r}{p'} + \dfrac{s}{q'} = 1$, where p', q' are the conjugate indices of p, q respectively. If $\underline{a} = \{a_i, i \in \mathbb{Z}\}$, $\underline{b} = \{a_j, j \in \mathbb{Z}\}$, $\underline{c} = \{c_{ij}, i, j \in \mathbb{Z}\}$ are non-negative with $\|\underline{a}\|_p \leq A$, $\|\underline{b}\|_q \leq B$, $\sum_{i\in\mathbb{Z}} c_{ij}^r < C$, $\sum_{j\in\mathbb{Z}} c_{ij}^s < D$, then

$$\sum_{i,j\in\mathbb{Z}} c_{ij} a_i b_j \leq C^{1/p'} D^{1/q'} AB.$$

COMMENTS (ii) Another special case is **Young's Convolution Inequality** (b).

(iii) Another important inequality for multilinear forms is the Riesz convexity theorem; see **Riesz-Thorin Theorem** COMMENTS (ii).

(iv) See also **Bilinear Form Inequalities, Quadratic Form Inequalities**.

REFERENCES [HLP, pp. 196–225]; [General Inequalities, vol. 3, pp. 205–218].

Myers Inequality *If $\underline{a}$ is a real n-tuple with $A_n = 0$ then*

$$\prod_{i=1}^{n}(1+a_1) \leq 1$$

with equality if and only if $\underline{a}$ is null.

COMMENTS (i) The proof is by induction; or the result follows from **Weierstrass's Inequalities** (b).

(ii) This inequality has been used to prove (GA).

REFERENCES [MI, p. 82].

N

Nanjundiah Inequalities (a) If $0 < a < b$ and

$$\underline{a} = \left\{ a, a + \frac{b-a}{n-1}, a + 2\frac{b-a}{n-1}, \ldots, a + (n-2)\frac{b-a}{n-1}, b \right\},$$

$$\underline{g} = \left\{ a, a\left(\frac{b}{a}\right)^{1/(n-1)}, a\left(\frac{b}{a}\right)^{2/n-1}, \ldots, a\left(\frac{b}{a}\right)^{(n-2)/(n-1)}, b \right\},$$

$$\underline{h} = \left\{ a, \frac{ab}{b - \frac{(b-a)}{n-1}}, \frac{ab}{b - \frac{2(b-a)}{n-1}}, \ldots, \frac{ab}{b - \frac{(n-2)(b-a)}{n-1}}, b \right\};$$

then

$$\mathfrak{H}_n(\underline{a}) > \mathfrak{A}_n(\underline{g}), \quad \text{and} \quad \mathfrak{H}_n(\underline{g}) > \mathfrak{A}_n(\underline{h}) \tag{1}$$

(b) If $r > 1$, $q > 1$, $r > q$, $ra > (r-1)c > 0$, $qc > (q-1)b > 0$ then

$$\frac{(ra - (r-1)c)^r}{(qc - (q-1)b)^{r-1}} \geq r\frac{a^r}{c^{r-1}} - (r-1)\frac{c^q}{b^{q-1}}. \tag{2}$$

COMMENTS (i) The n-tuples $\underline{a}, \underline{g}, \underline{h}$ are said to interpolate $(n-2)$ arithmetic, geometric, harmonic means between a and b, respectively.

(ii) Inequalities (1) follow from the convexity of e^x and $1/x$ and an application of **Hermite-Hadamard's Inequality** (2).

(iii) Taking limits in (1) leads to special cases of **Logarithmic Mean Inequalities** (1),

$$\mathfrak{L}^{[-2]}(a,b) \leq \mathfrak{L}^{[-1]}(a,b) \leq \mathfrak{L}^{[0]}(a,b) \leq \mathfrak{L}^{[1]}(a,b).$$

(iii) Inequality (2) follows by an application of **Nanjundiah's Inverse Mean Inequalities** (b) and (GA).

REFERENCES [MI, pp. 128–130].

Nanjundiah's Inverse Mean Inequalities

The *inverse means* of Nanjundiah are defined by

$$(\mathfrak{M}_n^{[p]})^{-1}(\underline{a};\underline{w}) = \begin{cases} \left(\left(\frac{W_n}{w_n}\right)a_n^p - \left(\frac{W_{n-1}}{w_n}\right)a_{n-1}^p\right)^{1/p}, & \text{if } -\infty < p < \infty,\ p \neq 0; \\ (a_n^{W_n/w_n}) / (a_{n-1}^{W_{n-1}/w_n}), & \text{if } p = 0. \end{cases}$$

The cases $p = -1, 0, 1$ are also written $\mathfrak{H}_n^{-1}(\underline{a}; \underline{w}), \mathfrak{G}_n^{-1}(\underline{a}; \underline{w}), \mathfrak{A}_n^{-1}(\underline{a}; \underline{w})$ respectively. The name is suggested by the fact that applying these means to the corresponding power mean sequence gives the original sequence back.

Let $\underline{a}$ be a positive n-tuple, $n \geq 2$.
(a)
$$\mathfrak{G}_n^{-1}(\underline{a}; \underline{w}) \geq \mathfrak{A}_n^{-1}(\underline{a}; \underline{w}),$$
with equality if and only if $a_{n-1} = a_n$.
(b)
$$\mathfrak{G}_n^{-1}(\underline{a}; \underline{w}) + \mathfrak{G}_n^{-1}(\underline{b}; \underline{w}) \geq \mathfrak{G}_n^{-1}(\underline{a} + \underline{b}; \underline{w})),$$
with equality if and only if $a_{n-1}b_n = a_n b_{n-1}$.
(c) If $(a_{n-1}, a_n), (b_{n-1}, b_n)$ are similarly ordered then
$$\mathfrak{A}_n^{-1}(\underline{a}; \underline{w})\mathfrak{A}_n^{-1}(\underline{b}; \underline{w}) \geq \mathfrak{A}_n^{-1}(\underline{a}\,\underline{b}; \underline{w}),$$
with equality if and only if $a_n = a_{n-1}$ and $b_n = b_{n-1}$.
(d) If $W_1 a_1 < W_2 a_2 < \cdots, W_1/w_1 < W_2/w_2 < \cdots$ then
$$\mathfrak{G}_n(\mathfrak{A}^{-1}; \underline{w}) \geq \mathfrak{A}_n(\mathfrak{G}^{-1}; \underline{w}),$$
with equality if and only if $a_{n-2} = a_{n-1} = a_n$.

COMMENTS (i) Inequalities (a), (b) and (c) are analogous to (GA), **Geometric Mean Inequalities** (1) and (Č) respectively.
(ii) (a) is easily seen to be equivalent to (B), and was used by Nanjundiah to give a simultaneous proof of (GA), **Popoviciu's' Geometric-Arithmetic Mean Inequality Extension** (1) and **Rado's Geometric-Arithmetic Mean Inequality Extension** (1). .
(iii) (b) has been used to prove a Popoviciu-type extension of **Power Mean Inequalities** (4).
(iv) (c) can be used to give a Rado-type extension of (Č).
(v) (d) is **Nanjundiah's Inequalities** (2) with a change of notation

REFERENCES [Bullen, 1997].

Nanjundiah's Mixed Mean Inequalities

If $\mathfrak{A}, \mathfrak{G}$ are, respectively, the sequences of arithmetic, geometric, means of a sequence $\underline{a}$ with weight sequence $\underline{w}$ then
$$\mathfrak{G}_n(\mathfrak{A}; \underline{w}) \geq \mathfrak{A}_n(\mathfrak{G}; \underline{w}),$$

with equality if and only if $\underline{a}$ is constant.

COMMENTS (i) This follows from **Nanjundiah's Inverse Mean Inequalities** (d) applied to the sequence $\mathfrak{A}$ in place of $\underline{a}$.
(ii) The equal weight case of this result was shown by Nanjundiah to imply **Carleman's Inequality** (1).

EXTENSIONS If $-\infty < r < s < \infty$ and $\mathfrak{M}^{[r]}$, $\mathfrak{M}^{[s]}$ are, respectively, the sequences of r-th, s-th, means of a sequence $\underline{a}$ with weight sequence $\underline{w}$, and if $\{W_1 a_1, W_2 a_2, \ldots\}$ and $\{w_1^{-1} W_1, w_2^{-1} W_2, \ldots\}$ are strictly increasing then

$$\mathfrak{M}_n^{[r]}(\mathfrak{M}^{[s]}; \underline{w}) \geq \mathfrak{M}_n^{[s]}(\mathfrak{M}^{[r]}; \underline{w}).$$

COMMENTS Although these results were proved by Nanjundiah he has not published his proofs. An incorrect proof is given in [MI, pp. 121-122]; the correct proof of Nanjundiah is given in [Bullen, 1997]; other proofs can be found in the last three references.

REFERENCES [MI, pp.67, 96], [Bullen, 1997], [Kedlaya], [Matsuda], [Mond & Pečarić].

Nanson's Inequality If $\underline{a}$ is a convex positive $(2n+1)$-tuple and if

$$\underline{b} = \{a_2, a_4, \ldots, a_{2n}\}, \qquad \underline{c} = \{a_1, a_3, \ldots, a_{2n+1}\},$$

then

$$\mathfrak{A}_n(\underline{b}) \leq \mathfrak{A}_{n+1}(\underline{c}), \qquad (1)$$

with equality if and only if the elements of $\underline{a}$ are in arithmetic progression

COMMENTS (i) By taking $a_i = x^{i-1}, 1 \leq i \leq 2n+1, 0 < x < 1$, we get as a particular case Wilson's inequalities:

$$\frac{1 + x^2 + \cdots + x^{2n}}{x + x^3 + \cdots + x^{2n-1}} > \frac{n+1}{n}, \quad \text{equivalently} \quad \frac{1 - x^{n+1}}{n+1} > \frac{1 - x^n}{n} \sqrt{x}. \qquad (2)$$

(ii) Inequality (2) has been improved; see [AI].

EXTENSIONS (a) [STEINIG] Under the above assumptions

$$\mathfrak{A}_n(\underline{b}) \leq \mathfrak{A}_{2n+1}(\underline{a}) \leq \mathfrak{A}_{n+1}(\underline{c}) \leq \sum_{i=1}^{2n+1} (-1)^{i+1} a_i.$$

(b) [ANDRICA, RAŞA & TOADER] If $m \leq \Delta^2 \underline{a} \leq M$ then

$$\frac{2n+1}{6} m \leq \mathfrak{A}_{n+1}(\underline{c}) - \mathfrak{A}_n(\underline{b}) \leq \frac{2n+1}{6} M.$$

COMMENTS (iii) Extension (b) follows because the hypothesis implies the convexity of the sequences $\alpha_i = a_i - mi^2/2$, $\beta_i = Mi^2/2 - a_i, i = 1, 2, \ldots$.

REFERENCES [AI, pp. 205–206], [HLP, p. 99], [MI pp. 130–131], [PPT, pp. 247–251].

Nash's Inequality Let $f : \mathbb{R}^n \to \mathbb{R}$ be continuously differentiable then

$$\left(\int_{\mathbb{R}^n} |f|^2\right)^{(n+2)/n} \leq A_n \int_{\mathbb{R}^n} |\nabla f|^2 \left(\int_{\mathbb{R}^n} |f|\right)^{4/n}.$$

COMMENTS A sharp value for the constant A_n has been determined.

REFERENCES [Carlen & Loss].

n-Convex Function Inequalities (a) If f is n-convex on interval $[a, b]$ and if $x_0, \ldots, x_n$ are any $(n+1)$ distinct points from that interval then

$$[x; f] = [x_0, \ldots, x_n; f] = \sum_{i=0}^{n} \frac{f(x_i)}{w'(x_i)} \geq 0; \tag{1}$$

where

$$w(x) = w_n(x; \underline{x}) = \prod_{i=0}^{n} (x - x_i);$$

if f is strictly n-convex then (1) is strict for all choices of such $\underline{x}$.

(b) If f is n-convex on $[a, b]$ and if $x_0, \ldots, x_{n-1}, y_0, \ldots, y_{n-1}$ are two sets of distinct points in $[a, b]$ with $x_i \leq y_i, 0 \leq i \leq n-1$, then

$$[x_0, \ldots, x_{n-1}; f] \leq [y_0, \ldots, y_{n-1}; f].$$

(c) [PEČARIĆ & ZWICK] If f is $n+2$-convex on $[a, b]$, if $\underline{a} \prec \underline{b}$, where $\underline{a}, \underline{b}$ are $n+1$-tuples with elements in $[a, b]$ then

$$[\underline{a}; f] \leq [\underline{b}; f].$$

COMMENTS (i) (a) is just the definition of n-convexity and strict n-convexity. In addition the definition of a n-concave function, strictly n-concave function, is one for which (~ 1) holds, strictly.

(ii) The quantity on the left-hand side of (1) is called the n-th divided difference of f at the points $x_0, \ldots, x_n$. It is sometimnes written $[\underline{x}]f = [x_0, \ldots, x_n]f$.
If $n = 2$, $[\underline{x}; f]$ is just the left-hand side of **Convex Function Inequalities** (4), and if $n = 1$ it is the elementary Newton ratio, $(f(x_0) - f(x_1))/(x_0 - x_1)$.

(iii) Clearly 2-convex functions are just convex functions; 1-convex functions are increasing functions and 0-convex functions are just non-negative functions.

(iv) If $n \geq 1$ and $f^{(n)} \geq 0$ then f is n-convex, and if $f^{(n)} > 0$, except possibly at a finite number of points, then f is strictly n-convex. So, for instance, the exponential function is strictly n-convex for all n, the logarithmic function is strictly n-convex if n is odd, but is strictly n-concave if n is even, while x^α is strictly n-convex if $\alpha > n-1$, $\alpha < 0$ and n is even, or if α is positive, not an integer an $k - [\alpha]$ is odd.

(v) The inequality in (b) just says that if f is n-convex then the $n-1$-th divided difference is an increasing function of each of its variables; in the case $n = 2$ this is just **Convex Function Inequalities** (5).

(vi) The results **Convex Function Inequalities** DERIVATIVE INEQUALITIES extend to n-convex functions if the derivative of order $n-1$ is interpreted in the appropriate way.

(vii) (c) is an example of **Schur Convex Function Inequalities** (b).

(viii) See also **Čakalov's Inequality** COMMENTS (iv), **Farwig & Zwick's Lemma, Hermite-Hadamard Inequality** COMMENTS (iii), **Levinson's Inequality, Quasi-Convex Function Inequalities** COMMENTS (v).

REFERENCES [MI, pp. 32–33], [MPF, pp. 4–5], [PPT, pp. 14–18, 76]; [General Inequalities, vol.3, pp. 379–384], [Roberts & Varberg, pp. 237–240].

n-Convex Sequence Inequalities (a) If $\underline{a}$ is an n-convex real sequence, $n \geq 2$, then
$$\Delta^n \underline{a} \geq 0; \tag{1}$$
and if $\underline{a}$ is bounded then
$$\Delta^k \underline{a} \geq 0, \quad 1 \leq k \leq n-1; \tag{2}$$
in particular $\underline{a} \geq \underline{0}$.

(b) [OZEKI] If $\underline{a}$ in a real n-convex sequence so is $\mathfrak{A}_n(\underline{a})$.

COMMENTS (i) (1) is just the definition of an n-convex sequence.

(ii) In particular a sequence that is 1-convex is decreasing, and 2-convex is just convex, see **Convex Sequence Inequalities**. As usual if (1) holds strictly we say that $\underline{a}$ is *strictly n-convex*, while if (~ 1) holds, strictly, we say that *the sequence is n- concave, strictly n-concave*.

(iii) If the function f is such that $(-1)^n f$ is n-convex then if $a_i = f(i), i = 1, 2, \ldots$ the sequence $\underline{a}$ is n-convex.

(iv) (b) is a generalization of **Convex Sequence Inequalities** (d).

EXTENSIONS [VASIĆ, KEČKIĆ, LACKOVIĆ & MITROVIĆ, SIMIĆ] If $\underline{a}$ in a real n-convex sequence then so is the sequence $\mathfrak{A}_n(\underline{a}; \underline{w})$ if and only if $\underline{w}$ is given by $w_n = w_0 \binom{\alpha+n+1}{n}, n \geq 1$, for some positive real numbers w_0, α.

COMMENTS (v) See also **Čebišev's Inequality** COMMENTS (v).

REFERENCES [MI, pp. 8–12], [PPT, pp. 21, 253–257, 277–279].

Newton's Inequalities If $n \geq 2$, $c_0 c_n \neq 0$ and if all the zeros of $\sum_{i=1}^n c_i x^i = \sum_{i=1}^n \binom{n}{i} d_i x^i$ are real then for $1 \leq i \leq n-1$,

$$d_i^2 \geq d_{i-1} d_{i+1} \quad \text{and} \quad c_i^2 > c_{i-1} c_{i+1}; \tag{1}$$

$$d_i^{1/i} \geq d_{i+1}^{1/i+1} \quad \text{and} \quad c_i^{1/i} > c_{i+1}^{1/i+1}. \tag{2}$$

The inequalities on the left are strict unless the zeros are all equal.

COMMENTS (i) The right inequalities are weaker than the corresponding left inequalities.
(ii) The first inequality in (2) follows by writing the first inequality in (1) as $d_k^{2k} \geq d_{k-1}^k d_{k+1}^k$ for all k, $1 \leq k \leq i < n$, and multiplying; the second inequality in (2) follows similarly.
(iii) By writing the c_i in terms of the roots of the polynomial in the above results leads to an important inequality for elementary symmetric functions and means; see **Notations 3** (1), and **Elementary Symmetric Function Inequalities** (1), (2).
(iv) The particular cases of (2) $d_1 \geq d_n^{1/n}$, $c_1 > c_n^{1/n}$ are the equal weight case of (GA).
(v) Various extensions of these results have been given by Mitrinović, see [AI].

REFERENCES [AI, pp. 95–96], [HLP, pp. 51–55, 104–105], [MI, pp. 3–4].

N-function Inequalities
If $p : [0, \infty[\to [0, \infty[$ is right continuous, increasing, $p(0) = 0, p(t) > 0, t > 0$, with $\lim_{t \to \infty} p(t) = \infty$ we will say that $p \in \mathcal{P}$. Then if $q(s) = \sup_{p(t) \leq s} t$, $s \geq 0$, $q \in \mathcal{P}$ and is called the *right inverse of p*.
A function M is called an *N-function* if it can be written as

$$M(x) = \int_0^{|x|} p, \quad x \in \mathbb{R},$$

for some $p \in \mathcal{P}$; p is the right derivative of M. If then q is the right inverse of p the N-function

$$N(x) = \int_0^{|x|} q, \, x \in \mathbb{R},$$

is called the *complementary N- function* to M.

YOUNG'S[1] INEQUALITY If M, N are complementary N-functions then for all $u, v \in \mathbb{R}$

$$uv \leq M(u) + N(u),$$

---
[1] This is W H Young.

with equality if and only if either $u = q(v)$ or $v = p(u)$, for $u, v \geq 0$, where p, q are the right derivatives of M, N respectively.

COMMENTS   (i) If M, N are complementary N-functions and K is a compact set in $\mathbb{R}^n$, define
$$||f||_M = \sup \left| \int_K fg \right|,$$
where the sup is over all functions g for which
$$\int_K N \circ g \leq 1.$$
Then $|| \cdot ||_M$ satisfies the **Norm Inequalities** (1).
(ii) Particular cases of N-functions give the standard $|| \cdot ||_p$.

REFERENCES   [EM, vol.7, pp. 19–20], [PPT, pp. 241–242], [MPF, p. 382]; [Krasnocel'skiĭ & Rutickiĭ].

Nikol'skiĭ's[2] Inequality

If T_n is a trigonometric polynomial of degree at most n, and if $1 \leq p < q < \infty$ then
$$||T_n||_{q,[-\pi,\pi]} \leq A n^{(p^{-1}-q^{-1})} ||T_n||_{p,[-\pi,\pi]},$$
where A is a constant.

COMMENTS   For a definition of trigonometric polynomial of degree at most n see **Trigonometric Polynomial Inequalities**.

REFERENCES   [Zygmund, vol.I, p. 154].

Nirenberg's Inequality
See **Friederichs's Inequality** COMMENTS (ii), **Sobolev's Inequalities** (b).

Normal Distribution Function Inequalities
See **Error Function Inequalities**.

Norm Inequalities
If X is any vector space and $|| \cdot || : X \to \mathbb{R}$ is a norm on X then for all x, y in X,

$$||x + y|| \leq ||x|| + ||y||; \qquad (1)$$
$$||x - z|| \leq ||x - y|| + ||y - z||. \qquad (2)$$

---

[2] С М Никольский. Also transliterated as Nikolsky.

COMMENTS (i) A *norm* on a real, (complex), vector space X is a function $||\cdot|| : X \to [0, \infty[$ that is positive except for $x = 0$, $||0|| = 0$, and satisfies $||\lambda x|| = |\lambda| \, ||x||$, $\lambda \in \mathbb{R}$, ($\mathbb{C}$), and (1). Then X is called a *normed space*. A complete normed space is called a *Banach space*. If only $\lambda \geq 0$ required above then $||\cdot||$ is a *generalized norm*. See also **Banach Algebra Inequalities**.

(ii) Given that $||0|| = 0$, (1) and (2) are equivalent; (2) is a generalization of (T).

(iii) It follows from **Absolute Value Inequalities**, or (T), that $|\cdot|$ is a norm on $\mathbb{R}^n$. More generally if $p > 1$ then by (M), $||\cdot||_p$ is also a norm on $\mathbb{R}^n$; of course $||\underline{a}||_2 = |\underline{a}|$.

(iv) Other examples of a normed spaces are inner product spaces, see **Inner Product Inequalities** COMMENTS (iv), and the *Hardy spaces*, **Analytic Function Inequalities** COMMENTS (i).

(v) If zero values are allowed for non-zero elements we have what is called a *semi-norm*; so for instance the space $\mathcal{L}^p([a,b]), p \geq 1$ has $||\cdot||_p$ as a semi-norm; equivalently it is a normed space if we identify functions that are equal almost everywhere.

(vi) If the norm satisfies the parallelogram identity, see **Inner product Inequalities** (3), then the Banach sapce is a *Hilbert space*.

EXTENSIONS (a) If X is a normed space with norm $||\cdot||$ and if $x_i \in X, 1 \leq i \leq n$, and if $\underline{w}$ is a positive n-tuple then

$$\left\|\sum_{i=1}^n w_i x_i\right\| \leq \sum_{i=1}^n w_i ||x_i||. \tag{3}$$

If $\underline{w}$ has $w_1 > 0$ and $w_i < 0, 2 \leq i \leq n$ then ($\sim$3) holds.

(b) [PEČARIĆ & DRAGOMIR] If X is a normed space with norm $||\cdot||$ and if $x, y \in X$, and if $0 \leq t \leq 1$ then

$$\left\|\frac{x+y}{2}\right\| \leq \int_0^1 ||\overline{1-t}x + ty|| \, dt \leq \frac{||x|| + ||y||}{2}.$$

(c) If X is a normed space with norm $||\cdot||$ and if $x, y \in X$, and if $p, q, r \in \mathbb{R}$ with $pq(p+q) > 0, r \geq 1$ then

$$\frac{||x+y||^r}{p+q} \leq \frac{||x||^r}{p} + \frac{||y||^r}{q}; \tag{4}$$

if $pq(p+q) < 0$ then ($\sim$4) holds.

COMMENTS (vii) Inequalities (3) and ($\sim$3) should be compared to (J) and ($\sim$J); see **Jensen's Inequality** (a), (b). In particular by taking $n = 2, W_2 = 1$ in (3) we see that $||\cdot|| : X \to \mathbb{R}$ is convex. So (b) is a special case of **Hermite-Hadamard Inequality** (1).

(viii) For a particular case of (4) see **Complex Number Inequalities** (4).

(ix) See also **Dunkl & Williams Inequality, Hajela's Inequality, Matrix Norm Inequalities, Metric Inequalities, N-functions Inequalities** COMMENTS (i).

REFERENCES   [*EM, vol. 6, pp. 459–460*], [*MPF, pp. 483–594*].

n-Simplex Inequality

If R is the outer radius, and r the inner radius of an n-simplex then

$$R \geq nr.$$

COMMENTS   For definitions of the radii see **Isodiametric Inequalities** COMMENTS (ii).

REFERENCES   [*EM, Supp., p. 469*]; [*Mitrinović, Pečarić & Volenec*].

Number Theory Inequalities

If $x > 0$ we define: $\psi(x) = \sum_{n \leq x} \Lambda(n)$, where $\Lambda(n) = \log p$ if n is a power of the prime p, and otherwise it is 0; and $\pi(x) =$ the number of primes p with $p \leq x$. Then

$$\frac{\psi(x)}{x} \leq \frac{\pi(x) \log x}{x} < \frac{1}{\log x} + \frac{\psi(x) \log x}{x \log(x/\log^2 x)}.$$

COMMENTS   (i) As usual $\log^2 x = \log \circ \log x$.

(ii) This inequality is fundamental in proving the *Prime Number Theorem*,

$$\lim_{x \to \infty} \frac{\pi(x) \log x}{x} = 1.$$

REFERENCES   [*Rudin, 1973, pp. 212–213*]; [*Mitrinović & Popadić*].

- end of N -

O

Opial's Inequalities (a) If f is absolutely continuous on $[0, h]$ with $f(0) = 0$ then

$$\int_0^h |ff'| \le \frac{h}{2} \int_0^h f'^2, \tag{1}$$

with equality if and only if $f(x) = cx$.
The constant is best possible.

(b) If $\underline{a}$ is a non-negative $(2n+1)$-tuple satisfying

$$a_{2k} \le \min\{a_{2k-1}, a_{2k+1}\}, \quad 1 \le k \le n,$$

and if $a_0 = 0$ then

$$\left(\sum_{k=0}^{n} (\Delta a_{2k})\right)^2 \ge \sum_{k=0}^{2n+1} (-1)^{k+1} a_k.$$

COMMENTS (i) Inequality (1), usually referred to as *Opial's inequality*, has been the object of much study, and many extensions can be found in the references. In particular there are extensions that allow higher derivatives, and to higher dimensions.

EXTENSIONS (a) [YANG] If f is absolutely continuous on $[a, b]$ with $f(a) = 0$ and if $r \ge 0, s \ge 1$ then

$$\int_a^b |f|^r |f'|^s \le \frac{s}{r+s} (b-a)^r \int_a^b |f'|^{r+s}.$$

(b) [PACHPATTE] Let $f \in C^1(I)$, where $I = \{\underline{x}; \underline{a} \le \underline{x} \le \underline{b}\}$ is an interval in $\mathbb{R}^n$, with f zero on ∂I, then if $r, s \ge 1$,

$$\int_I |f|^r |\nabla f|^s \le M \int_I |\nabla f|^{r+s},$$

where

$$M = \frac{1}{n2^r} \left(\sum_{i=1}^{n} (b_i - a_i)^{r(r+s)/s}\right)^{s/(r+s)}.$$

(c) [FITZGERALD] If $f \in C^2([0, h])$ with $f(0) = f'(0) = 0$ then

$$\int_0^h |ff'| \le \frac{h^3}{4\pi^2} \int_0^h f''^2. \tag{1}$$

COMMENTS (ii) Extensions of (a) in which different weight functions are allowed in each integral have been given by Beesack & Das.

(iii) (c) can be obtained by combining (1) and **Wirtinger's Inequality**; as a result the constant is not best possible. A sharp result involving even higher order derivatives can be found in [General Inequalities].

DISCRETE ANALOGUES (a) If $\underline{a}$ is a real n-tuple and if $a_0 = 0$ then

$$\sum_{i=0}^{n-1} a_{i+1}|\Delta a_i| \leq \frac{n+1}{2} \sum_{i=0}^{n-1} (\Delta a_i)^2.$$

(b) [LEE[1]] If $\underline{a}$ is a non-negative, increasing n-tuple with $a_0 = 0$, $r, s > 0$, $r + s \geq 1$ or $r, s < 0$ then

$$\sum_{i=1}^{n} a_i^r (\tilde{\Delta} a_i)^s \leq K_{n,r,s} \sum_{i=1}^{n} (\tilde{\Delta} a_i)^{r+s},$$

where

$$K_{0,r,s} = \frac{s}{r+s}, \quad K_{n,r,s} = \max\{K_{n-1,r,s} + \frac{rn^{r-1}}{r+s}, \frac{s(n+1)^r}{r+s}\}.$$

COMMENTS (iv) Other discrete analogues of (1) have been given by Pachpatte.

REFERENCES [AI, pp. 154–162, 351]; [General Inequalities, vol. 4, pp. 25–36; vol. 7, pp. 157–178], [PPT, p. 162]; [Lee], [Pachpatte 1987, 1987[(2),(3)]].

Oppenheim's Inequality Let $\underline{a}, \underline{b}, \underline{w}$ be three positive n-tuples with $\underline{a}, \underline{b}$ increasing and satisfying

$$a_i \leq b_i, \ 1 \leq i \leq n-m, \qquad a_i \geq b_i, \ n-m+2 \leq i \leq n,$$

for some $m, 1 < m < n$. Then if $t < s$, $s, t \in \mathbb{R}$,

$$\mathfrak{M}_n^s(\underline{a}, \underline{w}) \leq \mathfrak{M}_n^s(\underline{b}, \underline{w}) \implies \mathfrak{M}_n^t(\underline{a}, \underline{w}) \leq \mathfrak{M}_n^t(\underline{b}, \underline{w}).$$

COMMENTS The case $n = 3, s = 1, t = 0, w_1 = w_2 = w_3$ is the original result of Oppenheim.

REFERENCES [AI, pp. 309–310], [MI, pp. 273–279].

---

[1] This is C M Lee.

Order Inequalities (a) [RADO] $\underline{a} \prec \underline{b}$ if and only if there is a doubly stochastic matrix S such that $\underline{a} = \underline{b}S$.

(b) [KARAMATA] If $\underline{a}, \underline{b} \in I^n$, I an interval in $\mathbb{R}$, then $\underline{a} \prec \underline{b}$ if and only if for all functions f, convex on I,
$$\sum_{i=1}^{n} f(a_i) \leq \sum_{i=1}^{n} f(b_i).$$

(c) [WARD] If $\underline{a}, \underline{b}$ are n-tuples of non-negative integers, with $\underline{a} \prec \underline{b}$ and if $\underline{x}$ is a positive n-tuple then
$$\sum ! x_{i_1}^{a_1} x_{i_2}^{a_2} \ldots x_{i_n}^{a_n} \leq \sum ! x_{i_1}^{b_1} x_{i_2}^{b_2} \ldots x_{i_n}^{b_n}.$$

COMMENTS (i) A matrix is *doubly stochastic* if its entries are non-negative, and have all row and column sums equal to 1.

(ii) (b) is a fundamental property both of the order and of convex functions.

(iii) If $a_i = \cdots = a_n = \mathfrak{A}_n(\underline{b})$ then (b) reduces to the equal weight case of (J).

(iv) When for some doubly stochastic S, $\underline{a} = \underline{b}S$, as in (a), we sometimes say that $\underline{a}$ is an *average of* $\underline{b}$. For this equality it is both necessary and sufficient that $\underline{a}$ lie in the convex hull of the $n!$ points obtained by permuting the elements of $\underline{b}$.

(v) The expressions on both sides of Ward's inequality are symmetric polynomials that are homogeneous of order $A_n (= B_n)$; (for definitions of these terms see **Segre's Inequalities**). If $\underline{a} = \{1, 1, \ldots, 1\}, \underline{b} = \{n, 0, \ldots, 0\}$ then the inequality is just the equal weight case of (GA).

EXTENSIONS [FUCHS] If $\underline{a}, \underline{b} \in I^n$, I an interval in $\mathbb{R}$, and $\underline{w}$ is a real n-tuple then
$$\sum_{i=0}^{n} w_i f(a_i) \leq \sum_{i=0}^{n} w_i f(b_i)$$
for every convex function f if and only if $\underline{a}, \underline{b}$ are decreasing and
$$\sum_{i=0}^{k} w_i a_i \leq \sum_{i=0}^{k} w_i b_i, 1 \leq k < n, \quad \sum_{i=0}^{n} w_i a_i = \sum_{i=0}^{n} w_i b_i.$$

INTEGRAL ANALOGUES [KY FAN & LORENTZ] $\alpha \prec \beta$ on $[a, b]$ if and only if for all convex functions f
$$\int_a^b f \circ \alpha \leq \int_a^b f \circ \beta.$$

COMMENTS (vi) This concept is used to prove many particular inequalities; see **Absolutely and Completely Monotonic Function Inequalities** (c), **Barnes's Inequalities** (b), **Chong's Inequalities** (b), **Muirhead Symmetric Function and Mean Inequalities** (1), **Permanent Inequalities** (c), **Schur Convex Function Inequalities** (b), **Shannon's Inequality** (b), **Steffensen's Inequalities** COMMENTS (iii), **Walker's Inequality** COMMENTS, **Weierstrass's Inequality** COMMENTS (ii).

REFERENCES [AI, pp. 162–170], [BB, pp. 30–33], [EM, vol. 6, pp. 74–76], [MI, pp. 18–21], [MO, pp. 21–23, 108–115], [PPT, pp. 319–332]; [General Inequalities, vol. 4, pp. 41–46]; [Ward].

O'Shea's Inequality If $a_1 \geq \cdots \geq a_n > 0$, $a_1 \neq a_n$, and if $a_{n+k} = a_k, 1 \leq k \leq n$, then

$$\sum_{i=1}^{n} \left(a_i^k - \prod_{j=1}^{k} a_{i+j} \right) \geq 0; \tag{1}$$

further (1) is strict if $k > 1$.

COMMENTS   With $k = n$ this inequality is just (GA).

REFERENCES   [MI, pp. 78–79].

Ostrowski's Inequalities   (a) If $\underline{a}, \underline{b}, \underline{c}$ are real n-tuples with $\underline{a} \not\sim \underline{b}$, $\underline{a}.\underline{c} = 0$ and $\underline{b}.\underline{c} = 1$, then

$$\frac{|\underline{a}|^2}{|\underline{a}|^2 |\underline{b}|^2 - |\underline{a}.\underline{b}|^2} \leq |\underline{c}|^2. \tag{1}$$

with equality if and only if

$$c_i = \frac{b_i |\underline{a}|^2 - a_i \, \underline{a}.\underline{b}}{|\underline{a}|^2 |\underline{b}|^2 - |\underline{a}.\underline{b}|^2}, \ 1 \leq i \leq n.$$

(b) Let f be bounded on $[a,b]$ with $\alpha \leq f \leq A$, and assume that g has a bounded derivative on $[a,b]$ then

$$\left| \mathfrak{A}_{[a,b]}(fg) - \mathfrak{A}_{[a,b]}(f) \mathfrak{A}_{[a,b]}(g) \right| \leq \frac{1}{8}(b-a)(A-\alpha) \|g'\|_{\infty, [a,b]}.$$

The constant $1/8$ is best possible.

(c) If $f \in \mathcal{C}^1([a,b])$ then

$$\left| f(x) - \mathfrak{A}_{[a,b]}(f) \right| \leq \frac{(x-a)^2 + (b-x)^2}{2(b-a)} \|f'\|_{\infty, [a,b]}.$$

The function on the right-hand side cannot be replaced by a smaller function.

COMMENTS   (i) Inequality (1) can be regarded as a special case of **Bessel's Inequality** for non-orthonormal vectors. The result also holds for complex n-tuples if throughout the inner product is replaced by $\underline{u}.\overline{\underline{v}}$.

(ii) (b) is converse of (Č), or of **Power Mean Inequalities** (2), in the case $q = r = s = 1$; see also **Grüsses' Inequalities** (a).

EXTENSIONS (a) [KY FAN & TODD] If $\underline{a}, \underline{b}$ are real n-tuples with $a_i b_j \neq a_j b_i$, $i \neq j$, $1 \leq i, j \leq n$, then

$$\frac{|\underline{a}|^2}{|\underline{a}|^2 |\underline{b}|^2 - |\underline{a}.\underline{b}|^2} \leq \binom{n}{2}^{-2} \sum_{i=1}^n \left(\sum_{j \neq i, j=1}^n \frac{a_j}{a_j b_i - a_i b_j} \right)^2. \tag{2}$$

(b) [ANASTASSIOU] If $f \in C^n([a,b])$, $n \geq 2$ with $f^{(k)}(x) = 0, 1 \leq k \leq n-1$, then

$$|f(x) - \mathfrak{A}_{[a,b]}(f)| \leq \frac{(x-a)^{n+1} + (b-x)^{n+1}}{(n+1)!(b-a)} \sup_{a \leq x \leq b} |f^{(n)}(x)|.$$

The function on the right-hand side cannot be replaced by a smaller function.

COMMENTS (iii) Inequality (2) gives Chassan's inequality on putting $a_i = \sin \alpha_i$ and $b_i = \cos \alpha_i$, $0 \leq \alpha_i \leq \pi$, $\alpha_i \neq \alpha_j$, $i \neq j$, $1 \leq i, j \leq n$.

(iv) Further extensions can be found in the references. Other inequalities due to Ostrowski are in **Matrix Norm Inequalities, Schur Convex Function Inequalities** (b), **Trigonometric Integral Inequalities**.

REFERENCES [AI, pp. 66–70], [MI, pp. 157–158], [MPF, pp. 92–95], [PPT, pp. 209–210]; [Anastassiou], [Fink].

Ozeki's Inequalities

(a) If $\underline{a}$ is a real sequence then

$$\sum_{i=1}^{n-1} a_i a_{i+1} \leq a_n a_1 + \cos\frac{\pi}{n} \sum_{i=1}^n a_i^2. \tag{1}$$

(b) If $\underline{a}$ is a real n-tuple and $p > 0$ then

$$\min_{i \neq j} |a_i - a_j|^p \leq C_{n,p} \min_{x \in \mathbb{R}} \sum_{i=1}^n |a_i - x|^p$$

where

$$C_{n,p}^{-1} = \begin{cases} 2 \sum_{i=1}^{(n-1)/2} i^p, & \text{if } n \text{ is odd}, \\ \min\{1, 2^{1-p}\} \sum_{i=1}^{n/2} (2i-1)^p, & \text{if } n \text{ is even}. \end{cases}$$

COMMENTS (i) (1) can be used to obtain a discrete version of **Wirtinger's Inequality**.

(ii) For other inequalities by Ozeki see **Complete Symmetric Function Inequalities** (c), **Convex Sequence Inequalities** (d), **Elementary Symmetric Function Inequalities** EXTENSIONS (f), **n-Convex Sequence Inequalities** (b).

REFERENCES [AI, pp. 202, 340–341], [MI, pp. 127–128, 212], [MPF, pp. 438–439], [PPT, p. 277]; [General Inequalities, vol. 4, pp. 83–86].

- end of O -

P

Pachpatte's Series Inequalities If $\underline{a}$ is a non-negative sequence, and $p, q, r \geq 1$ then

$$\sum_{k=1}^{n} a_k A_k \leq \frac{n+1}{2} \sum_{k=1}^{n} a_k^2,$$

$$\sum_{k=1}^{n} A_k^{p+q} \leq ((p+q)(n+1))^q \sum_{k=1}^{n} a_k^q A_k^p,$$

$$\sum_{k=1}^{n} a_k^r A_k^{p+q} \leq ((p+q+r)(n+1))^q \sum_{k=1}^{n} a_k^{q+r} A_k^p.$$

COMMENTS These are proved using (H) and **Davies & Petersen's Inequality**.

REFERENCES [Pachpatte, 1996].

Padoa's Inequality
See **Adamović's Inequality** COMMENTS (ii).

Paley's Inequalities (a) If $1 < p \leq 2$, $f \in \mathcal{L}^p(a,b)$ and if $\phi_n, n \in \mathbb{N}$ is a uniformly bounded orthonormal sequence of complex valued functions defined on an interval $[a,b]$, with $\sup_{a \leq x \leq b} |\phi_n(x)| \leq M, n \in \mathbb{N}$, and if $\underline{c}$ is the sequence of Fourier coefficients of f with respect to $\phi_n, n \in \mathbb{N}$, then

$$\left(\sum_{n \in \mathbb{N}} |c_n|^p n^{p-2} \right)^{1/p} \leq A_p M^{(2-p)/p} \|f\|_{p,[a,b]}.$$

(b) If $\phi_n, n \in \mathbb{N}$, is as in (a), $q \geq 2$ and $\left(\sum_{n \in \mathbb{N}} |c_n|^q n^{q-2} \right)^{1/q} < \infty$, $\underline{c} = \{c_n, n \in \mathbb{N}\}$ a complex sequence, then there is a function $f \in \mathcal{L}^q(a,b)$ having $\underline{c}$ as its sequence of Fourier coefficients with respect to $\phi_n, n \in \mathbb{N}$, and,

$$\|f\|_{q,[a,b]} \leq B_q M^{(q-2)/q} \left(\sum_{n \in \mathbb{N}} |c_n|^q n^{q-2} \right)^{1/q}.$$

COMMENTS (i) The constants A_p, B_q can be taken so that if p, q are conjugate indices then $A_p = B_q$.

(ii) These results can be even further extended by replacing the sequence $\underline{c}$ by the decreasing rearrangement $\underline{c}^*$. That this strengthens the results follows from **Rearrangement Inequalities** (1). The proof consists in simultaneously rearranging the orthonormal sequence.
(iii) The two parts of all these results are equivalent in that either implies the other.
(iv) The case $p = 2$ of part (a) of the above results is Bessel's Inequality, see **Bessel's Inequalities** (b).
(v) These results extend the **Hausdorff-Young Inequalities**. Integral analogues are given in **Fourier Transform Inequalities** EXTENSIONS (a).
REFERENCES   [MPF, p. 398]; [Zygmund, vol. II, pp. 120–127].

Paley-Titchmarsh Inequality
See **Fourier Transform Inequalities** EXTENSIONS (a).

Parallelogram Inequality
If $\underline{a}, \underline{b}, \underline{c}, \underline{d}$ are real n-tuples then

$$|\underline{a} - \underline{b}|^2 + |\underline{b} - \underline{c}|^2 + |\underline{c} - \underline{d}|^2 + |\underline{d} - \underline{a}|^2 \geq |\underline{a} - \underline{c}|^2 + |\underline{d} - \underline{b}|^2, \tag{1}$$

with equality if and only if the n-tuples are co-planar and form a parallelogram in the correct order.

COMMENTS   (i) In the case that they form a parallelogram (1) becomes, putting $\underline{a} = \underline{0}$, $\underline{b} = \underline{u}$ and $\underline{d} = \underline{v}$,

$$2(|\underline{u}|^2 + |\underline{v}|^2) = |\underline{u} + \underline{v}|^2 + |\underline{u} - \underline{v}|^2.$$

This is called the *parallelogram identity*, see **Inner Product Inequalities** COMMENTS (iii).
(ii) The geometrical interpretation of (1) is:

the sum of the squares of the lengths of the sides of the quadrilateral formed by the n-tuples exceed the squares of the lengths of the diagonals.

(iii) This result can be extended to 2^n n-tuples.
REFERENCES   [Gerber].

Permanent Inequalities
(a) [MARCUS & NEWMAN] If A is an $m \times n$ complex matrix, B an $n \times m$ complex matrix then

$$|\text{per}(AB)|^2 \leq \text{per}(AA^*)\text{per}(B^*B).$$

There is equality if and only if either A has a zero row, B has a zero column or $A = DPB^*$, where D is diagonal and P is a permutation matrix.
In particular if A is a complex square matrix

$$|\text{per}(A)|^2 \leq \text{per}(AA^*).$$

(b) [SCHUR] If A is a positive semi-definite Hermitian matrix then

$$\det(A) \leq \text{per}(A),$$

with equality if and only if either A is diagonal or A has a zero row.

(c) Let $\underline{w}$ be a positive n-tuple, $\underline{a}, \underline{b}$ be n-tuples of positive integers, and let A, B be $n \times n$ matrices with (i,j) entries $w_i^{a_j}, w_i^{b_j}$ respectively. Then

$$\mathrm{per}(A) \leq \mathrm{per}(B)$$

if and only if $\underline{a} \prec \underline{b}$. There is equality if and only if either $\underline{a} = \underline{b}$, or $\underline{w}$ is constant.

COMMENTS (i) (c) is a form of Muirhead's inequality, **Muirhead Symmetric Function and Mean Inequalities** (1).

(ii) See also **van der Waerden's Conjecture**.

REFERENCES [MPF, pp. 225–227]; [Marcus & Minc, p. 118], [Minc, pp. 9–10, 25].

Petrović's[1] Inequality Let f be convex on $[0, a]$ and $a_i \in [0, a], 0 \leq i \leq n$, with $\sum_{i=1}^{n} a_i \in [0, a]$, then

$$\sum_{i=1}^{n} f(a_i) \leq f\left(\sum_{i=1}^{n} a_i\right) + (n-1)f(0).$$

EXTENSIONS [VASIĆ & PEČARIĆ] Let f be convex on $[0, a]$, $\underline{w}$ be a non-negative n-tuple, $a_i \in [0, a[, 1 \leq i \leq n$, with $\sum_{i=1}^{n} w_i a_i \in [0, a]$ and $\sum_{i=1}^{n} w_i a_i \geq \max \underline{a}$; then

$$\sum_{i=1}^{n} w_i f(a_i) \leq f\left(\sum_{i=1}^{n} w_i a_i\right) + (W_n - 1)f(0).$$

COMMENTS See also **Szegö's Inequality**.

REFERENCES [AI, pp. 22–23], [MPF, pp. 11, 715]; [PPT, pp. 151–169].

Petschke's Inequality Let $f, g : [0,1] \to \mathbb{R}$ be concave, increasing and non-negative functions, and let $p, q \geq \lambda \geq 1$, where λ is a solution of $r + 1 = (3/2)^r$, then

$$\|f\|_{p,[0,1]} \|g\|_{q,[0,1]} \leq \frac{3}{(1+p)^{1/p}(1+q)^{1/q}} \|fg\|_{[0,1]}.$$

There is equality if and only if $f(x) = g(x) = x$.

COMMENTS (i) $\lambda \approx 3.939$.

(ii) Various other possible relations between p, q, λ are considered in the reference.

(iii) This is a converse for (H); for a similar result related to (M) see **Rahmail's Inequality**.

(iv) For another inequality of Petschke see **Moment Inequalities**.

REFERENCES [MPF, pp. 149–150].

---

[1] Also spelt Petrovich, Petrovitch.

Phragmén-Lindelöf Inequality Let $f : \mathbb{C} \to \mathbb{C}$ be continuous and bounded on $R = \{\alpha \leq \Re z \leq \beta\}$, and analytic on $\overset{\circ}{R}$, with

$$|f(\alpha + iy)| \leq A, \qquad |f(\beta + iy)| \leq B, \qquad y \in \mathbb{R}.$$

If then $z = x + iy \in \overset{\circ}{R}$

$$|f(z)| \leq A^{\ell(x)} B^{1-\ell(x)},$$

where

$$\ell(x) = \frac{\beta - x}{\beta - \alpha}.$$

There is equality if and only if $f(z) = A^{\ell(z)} B^{1-\ell(z)} e^{i\theta}$.

COMMENTS (i) Since this result connects the upper bounds of f on the three lines $\Re z = \alpha, x, \beta$ it is often called the *Three Lines Theorem*; it is a statement that this upper bound function is log-convex. See also **Hadamard's Three Circles Theorem** and **Hardy's Analytic Function Inequality**.

(ii) This result was used by Thorin to prove the **Riesz-Thorin Theorem**.

REFERENCES [EM, vol. 7, pp. 152–153]; [Conway, vol. I, pp. 131–133], [Pólya & Szegö, 1972, pp. 166-172], [Rudin, 1966, pp. 243–244], [Titchmarsh, 1939, pp. 176–187], [Zygmund, vol. II, pp. 93–94].

Picard-Schottky Theorem If f is analytic in D and if f does not take the values $0, 1$ then

$$\log |f(z)| \leq \left(7 + \log^+ |f(0)|\right) \frac{1 + |z|}{1 - |z|}.$$

REFERENCES [Ahlfors, 1973, pp. 19–20]

Pittenger's Inequalities If $0 < a \leq b$ and $r \in \mathbb{R}$, then

$$\mathfrak{M}_2^{[r_1]}(a, b) \leq \mathfrak{L}^{[r]}(a, b) \leq \mathfrak{M}_2^{[r_2]}(a, b),$$

where

$$r_1 = \begin{cases} \min \left\{ \dfrac{r+2}{3}, \dfrac{r \log 2}{\log(r+1)} \right\}, & \text{if } r > -1,\ r \neq 0, \\ \min \left\{ \dfrac{2}{3}, \log 2 \right\}, & \text{if } r = 0, \\ \min \left\{ \dfrac{r+2}{3}, 0 \right\}, & \text{if } r \leq -1, \end{cases}$$

and r_2 is defined as r_1 but with min replaced by max. There is equality if and only if $a = b$, or $r = 2, 1$ or $1/2$. The exponents r_1, r_2 are best possible.

COMMENTS (i) These inequalities follow from (r;s). Taking $r = -1$ gives Lin's inequality; see **Logarithmic Mean Inequalities** (2).

(ii) See also **Rado's Inequality**.

REFERENCES [MI, pp. 349-350].

Poincaré's Inequalities (a) If $1 \leq p < n$, $B = B_{\underline{x},r} = \{\underline{u}; |\underline{u} - \underline{x}| < r\} \subset \mathbb{R}^n$, and $f \in \mathcal{W}^{1,p}(B)$, then

$$\left(\frac{1}{|B|}\int_B |f - \tilde{f}|^{p^*}\right)^{1/p^*} \leq C_{n,p} r \left(\frac{1}{|B|}\int_B |\nabla f|^p\right)^{1/p},$$

where $\tilde{f} = \frac{1}{|B|}\int_B f$, and p^* is the Sobolev conjugate.
(b) If $f \in C^1(Q), Q = [0,a]^n$, then

$$\int_Q f^2 \leq \frac{1}{a^n}\left(\int_Q f\right)^2 + \frac{na^2}{2}\int_Q |\nabla f|^2.$$

COMMENTS (i) For a definition of $\mathcal{W}_p^1(B)$, and p^*, see **Sobolev's Inequalities**.
(ii) Both inequalities are related to **Friederichs's Inequality**; and a one dimensional analogue of (b) is **Wirtinger's Inequality**.

EXTENSIONS [PACHPATTE] If $f, g \in C^1(Q), Q = [0,a]^n$, then

$$\int_Q fg \leq \frac{1}{a^n}\left(\int_Q f\right)\left(\int_Q g\right) + \frac{na^2}{4}\int_Q (|\nabla f|^2 + |\nabla g|^2).$$

REFERENCES [Evans & Gariepy, pp. 141–142], [Mitrović & Žubrinić, pp. 184, 242], [Opic & Kufner, p. 2]; [Pachpatte, 1988[3]].

Poisson Kernel Inequalities
The quantity

$$P(r,x) = \frac{1}{2} + \sum_{i=1}^{\infty} r^i \cos it = \frac{1}{2}\frac{1-r^2}{1 - 2r\cos x + r^2},$$

is called the *Poisson kernel*.

(a)
$$P(r,x) > 0; \quad \frac{1}{2}\frac{1-r}{1+r} \leq P(r,x) \leq \frac{1}{2}\frac{1+r}{1-r}.$$

(b) If $0 \leq t \leq \pi$ then for some C,

$$P(x,r) \leq C\frac{1-r}{x^2 + (1-r)^2}.$$

In particular if $0 \leq r < 1$

$$P(r,x) \leq \frac{1}{1-r}; \quad P(r,x) \leq \frac{C(1-r)}{x^2}, \quad 0 < x \leq \pi.$$

COMMENTS These should be compared to the similar inequalities in **Dirichlet Kernel Inequalities, Fejér Kernel Inequalities**.
REFERENCES [Zygmund, vol. I, pp. 96–97].

Pólya's Inequality Let $f, g, h : [a, b] \to \mathbb{R}$, with f increasing, g, h continuously differentiable and $g(a) = h(a), g(b) = h(b)$ then

$$\int_a^b fg' \int_a^b fh' \leq \left(\int_a^b f\sqrt{(gh)'} \right)^2.$$

COMMENTS The case $a = 0, b = 1$, $g(x) = x^{2p+1}, h(x) = x^{2q+1}$ is due to Pólya; the generalization is by Alzer.

REFERENCES [Pólya & Szegö, 1972, p.72]; [Alzer, 1990$^{(2)}$].

Polyá & Szegö's Inequality If $0 < a \leq \underline{a} \leq A$, $0 < b \leq \underline{b} \leq B$ then

$$\left(\sum_{i=1}^n a_i^2 \right)^{1/2} \left(\sum_{i=1}^n b_i^2 \right)^{1/2} \leq \frac{1}{2} \left(\sqrt{\frac{AB}{ab}} + \sqrt{\frac{ab}{AB}} \right) \left(\sum_{i=1}^n a_i b_i \right),$$

with equality if and only if $\nu = nAb/(Ab + Ba)$ is an integer and if ν of the a_i are equal to a, and the others equal to A, with the corresponding b_i being B, b respectively.

COMMENTS (i) This inequality is a converse to (C) and is equivalent to **Kantorovič's Inequality**.

EXTENSIONS (a) [CASSELS] If $\underline{w}$ is a non-negative n-tuple with $W_n \neq 0$ then

$$\left(\sum_{i=1}^n w_i a_i^2 \right)^{1/2} \left(\sum_{i=1}^n w_i b_i^2 \right)^{1/2} \leq \max_{1 \leq i,j \leq n} \frac{1}{2} \left(\sqrt{\frac{a_i b_j}{a_j b_i}} + \sqrt{\frac{a_j b_i}{a_i b_j}} \right) \left(\sum_{i=1}^n w_i a_i b_i \right).$$

(b) [DIAZ & METCALF] If $\underline{a}, \underline{b}$ are real n-tuples with $a_i \neq 0, 1 \leq i \leq n$, and if $m \leq a_i/b_i \leq M, 1 \leq i \leq n$, then

$$\sum_{i=1}^n b_i^2 + mM \sum_{i=1}^n a_i^2 \leq (M + m) \sum_{i=1}^n a_i b_i,$$

with equality if and only if for all $i, 1 \leq i \leq n$, either $b_i = ma_i$ or $b_i = Ma_i$.

COMMENTS (ii) The inequalities of Cassels, Kantorovič and Schweitzer are all special cases of the very elementary inequality of Diaz & Metcalf, (b).
(iii) Cassel's inequality is often called the *Greub & Rheinboldt inequality*.
(iv) The result of Diaz & Metcalf is equivalent to **Rennie's Inequality**.

REFERENCES [AI, pp. 59–66], [HLP, pp. 62, 166], [BB, pp. 44–45], [MI, pp. 207–212], [PPT, pp. 114–115].

Polynomial Inequalities   (a) If $x \geq 0$, $x \neq 1$ and $n \geq 2$ then

$$x^n - nx + (n-1) > 0; \tag{1}$$
$$(x + n - 1)^n - n^n x > 0. \tag{2}$$

(b) If $0 \leq x \leq 1$ and $n \in \mathbb{N}$ then

$$0 \leq x^n(1-x)^n \leq 1. \tag{3}$$

(c) If $n \in \mathbb{N}$ and $x > 1$ then

$$x^n - 1 > n \left(\frac{x+1}{2}\right)^{n-1} (x-1), \tag{4}$$

while if $0 \leq x \leq 1$ then ($\sim$4) holds.
(d) [HERZOG] If $0 < x < 1$, $0 \leq \nu \leq n$, $n, \nu \in \mathbb{N}$, then

$$\binom{n}{\nu} x^\nu (1-x)^{n-\nu} < \frac{1}{2enx\sqrt{1-x}}.$$

COMMENTS   (i) The proofs of both (1) and (2) follow by noting that $x = 1$ is the only positive root of the polynomials involved. These inequalities imply certain cases of (B).
(ii) The polynomial in (3) and all of its derivative have integer valued derivatives at $0, 1$; it can be used to prove that e^n is irrational for all non-zero integers n.
(iii) Inequality (4) follows from **Haber's Inequality**.

EXTENSIONS   [HADŽIĬNOV& PRODANOV[2]] (A) IF n IS EVEN AND $x \neq 1$

$$x^n - nx + (n-1) > 0,$$

WHILE IF n IS ODD

$$x^n - nx + (n-1) > 0, \text{ if } x > x_n, \ x \neq 1,$$
$$x^n - nx + (n-1) < 0, \text{ if } x < x_n,$$

WHERE x_n, $-2 \leq x_n < -1 - (1/n)$, IS THE UNIQUE NON-ZERO ROOT OF $x^n - nx + (n-1) = 0$.
(B) [OSTROWSKI] IF $0 \leq x \leq 1$ AND $n \geq 1$ THEN

$$0 \leq x^{n-1}(1-x)^n \leq \frac{1}{2^{2(n-1)}}.$$

(C) [OSTROWSKI & REDHEFFER] IF $n, \nu \in \mathbb{N}$, $n \geq 2$, $0 \leq \nu \leq n$, $0 < x < 1$ AND $q = \nu/n$ THEN

$$\binom{n}{\nu} x^\nu (1-x)^{n-\nu} < \exp\left(-2n(x-q)^2\right). \tag{5}$$

---

[2] Н Хаджийнов, И Проданов.

COMMENTS (iv) Inequality (5) is sharp in the sense that the coefficient -2 in the exponential and the coefficient in front of the exponential cannot be improved. See, however, **Statistical Inequalities** (b).

(v) See also Bernšteĭn Polynomial Inequalities, Bernoulli's Inequality, Binomial Function Inequalities, Brown's Inequalities, Čebišev Polynomial Inequalities, Descartes' Rule of Signs, Erdós's Inequality, Erdös & Grünwald's Inequality, Integral Mean Value Theorems EXTENSIONS (c), Kneser's Inequality, Labelle's Inequality, Markov's Inequality, Newton's Inequalities, Polynomial Interpolation Inequalities, Pommerenke's Inequality, Shampine's Inequality, Turán's Inequalities.

REFERENCES [AI, pp. 35, 198, 200, 226], [BB, pp. 12–13], [HLP, pp. 40–42, 103], [MI, p. 4], [MPF, pp. 65, 581–582]; [Borwein & Borwein, p. 352], [General Inequalities, vol. 1, pp. 125–129, 307].

Polynomial Interpolation Inequalities

(a) Let $a \le a_i < a_2 < \cdots < a_n \le b$, and if $f \in C^n([a,b])$, has $f^{(j)}(a_i) = 0$, $1 \le j \le k_i$, $1 \le i \le r$, with $k_1 + k_2 + \cdots + k_r + r = n$, then

$$\|f^{(k)}\|_{\infty,[a,b]} \le (b-a)^{n-k} \alpha_{n,k} \|f^{(n)}\|_{\infty,[a,b]}, \quad 0 \le k \le n-1,$$

where $\alpha_{n,k} = 1/(n-k)!$, $0 \le k \le n-1$.

(b) [TUMURA] If as above, but $f \in C^n([a_1, a_r])$, then

$$\|f^{(k)}\|_{\infty,[a_1,a_r]} \le (a_r - a_1)^{n-k} \beta_{n,k} \|f^{(n)}\|_{\infty,[a_1,a_r]}, \quad 0 \le k \le n-1,$$

where $\beta_{n,0} = \dfrac{(n-1)^{n-1}}{n^n} \alpha_{n,0}$, and if $1 \le k \le n-1$, $\beta_{n,k} = \dfrac{k}{n} \alpha_{n,k}$.

COMMENTS (i) The constants $\beta_{n,k}$ are best possible being exact for the functions, $f(x) = (x - a_1)^{n-1}(a_r - x)$, and $f(x) = (x - a_1)(a_r - x)^{n-1}$. These functions are, up to a constant factor, the only functions for which (b) is exact.

(ii) These results have been generalized by Agarwal.

REFERENCES [General Inequalities, vol. 3, pp. 371–378].

Pommerenke's Inequality

If $p(z) = a_0 + a_1 z + \cdots + a_{n-1} z^{n-1} + z^n$ has some $a_k \ne 0$ and if $E = \{z; z \in \mathbb{C}, |f(z)| \le 1\}$ is connected then

$$\max\{|f'(z)|; z \in E\} \le \frac{en^2}{2}.$$

COMMENTS It was conjectured by Erdös that the right-hand side can be replaced by $n^2/2$.

REFERENCES [General Inequalities, vol. 7, pp. 401–402].

Popa's Inequality *If K is a convex set in $\mathbb{R}^3$ put $K_\epsilon = \{\underline{x}; d(\underline{x}, K) < \epsilon\}$, $K^{-\rho} = \{\underline{x}; B(\underline{x}, \rho) \subseteq K\}$. If then $f : [0, \infty[\to [0, \infty[$ is a C^1 function,*

$$\int_{K_\epsilon \setminus K^{-\rho}} f(|\underline{x}|)\,d\underline{x} \leq 4\pi(\epsilon + \rho) \int_0^\infty t^2 |f'(t)|\,dt.$$

COMMENTS $d(\underline{x}, K)$ and $B(\underline{x}, \rho)$ are defined by:

$$d(\underline{x}, K) = \inf\{t; t = |\underline{y} - \underline{x}|, \underline{y} \in K\}, \qquad B(\underline{x}, \rho) = \{\underline{y}; |\underline{y} - \underline{x}| < \rho\};$$

the *distance of $\underline{x}$ from K* and the *sphere of centre $\underline{x}$, radius ρ*, respectively.

REFERENCES   [Popa].

Popoviciu's Geometric-Arithmetic Mean Inequality Extension   *If $n \geq 2$ then*

$$\left(\frac{\mathfrak{A}_n(\underline{a}; \underline{w})}{\mathfrak{G}_n(\underline{a}; \underline{w})}\right)^{W_n} \geq \left(\frac{\mathfrak{A}_{n-1}(\underline{a}; \underline{w})}{\mathfrak{G}_{n-1}(\underline{a}; \underline{w})}\right)^{W_{n-1}}, \tag{1}$$

with equality if and only if $a_n = \mathfrak{A}_{n-1}(\underline{a}; \underline{w})$.

COMMENTS   (i) This is a multiplicative analogue of **Rado's Geometric Arithmetic Mean Inequality Extension**, and several proofs have been given.

(ii) Repeated application of (1) leads to

$$\left(\frac{\mathfrak{A}_n(\underline{a}; \underline{w})}{\mathfrak{G}_n(\underline{a}; \underline{w})}\right)^{W_n} \geq \cdots \geq \left(\frac{\mathfrak{A}_1(\underline{a}; \underline{w})}{\mathfrak{G}_1(\underline{a}; \underline{w})}\right)^{W_1} = 1,$$

which exhibits (1) as an extension of (GA), to which it is equivalent. Stopping the above applications one step earlier and remarking that the left-hand side is invariant under simultaneous permutations of $\underline{a}$ and $\underline{w}$ leads to an improvement of (GA):

$$\mathfrak{A}_n(\underline{a}; \underline{w}) \geq \max_{1 \leq i,j \leq n} \left\{\left(\frac{w_i a_i + w_j a_j}{w_i + w_j}\right)^{w_i + w_j} a_i^{w_i} a_j^{w_j}\right\}^{1/W_n} \mathfrak{G}_n(\underline{a}; \underline{w}).$$

In the case of equal weights this is,

$$\mathfrak{A}_n(\underline{a}) \geq \sqrt[n]{\max_{1 \leq i,j \leq n}\left\{\frac{1}{2} + \frac{1}{4}\left(\frac{\max \underline{a}}{\min \underline{a}} + \frac{\min \underline{a}}{\max \underline{a}}\right)\right\}} \mathfrak{G}_n(\underline{a}).$$

A similar discussion can be found in **Rado's Geometric Arithmetic Mean Inequality Extension** COMMENTS (i), **Jensen's Inequality** COROLLARIES.

(iii) Inequalities that extend $P/Q \geq 1$ to P/Q is a decreasing function of index sets are called *Popoviciu-type*, or *Rado-Popoviciu type*, inequalities; see for instance **Harmonic Mean Inequalities** (c), **Hölder's Inequality** EXTENSIONS (d), **Sierpinski's Inequalities** EXTENSIONS (a), **Symmetric Mean Inequalities** EXTENSIONS.

EXTENSIONS [FUNCTIONS OF INDEX SETS]   *Let π be the following function defined on the index sets,*

$$\pi(\mathcal{I}) = \left(\frac{\mathfrak{A}_\mathcal{I}(\underline{a}; \underline{w})}{\mathfrak{G}_\mathcal{I}(\underline{a}; \underline{w})}\right)^{W_\mathcal{I}};$$

then $\pi \geq 1$, is increasing, and $\log \circ \pi$ is super-additive.

REFERENCES   [MI, pp. 94–105].

Power Mean Inequalities (a) If $-\infty \leq r < s \leq \infty$ then:

$$\mathfrak{M}_n^{[r]}(\underline{a};\underline{w}) \leq \mathfrak{M}_n^{[s]}(\underline{a};\underline{w}), \qquad (r;s)$$

with equality if and only if $\underline{a}$ is constant.
(b) If $q, r, s \in \overline{\mathbb{R}}$ with $s \leq q$ and $s \leq r$ and if

$$\frac{1}{q} + \frac{1}{r} \leq \frac{1}{s}; \qquad (1)$$

then

$$\mathfrak{M}_n^{[s]}(\underline{a}\,\underline{b};\underline{w}) \leq \mathfrak{M}_n^{[q]}(\underline{a};\underline{w})\mathfrak{M}_n^{[r]}(\underline{b};\underline{w}). \qquad (2)$$

If $q \leq s$ and $r \leq s$ ($\sim$2) holds.
Inequality (2) is strict except under the following circumstances:
(i) $q, r, s \neq 0, \pm\infty$, (1) strict, $\underline{a}$ and $\underline{b}$ constant; (ii) $q, r, s \neq 0, \pm\infty$, (1) is equality, and, $\underline{a}^q \sim \underline{b}^r$; (iii) $s \neq 0, \pm\infty, q = \pm\infty$ and $\underline{a}$ constant; (iv) $s \neq 0, \pm\infty, r = \pm\infty$ and $\underline{b}$ constant; (v) $s = 0, q = 0$ or $r = 0$ and $\underline{b}$ constant; (vi) $q = r = s = 0$; (vii) $q = r = s = \infty$ and for some $i, 1 \leq i \leq n$, $\max \underline{a} = a_i$ and $\max \underline{b} = b_i$; (viii) $q = r = \pm\infty, s = \pm\infty$, $\underline{a}$ and $\underline{b}$ constant; (ix) $q = r = s = -\infty$ and for some $i, 1 \leq i \leq n$, $\min \underline{a} = a_i$ and $\min \underline{b} = b_i$.
(c) If $r \geq 1$ then

$$\mathfrak{M}_n^{[r]}(\underline{a}+\underline{b};\underline{w}) \leq \mathfrak{M}_n^{[r]}(\underline{a};\underline{w}) + \mathfrak{M}_n^{[r]}(\underline{b};\underline{w}). \qquad (3)$$

If $r < 1$ inequality ($\sim$3) holds; in particular

$$\mathfrak{G}_n(\underline{a}+\underline{b};\underline{w}) \geq \mathfrak{G}_n(\underline{a};\underline{w}) + \mathfrak{G}_n(\underline{b};\underline{w}). \qquad (4)$$

If $r \geq 1$ (3) is strict unless: (i) $r = 1$, (ii) $1 < r < \infty$ and $\underline{a} \sim \underline{b}$, or (iii) $r = \infty$ and for some $i, 1 \leq i \leq n$, $\max \underline{a} = a_i$ and $\max \underline{b} = b_i$.
(d) If $\underline{a}$ is not constant, $\left(\mathfrak{M}_n^{[r]}(\underline{a};\underline{w})\right)^r$ is strictly log-convex on $]-\infty, 0[$ and on $]0, \infty[$.
(e) $\mathfrak{M}_n^{[r]}(\underline{a};\underline{w})$ is log-convex, and so convex, in $1/r$.

COMMENTS (i) Simple algebraic arguments show that to prove (r;s) it is sufficient to consider cases (0;1), (r;1), $0 < r < 1$, and (1; s). The case (0;1) is just (GA), and the other cases follow from (H).
Many other proofs have been given. In particular (r;s) follows from (J), and from **Mixed Mean Inequalities** (1), (2).
(ii) While (H) can be used to prove (r;s), the case (r;1), $0 < r < 1$, is, on putting $r = 1/p$ and changing notation, just (H).
(iii) The cases when q, r or s of (b) are infinite are trivial. If q, r and s are finite but not zero the result follows from (H) and **Hölder's Inequality** (5); if one or more of q, r, s is zero then use (r;s). This is really a weighted form of (H).
(iv) If $r \neq 0$ (c) is either a weighted form of (M), or is trivial. The case $r = 0$, (4), which is **Geometric Mean Inequalities** (1), can be proved using (GA); this inequality is sometimes called Hölder's Inequality.
(v) The convexity results, (d), (e) follow from **Power Sums Inequalities** COMMENTS (ii).

EXTENSIONS (a) If $0 < r < s$ and if $W_n \leq 1$ then

$$\left(\sum_{i=1}^n w_i a_i^r\right)^{1/r} \leq \left(\sum_{i=1}^n w_i a_i^s\right)^{1/s}, \tag{5}$$

and if $W_n < 1$ the inequality is strict.

(b) If $0 < r < s < t$ and if $\underline{a}$ is not constant then

$$1 < \frac{\mathfrak{M}_n^{[t]}(\underline{a}) - \mathfrak{M}_n^{[r]}(\underline{a})}{\mathfrak{M}_n^{[t]}(\underline{a}) - \mathfrak{M}_n^{[s]}(\underline{a})} < \frac{s(t-r)}{r(t-s)}.$$

(c) If $0 < r$ and $\underline{a}$ is not constant then

$$1 < \frac{\mathfrak{M}_n^{[r]}(\underline{a};\underline{w}) - \mathfrak{M}_n^{[-r]}(\underline{a};\underline{w})}{\mathfrak{M}_n^{[r]}(\underline{a};\underline{w}) - \mathfrak{M}_n^{[0]}(\underline{a};\underline{w})} < \frac{W_n}{\min \underline{w}}.$$

(d) [POPOVICIU TYPE] If $-\infty < r \leq 0 \leq s < \infty$ then

$$\left(\frac{\mathfrak{M}_n^{[s]}(\underline{a};\underline{w})}{\mathfrak{M}_n^{[r]}(\underline{a};\underline{w})}\right)^{W_n} \geq \left(\frac{\mathfrak{M}_{n-1}^{[s]}(\underline{a};\underline{w})}{\mathfrak{M}_{n-1}^{[r]}(\underline{a};\underline{w})}\right)^{W_{n-1}},$$

with equality if and only if (i) when $s = 0$, $a_n = \mathfrak{M}_{n-1}^{[r]}(\underline{a};\underline{w})$, (ii) when $r = 0$, $a_n = \mathfrak{M}_{n-1}^{[s]}(\underline{a};\underline{w})$, (iii) if $r < 0 < s$ both conditions in (i) and (ii) hold.

(e) [RADO TYPE] If $-\infty \leq r \leq 1 \leq s \leq \infty, r \neq s$ then

$$W_n\left(\mathfrak{M}_n^{[s]}(\underline{a};\underline{w}) - \mathfrak{M}_n^{[r]}(\underline{a};\underline{w})\right) \geq W_{n-1}\left(\mathfrak{M}_{n-1}^{[s]}(\underline{a};\underline{w}) - \mathfrak{M}_{n-1}^{[r]}(\underline{a};\underline{w})\right)$$

with equality if and only if one of the following holds: (i) $s = 1$, $a_n = \mathfrak{M}_{n-1}^{[r]}(\underline{a};\underline{w})$; (ii) $r = 1$, $a_n = \mathfrak{M}_{n-1}^{[s]}(\underline{a};\underline{w})$; (iii) $r < 1 < s$ both conditions in (i) and (ii) hold.

(f) [FUNCTIONS OF INDEX SETS] If $\mathcal{I}$ is an index set define $\sigma(\mathcal{I}) = W_{\mathcal{I}} \mathfrak{M}_{\mathcal{I}}^{[s]}(\underline{a};\underline{w})$. If $s > 1$ and if $\mathcal{I} \cap \mathcal{J} = \emptyset$ then

$$\sigma(\mathcal{I} \cup \mathcal{J}) \geq \sigma(\mathcal{I}) + \sigma(\mathcal{J}), \tag{6}$$

with equality in (6) if and only if $\mathfrak{M}_{\mathcal{I}}(\underline{a};\underline{w}) = \mathfrak{M}_{\mathcal{J}}(\underline{a};\underline{w})$.
If $s < 1$ then (~ 6) holds, while if $s = 1$ (6) becomes an equality.

(g) If $k = 0, 1, 2, \ldots$

$$\left(\mathfrak{A}_n^k(\underline{a}) - \mathfrak{G}_n^k(\underline{a})\right) \geq n^{1-k}\left(\left(\mathfrak{M}_n^{[k]}\right)^k(\underline{a}) - \mathfrak{G}_n^k(\underline{a})\right), \tag{7}$$

with equality if and only if one of the following holds: (i) $k = 0, 1$; (ii) $n = 1$; (iii) $k = n = 2$; (iv) $\underline{a}$ is constant.

If $k < 0$ ($\sim$7) holds and is strict unless $\underline{a}$ is constant.
(h) [PEČARIĆ & JANIĆ] If $\underline{a}, \underline{u}, \underline{v}$ are real n-tuples with $\underline{a}$ increasing, and with

$$0 \leq V_n - V_{k-1} \leq U_n - U_{k-1} \leq U_n = V_n, 2 \leq k \leq n,$$

then

$$\mathfrak{M}_n^{[r]}(\underline{a}; \underline{v}) \leq \mathfrak{M}_n^{[r]}(\underline{a}; \underline{u}).$$

(j) [IZUMI, KOBAYASHI & TAKAHASHI] If $\underline{a}, \underline{b}, \underline{w}$ are positive n-tuples with $\underline{b}$ and $\underline{b}/\underline{a}$ similarly ordered and if $r \leq s$, then

$$\frac{\mathfrak{M}_n^{[r]}(\underline{a}; \underline{w})}{\mathfrak{M}_n^{[r]}(\underline{b}; \underline{w})} \leq \frac{\mathfrak{M}_n^{[s]}(\underline{a}; \underline{w})}{\mathfrak{M}_n^{[s]}(\underline{b}; \underline{w})}. \tag{8}$$

If $\underline{b}$ and $\underline{b}/\underline{a}$ are oppositely ordered then ($\sim$8) holds.

COMMENTS (vi) (5) is a consequence of (r;s).
(vii) If $n > 2$ and $0 < k < 1$ then neither (7) nor ($\sim$7) holds; (7) also holds if $k = 1, 2$ and $n = 2$.
(viii) Inequality (8) follows by a use of (r;s) and (Č).

CONVERSE INEQUALITIES (a) [CARGO & SHISHA]

$$\mathfrak{M}_n^{[s]}(\underline{a}; \underline{w}) - \mathfrak{M}_n^{[r]}(\underline{a}; \underline{w}) \leq \mathfrak{M}_2^{[s]}(m, M; 1 - t_0, t_0) - \mathfrak{M}_2^{[r]}(m, M; 1 - t_0, t_0)$$

where m, M are, respectively the smallest and the largest of the a_i with associated non-zero w_i, and where $\overline{1 - t_0}\, m + t_0 M$ is the mean-value point for f on $[m, M]$. There is equality if and only if either all the a_i with associated non-zero w_i are equal or if all the w_i are zero except those associated with m and M, and then these have weights $1 - t_0, t_0$ respectively.
(b) [SPECHT] If $0 < m \leq a_i \leq M, 1 \leq i \leq n, -\infty < r < s < \infty, rs \neq 0$ then

$$\frac{\mathfrak{M}_n^{[s]}(\underline{a}, \underline{w})}{\mathfrak{M}_n^{[r]}(\underline{a}, \underline{w})} \leq \left(\frac{r}{\mu^r - 1}\right)^{1/s} \left(\frac{\mu^s - 1}{s}\right)^{1/r} \left(\frac{\mu^s - \mu^r}{s - r}\right)^{1/s - 1/r},$$

where $\mu = M/m$.

COMMENTS (ix) A mean value point is defined in **Mean Value Theorem of Differential Calculus** COMMENTS.
(x) Specht's result, (b), has extensions to allow one or other of the powers to be zero; see [MI]. On putting $s = 1, r = -1$ this result reduces to **Kantorovič's Inequality**.
(xi) Other converse inequalities can be found in the references; see also **Ostrowski's Inequalities** (b).

INTEGRAL ANALOGUES (a) If f is defined and not zero almost everywhere on $[a,b]$ and $r < s$,
$$\mathfrak{M}^{[r]}_{[a,b]}(f,w) \leq \mathfrak{M}^{[s]}_{[a,b]}(f,w),$$
provided the integrals exist.
There is equality only if either $r \geq 0$ and $\mathfrak{M}^{[r]}_{[a,b]}(f,w) = \mathfrak{M}^{[s]}_{[a,b]}(f,w) = \infty$, or $s \leq 0$ and $\mathfrak{M}^{[r]}_{[a,b]}(f,w) = \mathfrak{M}^{[s]}_{[a,b]}(f,w) = 0$.
(b) If f, g, w are positive functions on $[a, b]$ and if $g, f/g$ are similarly ordered then if $r \leq s$
$$\frac{\mathfrak{M}^{[r]}_{[a,b]}(f;w)}{\mathfrak{M}^{[r]}_{[a,b]}(g;w)} \leq \frac{\mathfrak{M}^{[s]}_{[a,b]}(f;w)}{\mathfrak{M}^{[s]}_{[a,b]}(g;w)}.$$

COMMENTS (xii) The power means and their various inequalities have been extended by Kalman to allow the power to vary with the index; see [MPF].
(xiii) See also **Čebišev's Inequality** (1), **Gauss-Winckler Inequality** COMMENTS (i), **Hamy Mean Inequalities** COMMENTS, **Jessen's Inequalities**, **Ky Fan Inequalities** EXTENSIONS (a), **Levinson's Inequality** SPECIAL CASES, **Liapunov's Inequality**, **Logarithmic Mean Inequalities** (d), COROLLARIES (a), EXTENSIONS, **Love-Young Inequalities**, **Muirhead Symmetric Function and Mean Inequalities** COMMENTS (ii), **Rennie's Inequalities** EXTENSIONS, **Rado's Inequality**, **Thunsdorff's Inequality**.

REFERENCES [AI, pp. 76–80, 85–95], [BB, pp. 16–18], [HLP, pp. 26–28, 44, 134–145], [MI, pp. 135, 159–183, 198–207], [MO, pp. 130–131], [MPF, pp. 14–15, 48, 181], [PPT, pp. 108–112, 117]; [General Inequalities, vol. 3, pp. 43–68], [Polyá & Szegö, 1972, p. 69]; [Bullen, 1997], [Diananda].

Power Sums Inequalities

(a) If $r, s \in \mathbb{R}$, $r < s$ then,
$$\left(\sum_{i=1}^{n} a_i^s\right)^{1/s} < \left(\sum_{i=1}^{n} a_i^r\right)^{1/r}. \tag{1}$$

(b) If $r, s \in \mathbb{R}$ and $0 \leq \lambda \leq 1$ then
$$\left(\sum_{i=1}^{n} a_i^{(1-\lambda)r + \lambda s}\right) \leq \left(\sum_{i=1}^{n} a_i^r\right)^{1-\lambda} \left(\sum_{i=1}^{n} a_i^s\right)^{\lambda}, \tag{2}$$

and
$$\left(\sum_{i=1}^{n} a_i^{(1-\lambda)r + \lambda s}\right)^{1/\{(1-\lambda)r + \lambda s\}} \leq \left(\sum_{i=1}^{n} a_i^r\right)^{(1-\lambda)/r} \left(\sum_{i=1}^{n} a_i^s\right)^{\lambda/s}. \tag{3}$$

(c) If $\underline{a}$ is a decreasing non-negative n-tuple, $\underline{b}$ a real n-tuple then
$$\sum_{i=1}^{k} a_i \leq \sum_{i=1}^{k} b_i, 1 \leq k \leq n \quad \Longrightarrow \quad \sum_{i=1}^{n} a_i^2 \leq \sum_{i=1}^{n} b_i^2,$$

with equality if and only if $\underline{a} = \underline{b}$.

COMMENTS (i) In (1), that is often called *Jensen's inequality*, we can, without loss in generality, assume that $0 < r < s$ and that the left-hand side of (1) is equal to 1; then $\underline{a} < \underline{e}$ and so $\underline{a}^s < \underline{a}^r$.
Inequalities (2) and (3) follow by a simple application of (H).
(ii) While (1) says that the power sum is decreasing as a function of the power, (2) and (3) say that certain sums are log-convex functions of their parameters.

EXTENSIONS (a) If $r > 1$ then

$$\sum_{i=1}^{m}\sum_{j=1}^{n} a_{i,j}^r < \sum_{j=1}^{n}\left(\sum_{i=m}^{n} a_{i,j}\right)^r ;$$

with the opposite inequality if $0 < r < 1$.
(b)

$$\sum_{i=1}^{n} a_i^{\mathfrak{A}_n(r;\underline{w})} \leq \mathfrak{G}_n(\sum_{i=1}^{n} a_i^{r_1}, \ldots, \sum_{i=1}^{n} a_i^{r_n}; \underline{w}).$$

COMMENTS (iii) (b), an extension of (2), is proved by a simple induction. There is a similar extension of (3), but it can be improved by replacing the arithmetic mean on the left-hand side by the harmonic mean; it is an improvement because of (1) and (GA).
(iv) The power sums have been extended by Kalman to allow the power to vary with the index; see [MPF].
(v) See also **Bennett's Inequalities** (b)–(e), **Hölder's Inequality, Klamkin & Newman Inequalities, König's Inequality, Love-Young Inequalities** (b), **Minkowski's Inequality, Quasi- arithmetic Mean Inequalities** COMMENTS (iv), **Young's Convolution Inequality** DISCRETE ANALOGUES.
REFERENCES [AI, pp. 337–338], [BB, pp. 18–19], [HLP, pp. 28–30, 32, 72, 196–203],[MI, pp. 143–147], [MPF, p. 181], [PPT, pp. 164–169, 216–217].

Prékopa-Leindler Inequalities (a) If $p, q \geq 1$ are conjugate indices then

$$\int_{\mathbb{R}} \sup_{x+y=t} \{f(x)g(y)\}\, dt \geq p^{1/p} q^{1/q} \|f\|_{p,\mathbb{R}} \|g\|_{q,\mathbb{R}}. \tag{1}$$

(b) If $p, q, r \geq 1$ with $1/p + 1/q = 1 + 1/r$ then

$$\int_{\mathbb{R}} \left(\int_{\mathbb{R}} (f(x)g(t-x))^r \, dx \right)^{1/r} dt \geq \|f\|_{p,\mathbb{R}} \|g\|_{q,\mathbb{R}}. \tag{2}$$

(c) If $f, g : \mathbb{R}^n \to [0, \infty[$ and if $0 < \lambda < 1, \alpha \in \mathbb{R}$, define

$$h_\alpha(\underline{x}; f, g) = \sup_{\underline{y} \in \mathbb{R}^n} \left\{ \mathfrak{M}_2^{[\alpha]}\left(f(\tfrac{1}{\lambda}(\underline{x} - \underline{y})), g(\tfrac{1}{1-\lambda}\underline{y}); \lambda, 1-\lambda \right) \right\}.$$

Then if $\alpha \geq -1/n, \nu = \alpha/(1+n\alpha)$,

$$\int_{\mathbb{R}^n} h_\alpha \geq \mathfrak{M}_2^{[\nu]}\left(\int_{\mathbb{R}^n} f, \int_{\mathbb{R}^n} g; \lambda, 1-\lambda\right);$$

in particular

$$\int_{\mathbb{R}^n} h_0 \geq \left(\int_{\mathbb{R}^n} f\right)^\lambda \left(\int_{\mathbb{R}^n} g\right)^{1-\lambda}. \tag{3}$$

COMMENTS (i) In general the discrete analogue of (1) is not valid, but there is a discrete analogue of (2).
(ii) A probabilistic interpretation can be given to (1) in the case $p = q = 2$:

the marginal densities in each direction of a log-concave probability distribution on $\mathbb{R}^2$ are also log-concave.

(iii) Taking $f = 1_A, g = 1_B$ in (c), then $h_0 = 1_{\lambda A + \overline{1-\lambda} B}$ and (3) reduces to **Brunn-Minkowski Inequalities** (a). These inequalities have been further generalized by Brascamp & Lieb.

REFERENCES [MPF, pp. 168–173]; [General Inequalities, vol. 1, pp. 303–304]; [Brascamp & Lieb, 1976].

Probability Inequalities
If f is an even non-negative Borel function on $\mathbb{R}$, decreasing on $[0, \infty[$ then if $r \geq 0$

$$\frac{Ef \circ X - f(r)}{\|f\|_{\infty,P}} \leq P\{|X| \geq r\} \leq \frac{Ef \circ X}{f(r)}.$$

COMMENTS (i) If f is increasing we can replace the central term by $P\{X \geq r\}$.
(ii) Special cases of this result are **Markov's Probability Inequality** and, if $k > 0$,

$$\frac{Ee^{kX} - e^{kr}}{\|e^{kX}\|_{\infty,P}} \leq P\{X \geq r\} \leq \frac{Ee^{kX}}{e^{kr}}.$$

(iii) See also **Bernšteĭn's Probability Inequality, Berry-Esseen Inequality, Bonferroni's Inequalities, Čebišev's Probability Inequality, Entropy Inequalities, Gauss-Winkler Inequality, Kolmogorov's Probability Inequality, Lévy's Inequalities, Markov's Probability Inequality, Martingale Inequalities, Prékopa-Leindler Inequalities** COMMENTS (ii), **Statistical Inequalities**.

REFERENCES [Loève, pp. 157–158].

Pseudo Arithmetic and Geometric Mean Inequalities
See **Reverse Inequalities** COMMENTS (i).

Psi Function Inequalities
See **Digamma Function Inequalities**.

Q

Q-class Function Inequalities (a) If f is of class Q on $[a,b]$ then for all $x, y \in [a, b]$ and $\lambda, 0 < \lambda < 1$,

$$0 \leq f((1-\lambda)x + \lambda y) \leq \frac{f(x)}{1-\lambda} + \frac{f(y)}{\lambda}. \tag{1}$$

(b) If f is of class Q on $[a,b]$ and if $a \leq x_i \leq b, 1 \leq i \leq 3$ then

$$f(x_1)(x_1 - x_2)(x_1 - x_3) + f(x_2)(x_2 - x_3)(x_2 - x_1) + f(x_3)(x_3 - x_1)(x_3 - x_2) \geq 0.$$

(c) If f is of class Q on $[a,b]$ and if $a \leq x_1 < x_2 < x_3 \leq b$ then

$$f(x_2) \leq \frac{x_3 - x_1}{x_3 - x_2} f(x_1) + \frac{x_3 - x_1}{x_2 - x_1} f(x_3);$$

$$0 \leq \frac{f(x_1)}{x_3 - x_2} + \frac{f(x_2)}{x_1 - x_3} + \frac{f(x_3)}{x_2 - x_1}.$$

(d) If f is of class Q on $[a,b]$, $a \leq a_i \leq b, 1 \leq i \leq n, n \geq 2$, and if $\underline{w}$ is a positive n-tuple then

$$f\left(\frac{1}{W_n} \sum_{i=1}^{n} w_i a_i\right) \leq W_n \sum_{i=1}^{n} \frac{f(a_i)}{w_i}.$$

COMMENTS (i) (1) is just the definition of the Q-class, and the inequalities in (b), (c) are equivalent to (1).

(ii) (d) is an analogue of (J) for the Q-class. The various inequalities above should be compared to those in **Convex Function Inequalities**.

(iii) It is easy to see that if f is non-negative, and either monotonic or convex, then it is in the Q-class. As a result since $f(x) = x^r$ is a Q-class function on $]0, \infty[$ we get **Schur 's Inequality** (1) as a particular case of (b).

REFERENCES [MPF, pp. 410–413]; [Mitrinović & Pečarić, 1990[(3)]].

Quadratic Form Inequalities (a) If a, b, c are non-negative real numbers then

$$2bx \leq c + ax^2 \quad \text{for all} \quad x \geq 0 \quad \Longrightarrow \quad b^2 \leq ac.$$

(b) [POPOVICIU] *Let $\underline{a}, \underline{b}$ be increasing real n-tuples and let $w_{ij}, 1 \leq i, j \leq n$, be real numbers satisfying*

$$\sum_{i=r}^{n}\sum_{j=s}^{n} w_{ij} \geq 0, \quad 2 \leq r, s \leq n,$$

$$\sum_{i=r}^{n}\sum_{j=1}^{n} w_{ij} = 0, \quad 1 \leq r \leq n, \qquad \sum_{i=1}^{n}\sum_{j=s}^{n} w_{ij} = 0, \quad 1 \leq s \leq n; \tag{1}$$

then

$$\sum_{i,j=1}^{n} w_{ij} a_i b_j \geq 0. \tag{2}$$

Conversely if (2) holds for all such $\underline{a}, \underline{b}$ then (1) holds.

COMMENTS (i) Popoviciu's result has been extended to more general forms; see [MPF].

EXTENSIONS If a, b, c are positive real numbers, and if $0 < u < v$ then

$$bx^u \leq c + ax^v \quad \text{for all} \quad x > 0 \quad \implies \quad b^v \leq K_{u,v} c^{v-u} a^u.$$

COMMENTS (ii) This follows by taking $x = (bu/av)^{1/(v-u)}$.
(iii) See also **Aczel's Inequality, Bilinear Form Inequalities, Multilinear Form Inequalities**.
REFERENCES [BB, p. 177], [MPF, pp. 338, 340–344].

Quadrilateral Inequality *If $\underline{a}_i, 1 \leq i \leq 4$ are real n-tuples then*

$$||\underline{a}_1 - \underline{a}_4| - |\underline{a}_3 - \underline{a}_4|| \leq |\underline{a}_1 - \underline{a}_2| + |\underline{a}_3 - \underline{a}_2|,$$

with equality if and only if the four points are in order on the same line.

COMMENTS (i) This is an easy deduction from **Triangle Inequality** EXTENSIONS (a).
(ii) See **Metric Inequalities** EXTENSIONS (b).

Quasi-arithmetic Mean Inequalities
Let $[a, b] \subseteq \overline{\mathbb{R}}$ and suppose that $M : [a, b] \to \overline{\mathbb{R}}$ is continuous and strictly monotonic. Let $\underline{a}$ be an n-tuple with $a \leq a_i \leq b$, $1 \leq i \leq n$, $\underline{w}$ be a non-negative n-tuple with $W_n \neq 0$, then the quasi- arithmetic $\mathfrak{M}$-mean of $\underline{a}$ with weight $\underline{w}$ is

$$\mathfrak{M}_n(\underline{a}; \underline{w}) = M^{-1}\left(\frac{1}{W_n}\sum_{i=1}^{n} w_i M(a_i)\right),$$

$$= M^{-1}\left(A_n(M(\underline{a}); \underline{w})\right).$$

It is easy to see that if $M(x) = x^p, x > 0, -\infty < p < \infty$ then the $\mathfrak{M}$-mean is just the p-power mean, while if $M(x) = \log x, x > 0$ it is the geometric mean; many more special cases can be found in the references.

Different functions can define the same mean, as can easily be seen.

Simple restrictions on the properties of M can imply that the quasi-arithmetic mean reduces to a well known mean; see [MI].

(a)
$$\min \underline{a} \leq \mathfrak{M}_n(\underline{a};\underline{w}) \leq \max \underline{a}.$$

(b) If N is convex with respect to M then

$$\mathfrak{M}_n(\underline{a};\underline{w}) \leq \mathfrak{N}_n(\underline{a};\underline{w}); \tag{1}$$

further if N is strictly convex with respect to M this inequality is strict unless all the a_i with $w_i > 0$ are equal.

(c)
$$f\left(\mathfrak{M}_n(\underline{a};\underline{w})\right) \leq \mathfrak{N}_n\left(f(\underline{a});\underline{w}\right)$$

if and only if $N \circ f$ is convex with respect to M.

COMMENTS (i) To say that a function f is (strictly) convex with respect to a function g is to say that both f and g are strictly increasingc, and $f \circ g^{-1}$ is (strictly) convex. If both functions have continuous second derivatives and non-zero first derivatives this occurs if $g''/g' \leq f''/f'$.

(ii) (1) is just a restatement of (J), the basic property of convex functions.

(iii) (c) is an easy extension of (b); see **Log-convex Function Inequalities** (b) for a special case.

(iv) By omitting the weights in the definition of $\mathfrak{M}_n(\underline{a};\underline{w})$ we get analogues of the power sums. The results (M) and **Power Sum Inequalities** (1) have been extended to this situation.

(vi) See also **Čakalov's Inequality** COMMENTS (iv).

REFERENCES [HLP, pp. 65–101], [MI, pp. 31, 215–282], [MPF, pp. 193–194], [PPT, pp. 107–110, 165–169].

Quasi-conformal Function Inequalities

If Ω is a domain in $\mathbb{R}^n$, $n \geq 2$, and if $f \in \mathcal{W}^{1,n}(\Omega)$ is quasi-conformal then there is a $k \in \mathbb{R}$ such that

$$|\nabla f|^n \leq k n^{n/2} J(f'),$$

almost everywhere, where $J(f')$ is the Jacobian of f.

COMMENTS This is one of the definitions of quasi-conformality; the definition of $\mathcal{W}^{1,n}(\Omega)$ is given in **Sobolev's Inequalities**.

REFERENCES [EM, vol. 7, pp. 418–421].

Quasi-convex Function Inequalities

(a) If f is quasi-convex on a convex set X then for all λ, $0 \leq \lambda \leq 1$ and for all $x, y \in X$,

$$f\big((1-\lambda)x + \lambda y\big) \leq \max\{f(x), f(y)\}.$$

(b) If f is quasi-convex on a convex set X then for all $y \in X$ and if P, Q are two parallelopipeds, symmetric with respect to y, $P \subseteq Q$,

$$f(y) + \inf \{f(x); x \in P \setminus Q\} \leq \frac{2}{|P \setminus Q|} \int_{P \setminus Q} f.$$

(c) If f is continuous, differentiable and quasi-convex on the open convex set X then

$$f(x) \leq f(y) \implies f'(y)(x-y) \leq 0.$$

COMMENTS (i) Both (a) and (b) are definitions of quasi-convexity. If $-f$ is quasi-convex then f is said to be *quasi-concave*. Both convex and monotone functions are quasi-convex.
(ii) The converse of (c) is also valid; so within the class functions in (c) this is another definition of quasi-convexity.
(iv) A symmetric quasi-convex function is Schur convex; for definitions see **Segre's Inequalities, Schur Convex Function Inequalities**.
(v) The concept can be readily extended to higher dimensions; in addition, there is a higher order quasi-convexity.
(vi) See also **Minimax Theorems** (b); another definition of quasi-concavity is given in **Bergh's Inequality** COMMENTS (i).

REFERENCES [MO. pp. 68–69]; [Roberts & Varberg, pp. 104, 230]; [Longinetti], [Popoviciu, 1986].

Quaternion Inequalities

If $\mathfrak{a}_k, \mathfrak{b}_k, 1 \leq k \leq n$, are quaternions then

$$\left\| \sum_{k=1}^{n} \mathfrak{a}_k \bar{\mathfrak{b}}_k \right\|^2 \leq \left(\sum_{k=1}^{n} \|\mathfrak{a}_k\|^2 \right) \left(\sum_{k=1}^{n} \|\mathfrak{b}_k\|^2 \right),$$

with equality if and only if either $\mathfrak{a}_k = \mathfrak{o}, 1 \leq k \leq n$, $\mathfrak{b}_k = \mathfrak{o}, 1 \leq k \leq n$, or for some real λ, $\mathfrak{a}_k = \lambda \mathfrak{b}_k, 1 \leq k \leq n$.

COMMENTS (i) If $\mathfrak{c}$ is a quaternion then $\mathfrak{c} = c_1 + c_2 \mathfrak{i} + c_3 \mathfrak{j} + c_4 \mathfrak{k}$, where $c_m \in \mathbb{R}, 1 \leq m \leq 4$, and $\mathfrak{i}^2 = \mathfrak{j}^2 = \mathfrak{k}^2 = -1$, $\mathfrak{ij} = -\mathfrak{ji} = \mathfrak{k}$, $\mathfrak{jk} = -\mathfrak{kj} = \mathfrak{i}$, $\mathfrak{ki} = -\mathfrak{ik} = \mathfrak{j}$.
The zero quaternion is $\mathfrak{o} = 0 + 0\mathfrak{i} + 0\mathfrak{j} + 0\mathfrak{k}$.
In addition we define: $\bar{\mathfrak{c}} = c_1 - c_2 \mathfrak{i} - c_3 \mathfrak{j} - c_4 \mathfrak{k}$, and $\|\mathfrak{c}\|^2 = \mathfrak{c}\bar{\mathfrak{c}} = \sum_{m=1}^{4} c_m^2$.
An important property is that $\|\mathfrak{c}\mathfrak{d}\| = \|\mathfrak{c}\| \|\mathfrak{d}\|$.
(ii) The inequality above is a generalization of (C).

REFERENCES [AI, p. 43], [EM, vol. 7, pp. 440–442], [MPF, p. 90].

- end of Q -

R

Rado's Geometric-Arithmetic Mean Inequality Extension *If $n \geq 2$, then*

$$W_n \{\mathfrak{A}_n(\underline{a};\underline{w}) - \mathfrak{G}_n(\underline{a};\underline{w})\} \geq W_{n-1} \{\mathfrak{A}_{n-1}(\underline{a};\underline{w}) - \mathfrak{G}_{n-1}(\underline{a};\underline{w})\}, \tag{1}$$

with equality if and only if $a_n = \mathfrak{G}_{n-1}(\underline{a};\underline{w})$.

COMMENTS (i) Repeated application of (1) exhibits it as an extension of (GA), to which it is however equivalent. If these applications are stopped earlier we get an extension of (GA):

$$\mathfrak{A}_n(\underline{a};\underline{w}) - \mathfrak{G}_n(\underline{a};\underline{w}) \geq \frac{1}{W_n} \max_{1 \leq i,j \leq n} \left\{ w_i a_i + w_j a_j - (w_i + w_j)\left(a_i^{w_i} a_j^{w_j}\right)^{1/(w_i+w_j)} \right\}.$$

In the case of equal weights this is,

$$\mathfrak{A}_n(\underline{a}) - \mathfrak{G}_n(\underline{a}) \geq \frac{1}{n}\left(\sqrt{\max \underline{a}} - \sqrt{\min \underline{a}}\right)^2.$$

A similar discussion can be found in **Jensen's Inequality** COROLLARIES, **Popoviciu's Geometric-Arithmetic Mean Inequality Extension** COMMENTS (ii).

(ii) Inequalities that extend $P - Q \geq 0$ to $P - Q$ is a increasing function of index sets are called *Rado-type* or *Rado-Popoviciu inequalities*. See for instance **Power Mean Inequalities** EXTENSIONS (e).

(iii) Inequality (1) implies that the ratio

$$\frac{\mathfrak{A}_n(\underline{a};\underline{w}) - \mathfrak{G}_n(\underline{a};\underline{w})}{\mathfrak{A}_{n-1}(\underline{a};\underline{w}) - \mathfrak{G}_{n-1}(\underline{a};\underline{w})}$$

is bounded below as a function of $\underline{a}$ by W_{n-1}/W_n. Furthermore this bound is attained only when $a_n = \mathfrak{G}_{n-1}(\underline{a};\underline{w})$ which is not possible for non-constant increasing sequences. A natural question is whether this bound can be improved for such sequences; this is the basis of **Čakalov's Inequality**. No such extension exists for **Popoviciu's Geometric-Arithmetic Mean Inequality Extension** (1).

(iv) Inequality (1) implies that $\lim_{n \to \infty} n\left(\mathfrak{A}_n(\underline{a}) - \mathfrak{G}_n(\underline{a})\right)$ always exists, possibly infinite. The value of this limit has been obtained by Everitt.

(v) An easy deduction from Rado's inequality, (1), is

$$\sum_{i=m+1}^{n} a_i \geq n\mathfrak{G}_n(\underline{a}) - m\mathfrak{G}_m(\underline{a}); \tag{2}$$

an inequality that is sometimes also called *Rado's inequality*.

EXTENSIONS [FUNCTIONS OF INDEX SET] Let ρ be the following function defined on the index sets,
$$\rho(\mathcal{I}) = W_\mathcal{I}\{\mathfrak{A}_\mathcal{I}(\underline{a};\underline{w}) - \mathfrak{G}_\mathcal{I}(\underline{a};\underline{w})\}.$$
Then $\rho \geq 0$, is increasing and super-additive.

COMMENTS (vi) There are many other extensions in the references; see also **Harmonic Mean Inequalities** (c), **Henrici's Inequality** (c).

REFERENCES [AI, pp. 90–95], [HLP, p. 61], [MI, pp. 94–109].

Rado's Inequality If $a,b,r \in \mathbb{R}$, $a < b$ and if $f \in C([a,b])$ is positive and convex
$$\mathfrak{M}_2^{[r_1]}(f(a), f(b)) \leq \mathfrak{M}_{[a,b]}^{[r]}(f) \tag{1}$$
where
$$r_1 = \begin{cases} \min\left\{\dfrac{r+2}{3}, \dfrac{r\log 2}{\log(r+1)}\right\}, & \text{if } r > -1,\ r \neq 0, \\ \min\left\{\dfrac{2}{3}, \log 2\right\}, & \text{if } r = 0, \\ \min\left\{\dfrac{r+2}{3}, 0\right\}, & \text{if } r \leq -1. \end{cases}$$

The value r_1 is best possible.
If f is concave then ($\sim$1) holds if r_1 is replaced by r_2 where r_2 is defined as is r_1, but with min replaced by max.

COMMENTS (i) In particular (1) implies **Pittenger's Inequalities**.

EXTENSIONS [PEARCE & PEČARIĆ] With the above notation
$$\mathfrak{M}_2^{[r_1]}(f(a), f(b)) \leq \mathfrak{M}_{[a,b]}^{[r]}(f) \leq \mathfrak{L}^{[r]}(f(a), f(b)) \leq \mathfrak{M}_2^{[r_2]}(f(a), f(b)), \tag{2}$$
with equality if and only if f is linear. If f is concave then ($\sim$2) holds.

COMMENTS (ii) The case $r = 2$ of the central inequality is due to Sándor.
(iii) For other inequalities with the same name see **Rado's Geometric-Arithmetic Mean Inequality Extension** COMMENTS (v).

REFERENCES [Pearce & Pečaric, 1996].

Radon's Inequality If $\underline{a}, \underline{b}$ are positive n-tuples and $p > 1$ then
$$\frac{\left(\sum_{i=1}^n a_i\right)^p}{\left(\sum_{i=1}^n b_i\right)^{p-1}} \leq \sum_{i=1}^n \left(\frac{a_i^p}{b_i^{p-1}}\right), \tag{1}$$
while if $p < 1$ the inequality ($\sim$1) holds. The inequalities are strict unless $\underline{a} \sim \underline{b}$.

COMMENTS This is a deduction from (H).

REFERENCES [HLP, p. 61], [MI, p. 140]; [Redheffer & Voigt].

Rahmail's[1] Inequality Let $f, g : [a, b] \to \mathbb{R}$ be positive concave functions, then if $k < p \leq k+1, k \in \mathbb{N}, k > 0$,

$$\|f + g\|_{p,[a,b]} \geq \Lambda \left(\|f\|_{p,[a,b]} + \|g\|_{p,[a,b]} \right),$$

where

$$\Lambda = \left(\frac{\left(\frac{k+1}{2}\right)! \left(\frac{k+2}{2}\right)!}{(k+2)!} \right)^{1/k} (p+1)^{1/p} \left(\frac{2p-k}{p-k} \right)^{(p-k)/pk}.$$

COMMENTS This inequality is a converse to (M); for a similar result related to (H) see **Petschke's Inequality**.

REFERENCES [MPF, p. 150].

Rank Inequalities (a) [FRÖBENIUS] If A, B, C are matrices for which ABC is defined, then

$$\operatorname{rank}(AB) + \operatorname{rank}(BC) \leq \operatorname{rank} B + \operatorname{rank}(ABC).$$

(b) [SYLVESTER] If A is a $m \times n$ matrix and B an $n \times p$ matrix then

$$\operatorname{rank} A + \operatorname{rank} B - n \leq \operatorname{rank}(AB) \leq \min\{\operatorname{rank} A, \operatorname{rank} B\}.$$

REFERENCES [Marcus & Minc, pp. 27–28].

Rayleigh-Ritz Ratio If A is an $n \times n$ Hermitian matrix then for any complex n-tuple $\underline{a}$,

$$\lambda_{(1)} \leq \frac{\underline{a}^* A \underline{a}}{\underline{a}^* \underline{a}} \leq \lambda_{(n)}.$$

COMMENTS (i) The central term in this inequality is called *the Rayleigh-Ritz ratio*. The maximum of this ratio over all non-zero $\underline{a}$ of this ratio is $\lambda_{(n)}$, whilst the minimum is $\lambda_{(1)}$.
(ii) For a generalization see the [*Inequalities II*].

REFERENCES [BB, pp. 71–72]; [Mitrović & Žubrinić, p. 242], [Horn & Johnson, pp. 176–177], [General Inequalities, vol. 1, pp. 223–230], [Inequalities II, pp. 276–279].

Real Numbers Inequalities
See **Arithmetic Inequalities**.

Rearrangement Inequalities (a) If $\underline{a}, \underline{b}$ are real n-tuples then

$$\sum_{i=1}^n a_{(i)} b_{[i]} \leq \sum_{i=1}^n a_i b_i \leq \sum_{i=1}^n a_{(i)} b_{(i)}; \tag{1}$$

or $\underline{a}_* . \underline{b}^* = \underline{a}^* . \underline{b}_* \leq \underline{a} . \underline{b} \leq \underline{a}_* . \underline{b}_* = \underline{a}^* . \underline{b}^*.$

$$\prod_{i=1}^n (a_{(i)} + b_{(i)}) \leq \prod_{i=1}^n (a_i + b_i) \leq \prod_{i=1}^n (a_{[i]} + b_{(i)}).$$

---

[1] Р Т Рахмаил. Also transliterated as Rakhmail.

(b) If f is a positive increasing function then

$$\sum_{n=1}^{\infty} f(\mathfrak{A}_n(\underline{a})) \le \sum_{n=1}^{\infty} f(\mathfrak{A}_n(\underline{a}^*)).$$

EXTENSIONS (a) [HAUTUS] If f is strictly convex in the interval I, and if $\underline{a}, \underline{b}$ are two n-tuples with $a_i, b_i, a_i + b_i \in I, 1 \le i \le n$, then

$$\sum_{i=1}^{n} f\left(a_{(i)} + b_{[i]}\right) \le \sum_{i=1}^{n} f(a_i + b_i) \le \sum_{i=1}^{n} f\left(a_{(i)} + b_{(i)}\right).$$

(b) [HARDY & LITTLEWOOD] If $a_i, b_i, c_i, -n \le i \le n$, are real $(2n+1)$-tuples, with $\underline{c}$ symmetrical, then

$$\sum_{\substack{i,j,k=-n \\ i+j+k=0}}^{n} a_i b_j c_k \le \sum_{\substack{i,j,k=-n \\ i+j+k=0}}^{n} {}^+a_i b_j^+ c_k^{(*)} = \sum_{\substack{i,j,k=-n \\ i+j+k=0}}^{n} a_i^{+\,+} b_j c_k^{(*)}.$$

COMMENTS (i) Extension (b) is a generalization of the basic result (1). Extensions to more than two n-tuples are more complicated and more difficult to prove; see [HLP].

INTEGRAL ANALOGUES [HARDY-LITTLEWOOD] If I is an interval in $\mathbb{R}$, $fg \in \mathcal{L}(I)$ then

$$\int_I fg \le \int_I f^* g^*.$$

COMMENTS (ii) See also **Chong's Inequalities (a)**, **Duff's Inequality**, **Hardy-Littlewood Maximal Inequalities**, **Hardy-Littlewood-Sobolev Inequality**, **Modulus of Continuity Inequalities**, **Riesz's Rearrangement Inequality**, **Spherical Rearrangement Inequalities**, **Variation Inequalities**.

REFERENCES [AI, p. 284], [HLP, pp. 248, 261–276], [MI, pp. 20, 81], [MPF, p. 17]; [Kawohl, p. 23].

Recurrent Inequalities Let $f_k(a_1, \ldots, a_k), g_k(a_1, \ldots, a_k)$ be real-valued functions defined for a_k in the set D_k, $1 \le k \le n$ and for which there exist real-valued functions F_k such that for all k, $1 \le k \le n$,

$$\sup_{a_k \in D_k} \left(\mu f_k(a_1, \ldots, a_k) - g_k(a_1, \ldots, a_k)\right) = F_k(\mu) f_{k-1}(a_1, \ldots, a_{k-1}), \qquad (1)$$

where $f_0 = 1$. Then

$$\sum_{k=1}^{n} \mu_k f_k(a_1, \ldots, a_k) \le \sum_{k=1}^{n} g_k(a_1, \ldots, a_k), \qquad (2)$$

provided we can find real numbers δ_k, $1 \leq k \leq n+1$, such that

$$\delta_1 \leq 0, \quad \delta_{n+1} = 0$$
$$\mu_k = F_k^{-1}(\delta_k) - \delta_{k+1},$$

where $F_k^{-1}(y)$ denotes any x such that $F_k(x) = y$.

COMMENTS (i) The proof is by induction on n. Inequalities of this form are called *recurrent inequalities* and the functions F_k, $1 \leq k \leq n$, are called the *structural functions of the inequality*.
(ii) If we want the inequality ($\sim$2) then the sup in (1) is replaced by inf.
(iii) It is possible to replace the fixed sets $D_1, \ldots, D_n$ by

$$D_k = D_k(a_1, \ldots, a_{k-1}), \ 2 \leq k \leq n.$$

(iv) Many classical inequalities can be put into this form, and usually this not only gives a proof of the inequality but of a significant refinement.; see **Abel's Inequalities** EXTENSIONS (b), **Arithmetic Mean Inequalities** (d), **Carleman's Inequality** EXTENSIONS (b), **Hardy's Inequality** EXTENSIONS (a), **Ky Fan-Taussky-Todd** EXTENSIONS (b), **Redheffer's Inequalities**.
REFERENCES [AI, pp. 129–131], [MI, pp. 113–117]; [Redheffer].

Redheffer's Inequalities (a) If $\mathfrak{A}_n(\mathfrak{G}) = \mathfrak{A}_n(\mathfrak{G}_1(\underline{a}), \ldots, \mathfrak{G}_n(\underline{a}))$ then

$$\mathfrak{A}_n(\mathfrak{G}) \exp\left\{\frac{\mathfrak{G}_n(\underline{a})}{\mathfrak{A}_n(\mathfrak{G})}\right\} \leq e\mathfrak{A}_n(\underline{a}), \tag{1}$$

$$1 \leq \frac{1}{2}\left\{\frac{\mathfrak{A}_n(\mathfrak{G})}{\mathfrak{G}_n(\underline{a})} + \frac{\mathfrak{G}_n(\underline{a})}{\mathfrak{A}_n(\mathfrak{G})}\right\} \leq \frac{\mathfrak{A}_n(\underline{a})}{\mathfrak{G}_n(\underline{a})}; \tag{2}$$

there is equality if and only if $\underline{a}$ is constant.
(b) If $\underline{w}$ is a positive n-tuple then

$$\sum_{i=1}^n \frac{2i+1}{W_i} + \frac{n^2}{W_n} \leq 4\sum_{i=1}^n \frac{1}{w_i},$$

with equality if and only if $w_i = iw_1$, $1 \leq i \leq n$.
(c) If $\underline{a}$ is a positive sequence,

$$\sum_{i=1}^n \mathfrak{G}_i(\underline{a}) + n\mathfrak{G}_n(\underline{a}) < \sum_{i=1}^n \left(1 + \frac{1}{i}\right)^i a_i,$$

with equality if and only if $a_i = 2a_1 i^{i-1}(i+1)^{-i}$, $1 \leq i \leq n$.

COMMENTS (i) Inequality (2) follows from (1). The proofs are based on the method of **Recurrent Inequalities**.

(ii) Both (1) and (2) extend (GA). Further inequalities that involve the sequence $\mathfrak{G}$ can be found in **Nanjundiah Mixed Mean Inequalities**.
(iii) Since $\exp(\mathfrak{G}_n(\underline{a})/\mathfrak{A}_n(\mathfrak{G})) > 1$, (1) implies the finite form of **Carleman's Inequality** (1).
(iv) See also **Abel's Inequalities** EXTENSIONS (b), **Carleman's Inequality** EXTENSIONS (b).

REFERENCES [MI, pp. 113–117], [Redheffer].

Reisner's Inequality
See **Mahler's Inequalities** COMMENTS (ii).

Rellich's Inequality If $n \neq 2$, $f \in C_0^\infty(\mathbb{R}^n \setminus \{\underline{0}\})$ then

$$\int_{\mathbb{R}^n} |\nabla f|^2 \geq \frac{n^2(n-4)^2}{16} \int_{\mathbb{R}^n} \frac{|f|^2}{|\underline{x}|^4}\, d\underline{x}.$$

COMMENTS This result has been extended, under extra conditions, to the case $n = 2$. In addition the exponent 4 in the integral on the right-hand side can be replaced by α, $\alpha > 4$.

REFERENCES [Pachpatte, 1991, 1991$^{(2)}$].

Rennie's Inequality If $0 < m \leq \underline{a} \leq M$, and if $W_n = 1$ then

$$\sum_{i=1}^n w_i a_i + mM \sum_{i=1}^n \frac{w_i}{a_i} \leq M + m.$$

COMMENTS (i) This is equivalent to Diaz & Metcalf's inequality, **Polyá & Szegö's Inequality** EXTENSIONS (b).

EXTENSION [GOLDMAN] If $0 < m \leq \underline{a} \leq M$, and if $-\infty < r < s < \infty$, $rs < 0$ then

$$(M^s - m^s)\left(\mathfrak{M}_n^{[r]}(\underline{a}, \underline{w})\right)^r - (M^r - m^r)\left(\mathfrak{M}_n^{[s]}(\underline{a}, \underline{w})\right)^s \leq M^s m^r - M^r m^s.$$

The opposite inequality holds if $rs > 0$.

COMMENTS (ii) Putting $s = 1$, $r = -1$ Goldman's inequality reduces to Rennie's inequality. Goldman used his result to prove Specht's inequality, see **Power Mean Inequalities** CONVERSE INEQUALITIES (b).

REFERENCES [AI, pp. 62–63], [MI, p. 203], [MPF, pp. 123–125], [PPT, pp. 108–109].

Reverse Inequalities [ALZER] If $\underline{a}, \underline{w}$ are positive n-tuples define

$$\mathfrak{a}_n(\underline{a};\underline{w}) = \frac{W_n}{w_1}a_1 - \frac{1}{p_1}\sum_{i=2}^n w_1 a_1, \quad \mathfrak{g}_n(\underline{a};\underline{w}) = a_1^{W_n/w_1}\prod_{i=2}^n a_i^{w_i/w_1}; \quad (1)$$

then

$$\mathfrak{a}_n(\underline{a};\underline{w}) \leq \mathfrak{g}_n(\underline{a};\underline{w}), \quad (2)$$

with equality if and only if $\underline{a}$ is constant.

COMMENTS (i) The quantities in (1) are known as the the *pseudo arithmetic, and geometric means of order n of $\underline{a}$ with weight $\underline{w}$*, respectively. The inequality (2) is an example of what is called a *reverse inequality*; another example is **Aczel's Inequality**.
(ii) Inequality (2) is a an analogue, for these means, of (GA), and has extension of both Popoviciu- and Rado-type; there is also an analogue of **Ky Fan's Inequality**. The inequality of Aczel is an analogue of (C).

REFERENCES [*General Inequalities*, vol. 6, pp. 5–16].

Riesz's[2] Inequalities
See **Conjugate Harmonic Function Inequalities, Hausdorff-Young Inequalities, Riesz Mean Value Theorem, Riesz-Thorin Theorem**.

Riesz[2] Mean Value Theorem If $0 < p \leq 1$ and $y > 0$ then

$$\int_0^x (x-t)^{p-1} f(t)\,dt, \quad 0 \leq x \leq y, \implies \int_0^y (x-t)^{p-1} f(t)\,dt \geq 0, \quad y \leq x < \infty.$$

COMMENTS This result is important as the integral on the left-hand side is, when divided by $(p-1)!$, the fractional integral of order p of f; in addition it appears in various summability problems, and in the remainder of Taylor's theorem. The integral on the right-hand side is then its partial integral.

EXTENSIONS [TÜRKE & ZELLER] If $p > 1$ and $y > 0$ then

$$\int_0^x (x-t)^{p-1} f(t)\,dt, \ 0 \leq x \leq y, \implies \int_0^y ((y-t)(x-t))^{p-1} f(t)\,dt \geq 0, \ y \leq x < \infty.$$

DISCRETE ANALOGUES [ASKEY, GASPER & ISMAIL] If $p > 1$ and if for some integer $n \geq 0$,

$$\sum_{k=0}^n \binom{m-k+p-1}{m-k} a_k \geq 0, \quad 0 \leq m \leq n,$$

then

$$\sum_{k=0}^n \binom{n-k+p-1}{n-k}\binom{m-k+p-1}{m-1} a_k \geq 0, \quad n \leq m < \infty.$$

REFERENCES [*General Inequalities*, vol. 3, pp. 491–496].

---

[2] This is M Riesz.

Riesz's[3] Rearrangement Inequality If f, g, h are non-negative measurable functions on $\mathbb{R}^n$ then

$$\int_{\mathbb{R}^n \times \mathbb{R}^n} f(\underline{v})g(\underline{u}-\underline{v})h(\underline{u})\,d\underline{u}d\underline{v} \le \int_{\mathbb{R}^n \times \mathbb{R}^n} f^{(*)}(\underline{v})g^{(*)}(\underline{u}-\underline{v})h^{(*)}(\underline{u})\,d\underline{u}d\underline{v}.$$

COMMENTS (i) The case $n=1$ is due to Riesz; the general case was given by Sobolev.
(ii) Cases of equality have been given by Burchard; in addition the result has been extended to m functions in k variables; see [Lieb & Loss].

REFERENCES [HLP, pp. 279–291]; [Lieb & Loss, pp. 79–87].

Riesz[2]-Thorin Theorem Let $(X, \mu), (Y, \nu)$ be two measure spaces, $T: X \to Y$ a linear operation with the properties

$$\|Tf\|_{1/\beta_1} \le M_1 \|f\|_{1/\alpha_1}, \qquad \|Tf\|_{1/\beta_2} \le M_2 \|f\|_{1/\alpha_2},$$

where $P_i = (\alpha_i, \beta_i), i=1,2$ belong to the square $0 \le \alpha \le 1, 0 \le \beta \le 1$. Then for all (α, β) on the the segment joining the points P_1, P_2, $\alpha = (1-t)\alpha_1 + t\alpha_2$, $\beta = (1-t)\beta_1 + t\beta_2$, $0 \le t \le 1$, say,

$$\|Tf\|_{1/\beta} \le M_1^{1-t} M_2^t \|f\|_{1/\alpha}.$$

COMMENTS (i) This is the original example of what has become called an *Interpolation Theorem*. It is also called the *Riesz convexity theorem*.
(ii) Originally this theorem was a result about multilinear forms;

if S is the left-hand side of **Multilinear Form Inequalities** (1), and if the supremum of S subject to $\|\underline{a}_i\|_{p_i} \le 1, 1 \le i \le n$, is $M = M(p_1, \ldots, p_n)$, then M is log-convex in each variable in the region $p_i > 0, 1 \le i \le n$.

(iii) See also **Phragmén-Lindelöf Inequality**.

REFERENCES [EM, vol.5, pp. 146–148; vol.8, p. 154], [HLP, pp. 203–204], [MPF, pp. 205–206]; [Hirschman, pp. 170–173], [Zygmund, vol.II, pp.93–100].

Rotation Theorems If f is univalent in $\overline{D}$, with $f(0)=0, f'(0)=1$, then

$$|\arg f'(z)| \le \begin{cases} 4 \arcsin |z|, & \text{if } |z| \le 1/\sqrt{2}, \\ \pi + \log\left(\dfrac{|z|^2}{(1-|z|)^2}\right), & \text{if } 1/\sqrt{2} \le |z| < 1. \end{cases}$$

COMMENTS Rotation theorems give estimates for the rotation at a point under a conformal map.

REFERENCES [EM, vol.8, pp. 185–186].

- end of R -

---

[3] This is F Riesz.
[2] This is M Riesz.

S

Saffari's Inequality If $f : [0,1] \to \mathbb{R}$ is bounded and measurable with $H = \mathfrak{M}_{[0,1]}^{[\infty]}(f) > 0$, $h = -\mathfrak{M}_{[0,1]}^{[-\infty]}(f) > 0$, and if $p = 1$ or $p \geq 2$, then

$$\|f\|_p \leq \mathfrak{M}_2^{[p-2]}(H, h).$$

COMMENTS The result is false if $1 < p < 2$.
REFERENCES [General Inequalities, vol. 3, pp. 529–530].

Sándor's Inequality
See **Rado's Inequality** COMMENTS (ii).

Schlömilch-Lemonnier Inequality If $n \in \mathbb{N}$, $n > 1$, then

$$\log(n+1) < \sum_{i=1}^{n} \frac{1}{i} < 1 + \log n.$$

COMMENTS (i) This is a simple deduction from **Integral Test Inequality**, and many extensions are given by Alzer & Brenner.
(ii) See also **Euler's Constant Inequalities, Logarithmic Function Inequalities**.
REFERENCES [AI, pp.187–188]; [Alzer & Brenner, 1992]

Schur Convex Function Inequalities (a) If $F : \mathbb{R}^n \to \mathbb{R}$ is Schur convex then for all $i \neq j$, $1 \leq i, j \leq n$,

$$(x_i - x_j)\left(\frac{\partial F}{\partial x_i} - \frac{\partial F}{\partial x_j}\right) \geq 0.$$

(b) [OSTROWSKI] If F is a real valued function that is Schur convex on a set A in $\mathbb{R}^n$ and if $\underline{a}, \underline{b} \in A$ then

$$F(\underline{a}) \leq F(\underline{b}) \iff \underline{a} \prec \underline{b}.$$

(c) [MARSHALL & OLKIN] If f, g are both Schur convex on $\mathbb{R}^n$ so is $f \star g$.

COMMENTS (i) (a) is just the definition of a *Schur convex function on A*. If the inequality is always strict then F is said to be *strictly Schur convex on A*. It is worth noting that a function that is both convex, (even quasi-convex), and symmetric is Schur convex.
(ii) (b) identifies the class of order preserving functions from the pre-ordered set of real n-tuples to $\mathbb{R}$; see also **Muirhead Symmetric Function and Mean Inequalities** COMMENTS (iii).
(iii) Many inequalities can be obtained by using special Schur convex functions; for instance:

if $\underline{w}$ is a positive n-tuple with $W_n = 1$,

$$\sum_{i=1}^n \frac{w_i}{1-w_i} \geq \frac{n}{n-1}; \qquad \sum_{i=1}^n \frac{w_i}{\prod_{j=1}^n w_j} \geq n^{n-1};$$

$$\sum_r! \frac{w_1 + \cdots + w_r}{w_1 \cdots w_r} \geq r\binom{n}{r} n^{r-1}; \qquad \sum_{\substack{i,j=1 \\ i \neq j}}^n \frac{w_i w_j}{w_i + w_j} \leq \frac{n-1}{2}.$$

(iv) See also **Arithmetic Mean Inequalities** COMMENTS (iii), **Complete Symmetric Function Inequalities** COMMENTS (iii), **Elementary Symmetric Function Inequalities** COMMENTS (vii), **Geometric-Arithmetic Mean Inequality** COMMENTS (vi), **Geometric Mean Inequalities** COMMENTS (iii), **Mitrinović & Doković's Inequality** COMMENTS (i), **n-convex Function Inequalities** COMMENTS (vii).

REFERENCES   [AI, pp. 167–168, 209–210], [BB, p.32], [EM, vol.6, p. 75], [MO, pp.54–82, 90–91], [PPT, pp. 332–336].

Schur-Hardy Inequality
See **Hardy-Littlewood-Pólya-Schur Inequality**.

Schur's Inequality
If a, b, c are positive and $r \in \mathbb{R}$ then

$$a^r(a-b)(a-c) + b^r(b-c)(b-a) + c^r(c-a)(c-b) \geq 0, \tag{1}$$

with equality if and only if $a = b = c$.

COMMENTS   (i) Inequality (1) has been the subject of many generalizations. It can be deduced from properties of Q-class functions; see **Q-class Function Inequalities** COMMENTS (iii).

EXTENSIONS [GUHA]   If $a, b, c, u, v, w > 0, p \in \mathbb{R}$ are such that

$$a^{1/p} + c^{1/p} \leq b^{1/p}, \tag{2}$$
$$u^{1/p+1} + w^{1/p+1} \geq v^{1/p+1}; \tag{3}$$

then if $p > 0$

$$ubc - vca + wab \geq 0. \tag{4}$$

If $-1 < p < 0$ and we have (2), ($\sim$3) then ($\sim$4) holds; if $p < -1$ then (4) holds if we have ($\sim$2), ($\sim$3).
In each case there is equality in (4), or ($\sim$4) if and only if there is equality in (2), (3), or ($\sim$2), ($\sim$3), and also

$$\frac{a^{p+1}}{u^p} = \frac{b^{p+1}}{v^p} = \frac{c^{p+1}}{w^p}.$$

COMMENTS (ii) Guha's result follows from (H). It reduces to Schur's inequality (1) by putting $p = 1, a = y - z, b = x - z, c = x - y, u = x^\lambda, v = y^\lambda, w = z^\lambda$, assuming as we may that $0 \le z \le y \le x$.
(iii) This is a cyclic inequality; see **Cyclic Inequalities**.
(iv) For other inequalities by Schur see **Eigenvalue Inequalities**(a), **Hardy-Littlewood-Pólya-Schur Inequalities, Permanent Inequalities** (b).

REFERENCES [AI, pp. 119–121], [HLP, p. 64], [MPF, pp. 407–413].

Schwarz's Lemma If f is analytic in D, $|f| \le 1$, $f(0) = 0$, then

$$|f(z)| \le |z|, \ z \in D, \quad \text{and} \quad |f'(0)| \le 1.$$

There is equality if and only if $f(z) = e^{i\theta} z$.

COMMENTS (i) Schwarz's Lemma is an easy deduction from the **Maximum-Modulus Principle**.

EXTENSIONS (a) If f analytic in $D_R = |z| < R$, $|f| \le M$, and if $z_0 \in D_R$ then

$$\left| \frac{M(f(z) - f(z_0))}{M^2 - \overline{f(z_0)}f(z)} \right| \le \left| \frac{R(z - z_0)}{R^2 - \overline{z_0}z} \right|, \ z \in D_R. \tag{1}$$

(b) If f is analytic in D, $|f| \le 1$, then

$$\frac{|f'(z)|}{1 - |f(z)|^2} \le \frac{1}{1 - |z|^2}. \tag{2}$$

COMMENTS (ii) Inequality (1), with $M = R = 1$, and (2) can be expressed as follows:

an analytic function on the unit disk with values in the unit disk decreases the hyperbolic distance between two points, the hyperbolic arc length and the hyperbolic area.

In this form the result is known as the *Schwarz-Pick lemma*.
The definition of the hyperbolic metric, ρ_h, is given in **Metric Inequalities** COMMENTS (iv).

EXTENSIONS (c) [DIEUDONNÉ] If f is a analytic in D and takes values in D, and if $f(0) = 0$, then

$$|f'(z)| \le \begin{cases} 1, & \text{if } |z| \le \sqrt{2} - 1, \\ \frac{(1 + |z|^2)^2}{4|z|(1 - |z|^2)}, & \text{if } |z| \ge \sqrt{2} - 1. \end{cases}$$

(d) [BEARDON] If f is a analytic in D and takes values in D and if f is not a fractional linear transformation, then

$$\rho_h(f^*(0), f^*(z)) \le 2\rho_h(0, z),$$

where
$$f^*(z) = \frac{1-|z|^2}{1-|f(z)|^2}f'(z).$$

The inequality is strict unless $f(z) = zh(z)$ where h is a fractional linear transformation of the unit disk onto itself.

COMMENTS (iii) The quantity $f^*(z)$ is called the *hyperbolic derivative* of f at z. A fractional linear tranformation is a function of the form $(az+b)(cz+d)^{-1}$, with $ad-bc \neq 0$.

REFERENCES [EM, vol.8, pp. 224–225]; [Ahlfors, 1966, pp. 135–136, 164; 1973, pp. 1–3], [Inequalities III, pp. 9–21], [Pólya & Szegö, 1972, pp. 160–163], [Titchmarsh, p. 168]; [Beardon].

Schwarz-Pick Lemma
See **Schwarz's Lemma** COMMENTS (ii).

Schweitzer's Inequality
See **Kantorovič's Inequality** COMMENTS (i), and INTEGRAL ANALOGUES.

Segre's Inequalities
If $f(\underline{x}) = f(x_1, \ldots, x_n)$ is unchanged under all permutations of $x_1, \ldots, x_n$ then f is called a *symmetric function*.
If $f(\underline{x}) = f(\underline{x}; \underline{a}^1, \ldots, \underline{a}^m) = f(x_1, \ldots, x_n, a_1^1, \ldots, a_n^1, \ldots, a_1^m, \ldots, a_n^m)$ is unchanged under any simultaneous permutation of $x_1, \ldots, x_n$, and $a_1^1, \ldots, a_n^1$ and . $a_1^m, \ldots, a_n^m$ then f is called an *almost symmetric function*. Here f is regarded as a function of $\underline{x}$ and the $\underline{a}^1, \ldots \underline{a}^n$ are parameters. For example $\mathfrak{A}_n(\underline{a})$ is symmetric, $\mathfrak{A}_n(\underline{a}, \underline{w})$ is almost symmetric. If for all real λ, $f(\lambda \underline{x}) = \lambda^\alpha f(\underline{x})$ then f is called a *homogeneous function* (of degree α).

Hypotheses: $f(\underline{x})$ is almost symmetric on $\{\underline{x}; a < \underline{x} < b\}$ and satisfies the following conditions:

$$f(x_1, x_1, \ldots, x_1) = 0, \quad a < x_1 < b, \tag{1}$$
$$f'_1(\underline{x}) \leq \rho_1 f(\underline{x}), \quad a < x_1 \leq x_j < b, \quad 2 \leq j \leq n, \tag{2}$$
$$f'_j(x_j, x_2, \ldots, x_n) = \rho_j f'_1(x_j, x_2, \ldots, x_n), \quad 2 \leq j \leq n, \tag{3}$$

where all the ρ are non-negative, and there is equality in (2), if at all, only when $x_1 = \cdots = x_n$.
Conclusions:
$$f(\underline{x}) \geq 0$$
with equality if and only if $x_1 = \cdots = x_n$.

COMMENTS (i) If $a = 0, b = \infty$ and if f is also homogeneous then we need only assume that (1), (2) and (3) hold for $x_1 = 1$.
(ii) Many classical inequalities follow from this result.
REFERENCES [MI, pp. 369–372].

Semi-continuous Function Inequalities If $f : [a,b] \to \overline{\mathbb{R}}$ is lower semi-continuous at x then

$$\liminf_{y \to x} f(y) \geq f(x). \qquad (1)$$

COMMENTS This is sometimes taken as the definition of lower semi-continuity. As the opposite inequality is trivial the definition can also be taken as equality in (1).

REFERENCES [Bourbaki, 1960, p. 170].

Series' Inequalities
(a)
$$\frac{m}{n+m} < \sum_{i=1}^{m} \frac{1}{n+i} < \frac{m}{n}.$$

(b) If $0 \leq a < 1$ and $k \in \mathbb{N}, k > (3+a)/(1-a)$ then

$$\sum_{i=0}^{nk-(n+1)} \frac{1}{n+i} > 1 + a.$$

(c)
$$\frac{1}{2(2n+1)} < \left| \sum_{i=0}^{n-1} \frac{(-1)^i}{2i+1} - \frac{\pi}{4} \right| = \sum_{i=n}^{\infty} \frac{(-1)^{n-i}}{2i+1} < \frac{1}{2(2n-1)};$$

$$\frac{1}{2(n+1)} < \left| \sum_{i=1}^{n} \frac{(-1)^{i-1}}{i} - \log 2 \right| = \sum_{i=n+1}^{\infty} \frac{(-1)^{n+1-i}}{i} < \frac{1}{2(n-1)}.$$

(d) If $n \geq 1$ then

$$\left(1 + \frac{1}{n}\right)^n < \sum_{k=0}^{n} \frac{1}{k!} < \left(1 + \frac{1}{n}\right)^{n+1}.$$

(e) [TRIMBLE ET AL] If $p \geq 0, x, y > 0$ then

$$\frac{1}{\sum_{k \in \mathbb{N}}(x+k)^{-p}} + \frac{1}{\sum_{k \in \mathbb{N}}(y+k)^{-p}} \leq \frac{1}{\sum_{k \in \mathbb{N}}(x+y+k)^{-p}}. \qquad (1)$$

COMMENTS (i) All the terms in inequality (d) have e as limit when $n \to \infty$.
(ii) (1) follows from the log-convexity of the function $((r+s-1)!/r!s!)x^r(1-x)^s$ for $0 < x < 1, r, s > -1$.
(iv) See also **Binomial Function Inequalities** (b), **Enveloping Series Inequalities**, **Exponential Function Inequalities** INEQUALITIES INVOLVING THE REMAINDER OF THE TAYLOR SERIES, **Sums of Integer Powers Inequalities**, **Trigonometric Function Inequalities** (n)–(p).

REFERENCES [AI, pp. 187–188]; [Kazarinoff, pp. 40–42, 46–47]; [Trimble et al].

Shampine's Inequality If p is a polynomial of degree n with real coefficients, and $p(x) \geq 0$, $x \geq 0$, then

$$-\left[\frac{n}{2}\right] \int_0^\infty p(x) e^{-x}\, dx \leq \int_0^\infty p'(x) e^{-x}\, dx \leq \int_0^\infty p(x) e^{-x}\, dx.$$

REFERENCES   [Skalsky]

Shannon's Inequality (a) If $\underline{p}, \underline{q}$ are positive n-tuples with $P_n = Q_n = 1$ then

$$\sum_{i=1}^n p_i \log p_i \geq \sum_{i=1}^n p_i \log q_i, \tag{1}$$

with equality if and only if $p_i = q_i, 1 \leq i \leq n$.
(b) If $\underline{p} \prec \underline{q}$ then

$$\sum_{i=1}^n p_i \log p_i \leq \sum_{i=1}^n q_i \log q_i.$$

COMMENTS (i) The quantity $H(\underline{p}) = -\sum_{i=1}^n p_i \log_2 p_i$ is called the *entropy of* $\underline{p}$. See **Entropy Inequalities**.
(ii) The logarithmic function in (1) is essentially the only function for which an inequality of this type holds.
(iii) Inequality (1) can be deduced from **Log-convex Function Inequalities** (2) by taking $f(x) = x^x$; or directly using (J) and the convexity of $x \log x$.
(iv) There are other entropy functions with their own inequalities; see [Tong, ed.].

REFERENCES   [AI, pp. 382–383], [EM, vol.4, pp. 387–388; Supp., pp. 338–339], [MI, pp. 227–228], [MO, p. 71], [MPF, pp. 635–650]; [Tong, ed., pp. 68–77].

Shapiro's Inequality If the positive n-tuple $\underline{a}$ is extended to a sequence by defining $a_{n+r} = a_r, r \in \mathbb{N}$, then

$$\sum_{i=1}^n \frac{a_i}{a_{i+1} + a_{i+2}} \geq n\lambda(n);$$

where,

$$\lambda(n) = \begin{cases} = \frac{1}{2}, & \text{if } n \leq 12 \text{ or } n = 13, 15, 17, 19, 21, 23; \\ < \frac{1}{2}, & \text{for all other values of } n. \end{cases}$$

COMMENTS (i) It is known that the limit, $\lim_{n\to\infty} \lambda(n)$ exists, λ say. Its exact value $0.494566\ldots$, is given in (1) below.
(ii) This inequality was proposed as a problem by Shapiro, although special cases had been considered earlier. Shapiro conjectured that $\lambda(n) \geq 1/2$. As we see above Shapiro's conjecture is false; it is true if for some k, $a_{k-1}, \ldots, a_{k+n}$ is monotonic.
(iii) An expository article on this inequality has been given by Mitrinović.

EXTENSIONS (a) [DRINFELD[1]] *With the above notation*

$$\sum_{i=1}^{n} \frac{a_i}{a_{i+1} + a_{i+2}} \geq n \frac{0.989133\ldots}{2}. \quad (1)$$

(b) [DIANANDA] *If the positive n-tuple $\underline{a}$ is extended to a sequence by defining $a_{n+r} = a_r, r \in \mathbb{N}$ then*

$$\sum_{i=1}^{n} \frac{a_i}{a_{i+1} + \cdots + a_{i+m}} \geq \frac{n}{m},$$

if one of the following holds: (i) $\sin(r\pi)/n \geq \sin(2m+1)r\pi/n$, $1 \leq r \leq [\frac{n}{2}]$; (ii) $n|(m+2)$,; (iii) $n|2m+k$, $k = 0, 1,$ or 2; (iv) $n = 8, 9,$ or 12 and $n|m+3$; (v) $n = 12$ and $n|m+4$.

COMMENTS (iv) The constant in the numerator on the right-hand side of (1) is $g(0)$ where g is the lower convex hull of the functions e^{-x} and $2(e^x + e^{x/2})^{-1}$. The proof depends on an ingenious use of the **Rearrangement Inequalities** (1).
(v) These are examples of **Cyclic Inequalities**.
(vi) For another inequality of the same name see **Leindler's Inequality** COMMENTS.

REFERENCES [AI, pp. 132–137], [MPF, pp. 440–447]; [General Inequalities, vol. 6, pp. 17–31], [Mitrinović].

Shapiro-Kamaly Inequality

Let $\underline{a} = \{a_n, n \in \mathbb{Z}\}$ be a non-negative sequence such that the following series converge:

$$\sum_{n \in \mathbb{Z}} a_n, \quad \sum_{n \in \mathbb{Z}} a_n^2; \quad \sum_{n \in \mathbb{Z}} (\sum_{i+j=n} a_i a_j)^2; \quad \sum_{n \in \mathbb{Z}} (\sum_{i-j=n} a_i a_j)^2.$$

Then if $0 \leq \delta \leq 1$ or $\delta = 2$,

$$\sum_{n \in \mathbb{Z}} |n|^\delta (\sum_{i-j=n} a_i a_j) \leq \sum_{n \in \mathbb{Z}} |n|^\delta (\sum_{i+j=n} a_i a_j),$$

and

$$\sum_{n \in \mathbb{Z}} |n|^\delta (\sum_{i-j=n} a_i a_j)^2 \leq \sum_{n \in \mathbb{Z}} |n|^\delta (\sum_{i+j=n} a_i a_j)^2.$$

REFERENCES [MPF, pp. 552–553].

Siegel's Inequality

See **Geometric-Arithmetic Mean Inequality** EXTENSIONS (c).

---

[1] В Г Дринфельд.

Sierpinski's Inequalities *If $\underline{a}$ is a positive n-tuple then*

$$\mathfrak{A}_n(\underline{a}) \geq \mathfrak{G}_n(\underline{a}) \left(\frac{\mathfrak{G}_n(\underline{a})}{\mathfrak{H}_n(\underline{a})} \right)^{1/(n-1)} \quad ; \quad \mathfrak{G}_n(\underline{a}) \geq \mathfrak{H}_n(\underline{a}) \left(\frac{\mathfrak{A}_n(\underline{a})}{\mathfrak{G}_n(\underline{a})} \right)^{1/(n-1)}. \qquad (1)$$

COMMENTS (i) The first inequality in (1) extends (GA) and the second, implied by the first, extends (HG).
(ii) On rewriting the left hand inequality all we have to prove is that

$$\mathfrak{A}_n(\underline{a}) \geq \left(\frac{\mathfrak{G}_n^n(\underline{a})}{\mathfrak{H}_n(\underline{a})} \right)^{1/(n-1)},$$

which is just S(1; n-1).

EXTENSIONS (a) [POPOVICIU-TYPE] *If $\underline{a}, \underline{w}$ are positive n-tuples, $n \geq 2$, then*

$$\frac{\mathfrak{A}_n^{W_{n-1}}(\underline{a}; \underline{w}) \mathfrak{H}_n^{w_n}(\underline{a}; \underline{w})}{\mathfrak{G}_n^{W_n}(\underline{a}; \underline{w})} \geq \frac{\mathfrak{A}_{n-1}^{W_{n-2}}(\underline{a}; \underline{w}) \mathfrak{H}_{n-1}^{w_{n-1}}(\underline{a}; \underline{w})}{G_{n-1}^{W_{n-1}}(\underline{a}; \underline{w})},$$

with equality if and only if $\underline{a}$ is constant.
(b) [ADDITIVE ANALOGUES] (i) *If $\underline{a}$ is a positive n-tuple*

$$n \left(\mathfrak{G}_n(\underline{a}) - \mathfrak{A}_n(\underline{a}) \right) \leq \mathfrak{H}_n(\underline{a}) - \mathfrak{A}_n(\underline{a}),$$

with equality if and only if $\underline{a}$ is constant.
(ii) *If $\underline{a}$ is a positive n-tuple with $\mathfrak{A}_n(\underline{a}) > 1$ then*

$$\mathfrak{G}_n^n(\underline{a}) - \mathfrak{A}_n^n(\underline{a}) \leq \mathfrak{H}_n(\underline{a}) - \mathfrak{A}_n(\underline{a}),$$

with equality if and only if $\underline{a}$ is constant.

REFERENCES [MI, p. 118–119], [MPF, pp. 21–25].

Skarda's Inequalities *If $\underline{a}, \underline{b}$ are two real absolutely convergent sequences then*

$$\underline{a}.\underline{b} \geq \frac{1}{2} (|\underline{a} + \underline{b}| - |\underline{a}| - |\underline{b}|) (\max\{|\underline{a}|, |\underline{b}|\}); \qquad (1)$$

$$\underline{a}.\underline{b} \geq \frac{1}{4} (|\underline{b}| - |\underline{a}| - |\underline{a} - \underline{b}|) (|\underline{b}| - |\underline{a}| + |\underline{a} - \underline{b}|); \qquad (2)$$

$$\underline{a}.\underline{b} \leq \frac{1}{4} (|\underline{a} + \underline{b}| + |\underline{a}| - |\underline{b}|) (|\underline{a} + \underline{b}| - |\underline{a}| + |\underline{b}|); \qquad (3)$$

$$\underline{a}.\underline{b} \leq \frac{1}{2} (|\underline{b}| + |\underline{a}| - |\underline{a} - \underline{b}|) (\max\{|\underline{a}|, |\underline{b}|\}). \qquad (4)$$

There is equality in (1) or (4) if either $a_n b_n = 0$ for all n, or if $\underline{b}$ has one non-zero term of the opposite sign to the corresponding entry in $\underline{a}$ and $|\underline{b}| \geq |\underline{a}|$.
There is equality in the (2) or (3), if either of $\underline{a}, \underline{b}$ is the null sequence, or if $\underline{a}, \underline{b}$ have only one non-zero term that is the same entry and is of the same sign;

REFERENCES [General Inequalities, vol.2, pp. 137–142].

Slater's Inequality If f is convex and increasing on $[a,b]$, $\underline{a}$ and n-tuple with terms in $[a,b]$, and if $\underline{w}$ is a non-negative n-tuple with $W_n > 0$, then

$$\frac{1}{W_n}\sum_{i=1}^n w_i f(a_i) \le f\left(\frac{\sum_{i=1}^n w_i a_i f'_+(a_i)}{\sum_{i=1}^n w_i f'_+(a_i)}\right).$$

REFERENCES [PPT, pp. 63-67].

Sobolev's[2] Inequalities

A *Sobolev inequality* is one that estimates lower order derivatives in terms of higher order derivatives, as distinct from the **Hardy-Littlewood-Landau Derivative Inequalities, Hardy-Littlewood-Pólya Inequalities** and **Kolmogorov Inequalities** that estimate intermediate order derivatives in terms of a higher order and a lower order derivative. Another important difference is the spaces in which the derivatives are considered, and the fact that in general these derivatives are generalized derivatives defined using distributions or something equivalent.

The *Sobolev Space* $\mathcal{W}^{\kappa,p}(\Omega)$, $1 \le p \le \infty$, $\kappa \in \mathbb{N}$, is the class of functions defined on Ω, a domain in $\mathbb{R}^n$, with finite norm

$$\|f\|_{p,\kappa,\Omega} = \|f\|_{p,\Omega} + \sum_{1 \le |\underline{k}| \le \kappa} \|D^{\underline{k}} f\|_{p,\Omega};$$

where

$$D^{\underline{k}} f = \frac{\partial^{|\underline{k}|} f}{\partial x_1^{k_1} \cdots \partial x_n^{k_n}}; \quad \underline{k} = (k_1, \ldots, k_n), \quad |\underline{k}| = k_1 + \cdots + k_n.$$

The derivatives, of order $\kappa = |\underline{k}|$, are, as mentioned above, usually interpreted in the generalized or distribution sense; in case $\kappa = 1$ we write $\nabla f = (\partial f/\partial x_1, \ldots, \partial f/\partial x_n)$.

(a) If $f : \mathbb{R}^n \to \mathbb{R}$, $1 \le m \le n$, and if $0 \le k = \ell - n/p + m/q$ then

$$\|f\|_{q,[k],\mathbb{R}^m} \le C\|f\|_{p,\ell,\mathbb{R}^n},$$

where C is independent of f.
(b) [GAGLIARDO-NIRENBERG-SOBOLEV] If $f \in \mathcal{W}^{1,p}(\mathbb{R}^n)$ then

$$\|f\|_{p^*} \le C\| |\nabla f| \|_p,$$

where C is independent of f.

COMMENTS (i) p^*, the *Sobolev conjugate index*, is given by $1/p^* = 1/p - 1/n$.
(ii) (b) is essentially the case $k = 0$, $\ell = 1$, $m = n$ of (a) and is related to Hardy's inequality, see **Hardy's Inequality** (5), COMMENTS (viii). It is equivalent to **Isoperimetric Inequalities** (1).
(iii) In some particularly important cases exact values of the constants in these inequalities have been given; see [Lieb & Loss].

---

[2] С Л Соболев.

(iv) See also **Logarithmic Sobolev Inequality, Morrey's Inequality, Poincaré's Inequalities, Trudinger's Inequality**. In addition there are many other Sobolev inequalities on the references.

(v) Another inequality by Sobolev is the **Hardy-Littlewood-Sobolev Inequality**.

REFERENCES [EM, vol.5, pp. 16–21; vol.8, pp. 379–381]; [Adams, p. 95–113], [Bobkov & Houdré, p. 1], [Evans & Gariepy, pp. 138–140,189–190], [General Inequalities, vol. 4, pp. 401–408], [Lieb & Loss, pp. 183–199], [Mitrović & Žubrinić, pp. 160–213], [Opic & Kufner, pp. 1–3, 240–314].

Spherical Rearrangement Inequalities

(a) If $f : \mathbb{R}^n \to \mathbb{R}$ and $p \geq 1$ then

$$\|\nabla f^{(*)}\|_{p,\mathbb{R}^n} \leq \|\nabla f\|_{p,\mathbb{R}^n}.$$

(b) If $f : \Omega \to \mathbb{R}$, Ω a domain in $\mathbb{R}^n$, then

$$\int_{\Omega^*} \sqrt{1 + |\nabla u^{(*)}|^2} \leq \int_{\Omega} \sqrt{1 + |\nabla u|^2},$$

where Ω^* is the *n*-sphere centred at the origin having the same measure as Ω.

EXTENSIONS [BROTHERS & ZIEMER] If $f : [0, \infty[\to [0, \infty[$, $f(0) = 0$ be in $\mathcal{C}^2$ with $f^{1/p}$ convex, $p \geq 1$, and if $u \in W^{1,p}(\mathbb{R}^n)$ is non-negative then

$$\int_{\mathbb{R}^n} f\left(|\nabla u^{(*)}|\right) \leq \int_{\mathbb{R}^n} f(|\nabla u|).$$

COMMENTS (i) The case of equality has been given by Brothers & Ziemer.

(ii) See also **Hardy-Littlewood Maximal Inequalities, Hardy-Littlewood-Sobolev Inequality, Rearrangement Inequalities, Riesz Rearrangement Inequality**.

REFERENCES [Kawohl, pp. 94–95], [Lieb & Loss, pp. 174–176], [Pólya & Szegö, 1951, p. 194].

Starshaped Function Inequalities

(a) If $f : [0, a] \to \mathbb{R}$ is starshaped then

$$f(\lambda x) \leq \lambda f(x), \quad 0 \leq \lambda \leq 1, \quad 0 \leq x \leq a.$$

(b) If $\underline{a}$ is a positive *n*-tuple, and if $f :]0, \infty[\to \mathbb{R}$ is starshaped then

$$\sum_{i=1}^{n} f(a_i) \leq f\left(\sum_{i=1}^{n} a_i\right). \tag{1}$$

If f is strictly starshaped and $n > 1$ then (1) is strict.

COMMENTS (i) (a) is just the definition of a function being *starshaped* (with respect to origin); an equivalent formulation is that $f(x)/x$, the slope of the chord from the origin, is increasing. If the slope increases strictly, then the function is said to be *strictly starshaped*.

(ii) It follows from **Convex Function Inequalities** (5), and COMMENTS (vi), that if f is convex on $[0, a]$ then $f(x) - f(0)$ is starshaped.
(iii) Since (1) is satisfied by any negative function f such that $\inf f \geq 2 \sup f$, the converse of (b) does not hold.
(iv) (b) is a simple deduction from **Increasing Function Inequalities** (2); just apply that result to $f(x)/x$, taking $w_i = a_i, 1 \leq i \leq n$.
(v) The last inequality in the case $\underline{w} = \underline{e}$ just says that starshaped functions are super-additive; see **Subadditive Function Inequalities** COMMENTS (ii).

EXTENSIONS   If f is starshaped on $[0, a]$ and if $\underline{a}, \underline{v}$ are non-negative n-tuples satisfying

$$a_j \in [0, a], \quad \sum_{i=1}^{n} v_i a_i \geq a_j, \quad 1 \leq j \leq n; \quad \text{and} \quad \sum_{i=1}^{n} v_i a_i \in [0, a];$$

then

$$\sum_{i=1}^{n} v_i f(a_i) \leq f\left(\sum_{i=1}^{n} v_i a_i\right).$$

COMMENTS   (vi) This is an easy deduction from **Increasing Function Inequalities** EXTENSIONS; apply the result to $f(x)/x$, taking $w_i = v_i a_i, 1 \leq i \leq n$.
REFERENCES   [MO, pp. 452–453], [PPT, pp. 8, 152–154]; [Trimble et al].

Statistical Inequalities
(a) [KLAMKIN-MALLOWS] If $\underline{a}$ is a positive increasing n-tuple, and $\bar{a} = \mathfrak{A}_n(\underline{a})$, then

$$\min \underline{a} \leq \bar{a}\left(1 - \frac{CV}{\sqrt{n-1}}\right) \leq \bar{a}\left(1 + \frac{CV}{\sqrt{n-1}}\right) \leq \max \underline{a},$$

with equality on the right-hand side if and only if $a_2 = \cdots = a_n$, and on the left-hand side if and only if $a_1 = \cdots = a_{n-1}$.
(b) [REDHEFFER & OSTROWSKI] If $1 \leq \nu \leq n-1$, $q = \nu/n$, $n, \nu \in \mathbb{N}$, and $0 < x < 1$ then

$$\binom{n}{\nu} x^\nu (1-x)^{(n-\nu)} \leq \left(1 - \frac{1}{n}\right)^{n-1} e^{-2n(x-q)^2},$$

with equality if and only if $x = q$ and $\nu = 1$ or $n-1$; and

$$\binom{n}{\nu} x^\nu (1-x)^{(n-\nu)} \leq \sqrt{\frac{1}{2\pi n q(1-q)}} e^{-2n(x-q)^2},$$

where, for every q, 2π is sharp as $n \to \infty$.
(c) If $\underline{a}$ is a real n-tuple and if $\mathfrak{A}_n(\underline{a}) = 0$, $\mathfrak{A}_n(\underline{a}^2) = 1$ then:
   (i) [WILKINS, CHAKRABARTI]

$$\mathfrak{A}_n(\underline{a}^3) \leq \frac{n-2}{\sqrt{n-1}}, \quad \mathfrak{A}_n(\underline{a}^4) \leq n - 2 + \frac{1}{n-1},$$

with equality if and only if $a_1 = \sqrt{n-1}$, and $a_i = -\frac{1}{\sqrt{n-1}}, 2 \leq i \leq n$;

(ii) [PEARSON]
$$\mathfrak{A}_n(\underline{a}^4) \leq 1 + \mathfrak{A}_n(\underline{a}^3);$$

(iii) [LAKSHMANAMURTI]
$$\mathfrak{A}_n^2(\underline{a}^m) + \mathfrak{A}_n^2(\underline{a}^{m-1}) \leq \mathfrak{A}_n^2(\underline{a}^{2m}).$$

COMMENTS   See also **Bernšteĭn's Probability Inequality**, **Berry-Esseen Inequality**, **Bonferroni's Inequalities**, **Čebišev's Probability Inequality**, **Copula Inequalities**, **Correlation Inequalities**, **Entropy Inequalities**, **Error Function Inequalities**, **Gauss-Winkler Inequality**, **Lévy's Inequalities**, **Martingale Inequalities**, **Polynomial Function Inequalities** (c), **Probability Inequalities**, **Shannon's Inequality**, **Totally Positive Function Inequalities**.

REFERENCES   [MI, p. 158]; [General Inequalities, 1, pp. 125–139, 307];[Klamkin, 1975], [Mallows & Richter].

Steffensen's Inequalities   (a) If $f, g \in \mathcal{L}([a,b])$, with f decreasing and $0 \leq g \leq 1$, and if $\lambda = \int_a^b g$ then
$$\int_{b-\lambda}^b f \leq \int_a^b fg \leq \int_a^{a+\lambda} f. \tag{1}$$

(b) If $\underline{a}, \underline{b}$ are real n-tuples such that
$$\sum_{i=1}^k b_i \leq \sum_{i=1}^k a_i, \ 1 \leq k < n; \quad \sum_{i=1}^n b_i = \sum_{i=1}^n a_i; \tag{2}$$

then if $\underline{w}$ is an increasing n-tuple,
$$\sum_{i=1}^n w_i a_i \leq \sum_{i=1}^n w_i b_i. \tag{3}$$

Conversely if (3) holds for such n-tuples then (2) holds.

COMMENTS   (i) Inequality (1) is called *Steffensen's inequality* and the right-hand inequality is the basis of Steffensen's proof of **Jensen-Steffensen Inequality**, to which it is equivalent.
(ii) For another deduction from (1) see **Fourier Transform Inequalities** (a).
(iii) The result in (b), is related to **Abel's Inequality** RELATED RESULTS (b), and follows by two applications of **Abel's Inequality** (1). Condition (2) is similar to the definition of the order relation $\prec$, to which it reduces if the n-tuples are decreasing.

EXTENSION [GODUNOVA, LEVIN & ČEBAEVSKAYA[3]]   If $f \in \mathcal{L}^p[a,b], g \in \mathcal{L}^q[a,b]$ where $f \geq 0$ is decreasing, $g > 0$ is increasing with $\|g\|_q \leq 1$, and p, q are conjugate indices, then with $\lambda = \left(\int_a^b g\right)^p$,
$$\int_a^b fg \leq \left(\int_a^{a+\lambda} f^p\right)^{1/p}.$$

---

[3] Е К Годунова, В И Левин, И В Чебаевская.

INTEGRAL ANALOGUES If f, g are integrable functions on $[a, b]$ then

$$\int_a^b fh \le \int_a^b gh$$

holds for every increasing function h if and only if

$$\int_a^x g \le \int_a^x f, \ a \le x < b; \quad \int_a^b g = \int_a^b f.$$

COMMENTS (iv) See also **Gauss's Inequality. Moment Inequalities.**

REFERENCES [AI, pp. 107–112], [MPF, pp. 311–337], [PPT, pp. 181–195]; [Bourbaki, 1949, pp. 102–103].

Stirling's Formula

$$\sqrt{2n\pi}\, n^n e^{-n + \frac{1}{12n+1/4}} < n! < \sqrt{2n\pi}\, n^n e^{-n + \frac{1}{12n}}.$$

COMMENTS From this it follows that

$$\lim_{n \to \infty} \frac{n!}{\sqrt{2n\pi}\, n^n e^{-n}} = 1,$$

or as this is often written

$$n! \sim \sqrt{2n\pi}\, n^n e^{-n}.$$

This last is *Stirling's Formula*.

EXTENSIONS

$$\lim_{\Re z \to \infty} \frac{z!}{\sqrt{2z\pi}\, z^n e^{-z}} = 1, \quad z! \sim \sqrt{2z\pi}\, z^n e^{-z}, \text{ as } \Re z \to \infty.$$

REFERENCES [AI, pp. 181–185], [EM, vol.8, p. 540].

Stolarsky's Inequality

If $f : [0, 1] \to [0, 1]$ is decreasing and if $p, q > 0$ then

$$\int_0^1 f(x^{1/(p+q)})\, dx \ge \int_0^1 f(x^{1/p})\, dx \int_0^1 f(x^{1/q})\, dx.$$

COMMENTS This has been generalized in the reference.

REFERENCES [Maligranda, Pečarić & Persson, 1995], [Pečarić & Varošanec].

Strongly Convex Function Inequalities If f is a strongly convex function defined on the compact interval $[a,b]$ then for some $\alpha > 0$, all $x, y \in [a,b]$ and $0 \le \lambda \le 1$,

$$f((1-\lambda)x + \lambda y) \le (1-\lambda)f(x) + \lambda f(y) - \lambda(1-\lambda)\alpha(x-y)^2.$$

COMMENTS (i) This is just the definition of *strong convexity*.
(ii) If f has a second derivative then f is strongly convex if and only if $f'' \ge 2\alpha$.
REFERENCES [Roberts & Varberg, p. 268].

Subadditive Function Inequalities (a) If $f : \mathbb{R}^n \to \mathbb{R}$ is subadditive then for all $\underline{x}, \underline{y} \in \mathbb{R}^n$.

$$f(\underline{x} + \underline{y}) \le f(\underline{x}) + f(\underline{y}). \tag{1}$$

(b) [KY FAN] If X is a lattice and $f : X \to \mathbb{R}$ then for all $x, y \in X$,

$$f(x \wedge y) + f(x \vee y) \le f(x) + f(y).$$

COMMENTS (i) These are just the definitions of *subadditivity* in the two situations; to avoid confusion the second property is often called *L-subadditivity*; if $-f$ is subadditive then we say that f is *superadditive*. See **Measure Inequalities** COMMENTS for another usage of this term. For strong sub-additivity see **Capacity Inequalities** COMMENTS (ii).
(ii) In (a) the domain can be generalized, and if it is a convex cone and if f is also homogeneous then f is convex; for a definition see **Segre's Inequalities**. A starshaped function defined on $[0, \infty[$ is superadditive; , see **Starshaped Function Inequalities**.
REFERENCES [MO, pp. 150–151, 453], [PPT, pp. 175–176]; [General Inequalities, vol. 1, pp. 159–160].

Subharmonic Function Inequalities If Ω is a domain in $\mathbb{R}^n$, if u is subharmonic on Ω, if $B_{\underline{x}_0, r} = \{\underline{x}; |\underline{x} - \underline{x}_0| < r\}$ and $\overline{B}_{\underline{x}_0, r} \subset \Omega$ then

$$u(\underline{x}_0) \le \frac{1}{a_n r^{n-1}} \int_{\partial B_{\underline{x}_0, r}} u; \tag{1}$$

$$u(\underline{x}_0) \le \frac{1}{v_n r^n} \int_{B_{\underline{x}_0, r}} u. \tag{2}$$

COMMENTS (i) (1) says that the value of u at the centre of any ball is not greater than the mean value of u on the surface of the ball; while (2) says the same with the mean taken over the ball itself.
(ii) (1) is just the definition of a *subharmonic function*; such a function being an upper semi-continuous function, with values in $\overline{\mathbb{R}} \setminus \{\infty\}$, not identically equal to $-\infty$, that satisifies (1).

This class of functions is a natural extension of the idea of **Convex Functions**. The concept can be extended to complex valued functions, and to functions defined on more general spaces.
If both u and $-u$ are subharmonic then u is harmonic.
A function u on Ω with continuous second partial derivatives will be subharmonic if and only if $\nabla^2 u \geq 0$.
(iii) It is an immediate consequnce of the integral form of (J) that if h is harmonic and f is convex then $f \circ h$ is subharmonic; in particular if $p \geq 1$ then $|h|^p$ is subharmonic. Similarly if u is subharmonic and f is convex and increasing then $f \circ u$ is subharmonic.
(iv) See also **Maximum-Modulus Principle**.

REFERENCES [EM, vol. 9, pp. 59–61]; [Ahlfors,1966, pp. 237–239], [Conway, vol. II, pp. 220–223], [Helms, pp. 57–59], [Lieb & Loss, pp. 202–209], [Protter & Weinberger, pp. 54–55], [Rudin, 1966, pp. 328–330].

Subordination Inequalities

If g is analytic in D then any function $f = g \circ w$, for some w analytic in D, $w(0) = 0$ and $|w(z)| \leq |z|$, $z \in D$, is said to be *subordinate to g in D*. Clearly f is also analytic in D and $f(0) = g(0)$. We also say that f is subordinate to g in $\{z; |z| < R\}$ if $f(Rz)$ is subordinate to $g(Rz)$ in D.

(a) If f is subordinate to g in D then for $0 < r < 1$,

$$M_p(f;r) \leq M_p(g;r), \quad 0 \leq p \leq \infty;$$

$$\int_0^{2\pi} \log^+ f(re^{i\theta})\, d\theta \leq \int_0^{2\pi} \log^+ g(re^{i\theta})\, d\theta;$$

$$M_p(\Re f; r) \leq M_p(\Re g; r), \quad 1 \leq p \leq \infty.$$

(b) If f, g are as in (a) with $f(z) = \sum_{n \in \mathbb{N}} a_n z^n$, $g(z) = \sum_{n \in \mathbb{N}} b_n z^n$, $|z| < 1$ then

$$a_0 = b_0; \quad |a_1| \leq |b_1|; \quad |a_2| \leq \max\{|b_1|, |b_2|\}; \quad \sum_{n \in \mathbb{N}} |a_n|^2 \leq \sum_{n \in \mathbb{N}} |b_n|^2;$$

there is equality in the first inequality if and only if $f(z) = g(e^{i\theta} z)$, for some $\theta \in \mathbb{R}$.

REFERENCES [EM, vol.9, pp. 64–65]; [Conway, vol.II, pp. 133–136, 272], [General Inequalities, vol.3, pp. 340–348]. .

Sums of Integer Powers Inequalities

(a) If $p > 0$ then

$$\frac{n^{p+1}-1}{p+1} < \sum_{i=1}^n i^p < \frac{n^{p+1}-1}{p+1} + n^p.$$

(b) If $p < 0, p \neq -1$ then

$$\frac{n^{p+1}-1}{p+1} + n^p < \sum_{i=1}^n i^p < \frac{n^{p+1}-1}{p+1} + 1.$$

(c)
$$\frac{1}{n} - \frac{1}{2n+1} < \sum_{i=1}^{2n} \frac{1}{i^2} < \frac{1}{n-1} - \frac{1}{2n}, \ if\ n > 1;$$

$$\sum_{i=n}^{\infty} \frac{1}{i^2} < \frac{2}{2n-1}, \ n \geq 1.$$

(d)
$$2\left(\sqrt{n+1} - 1\right) < \sum_{i=1}^{n} \frac{1}{\sqrt{i}} < 2\sqrt{n} - 1.$$

COMMENTS (i) (a), (b) are easy deductions from the **Integral Test Inequality**.
(ii) See also **Bennett's Inequalities** (1) and (2), **Euler Constant Inequalities, Logarithmic Function Inequalities** (j), **Mathieu's Inequality, Schlömilch-Lemonnier Inequality, Zeta Function Inequalities**.

REFERENCES [AI, pp. 190–191], [PPT, p. 39]; [Milisavljević].

Sup and Inf Inequalities
See **Inf and Sup Inequalities**.

Superadditive Function Inequalities
See **Subadditive Function Inequalities**.

Symmetric Function Inequalities
See **Elementary Symmetric Function Inequalities, Segre's Inequalities**.

Symmetric Mean Inequalities
(a) If $1 \leq r \leq n$ then

$$\min \underline{a} \leq \mathfrak{P}_n^{[r]}(\underline{a}) \leq \max \underline{a},$$

with equality if and only if $\underline{a}$ is constant.
(b) If $1 \leq r < s \leq n$ then

$$\mathfrak{P}_n^{[s]}(\underline{a}) \leq \mathfrak{P}_n^{[r]}(\underline{a}), \qquad\qquad S(r;s)$$

with equality if and only if $\underline{a}$ is constant.
(c) [KU ET AL] If $1 \leq r < s \leq n$ and $0 < m \leq \underline{a} \leq M$ then

$$\mathfrak{A}_n(\underline{a}) \left(\frac{\mathfrak{P}_n^{[r]}(\underline{a})^r}{\mathfrak{P}_n^{[s]}(\underline{a})^s} \right)^{1/(s-r)} \leq \frac{(M+m)^2}{4Mm},$$

with equality if and only if $\underline{a}$ is constant.

COMMENTS (i) The case $s = r + 1$ of S(r;s) follows from **Elementary Symmetric Function Inequalities** (2) with $r = t$ and multiplying over t, $1 \leq t \leq r$

(ii) The basic inequality S(r;s) is yet another generalization of (GA) for if $1 \le r \le n$,
$$\mathfrak{P}_n^{[n]}(\underline{a}) = \mathfrak{G}_n(\underline{a}) \le \mathfrak{P}_n^{[r]}(\underline{a}) \le \mathfrak{P}_n^{[1]}(\underline{a}) = \mathfrak{A}_n(\underline{a}).$$
(iii) In the case $n = 3$ S(r;s) has a simple geometric interpretation; see **Geometric Inequalities** (b).
(iv) The inequality in (c) can be considered as a generalization of **Kantorovič's Inequality**.
(v) As was pointed out in **Mixed Means Inequalities**, symmetric means are special cases of mixed means; so consider the matrix $\mathbb{S}$ where

$$\mathbb{S} = \begin{pmatrix} \mathfrak{M}_n(0,1;1;\underline{a}) & \mathfrak{M}_n(0,1;2;\underline{a}) & \ldots & \mathfrak{M}_n(0,1;n-1;\underline{a}) & \mathfrak{M}_n(0,1;n;\underline{a}) \\ \mathfrak{M}_n(0,1;n;\underline{a}) & \mathfrak{M}_n(0,1;n-1;\underline{a}) & \ldots & \mathfrak{M}_n(0,1;2;\underline{a}) & \mathfrak{M}_n(0,1;1;\underline{a}) \\ \mathfrak{M}_n(n,0;n;\underline{a}) & \mathfrak{M}_n(n-1,0;n-1;\underline{a}) & \ldots & \mathfrak{M}_n(2,0;2;\underline{a}) & \mathfrak{M}_n(1,0;1;\underline{a}) \end{pmatrix}.$$

Then by S(r;s) and **Mixed Mean Inequalities** (1) the rows of $\mathbb{S}$ increase strictly to the right, and the entries in the second row are strictly less than those in the first row, and strictly greater than those in the last row, except for the first column all of whose entries are $\mathfrak{G}_n(\underline{a})$, and for the last column all of whose entries are $\mathfrak{A}_n(\underline{a})$. A similar matrix is discussed in **Mixed Mean Inequalities** COMMENTS (ii).
(vi) No general relations are known between the first and last rows of $\mathbb{S}$ except for the following special case: $\mathfrak{M}_3(2,0;2;\underline{a}) \le \mathfrak{M}_3(0,1;2;\underline{a})$ or
$$\sqrt{\frac{ab+bc+ca}{3}} \le \sqrt[3]{\frac{(a+b)(b+c)(c+a)}{8}}. \tag{1}$$
It is proved by using (GA) and S(r;s). This should be compared with **Mixed Means Inequalities** (3).

EXTENSIONS [POPVICIU TYPE] Let $\underline{a}$ be a positive $(n+m)$-tuple, r,k integers with $1 \le r < k \le n+m$, and put $u = \max\{r-n, 0\}$, $v = \min\{r, m\}$, $w = \max\{k-n, 0\}$, $x = \min\{k, m\}$.
(a) If $v \ge w$ and $r - u \le k - x$ then
$$\left(\frac{\mathfrak{P}_{n+m}^{[r]}(\underline{a})}{\mathfrak{P}_{n+m}^{[k]}(\underline{a})}\right)^k \ge \left(\frac{\mathfrak{P}_n^{[r-u]}(\underline{a})}{\mathfrak{P}_n^{[k-x]}(\underline{a})}\right)^{k-x} \left(\frac{\overline{\mathfrak{P}}_m^{[v]}(\underline{a})}{\overline{\mathfrak{P}}_m^{[w]}(\underline{a})}\right)^w.$$

(b) If $v \le w$
$$\left(\frac{\mathfrak{P}_{n+m}^{[r]}(\underline{a})}{\mathfrak{P}_{n+m}^{[k]}(\underline{a})}\right)^k \le \left(\frac{\overline{\mathfrak{P}}_m^{[v]}(\underline{a})}{\overline{\mathfrak{P}}_m^{[w]}(\underline{a})}\right)^w.$$

Here we use the notation $\overline{\mathfrak{P}}_m(\underline{a}) = \mathfrak{P}_m(a_{n+1}, \ldots, a_{n+m})$.

COMMENTS (vii) See also **Complete Symmetric Mean Inequalities** COMMENTS (ii), **Hamy Mean Inequalities**, **Ky Fan's Inequality** EXTENSIONS (b), **Marcus & Lopez's Inequality** (b), **Muirhead Symmetric Function and Mean Inequalities**.

REFERENCES [AI, pp. 95–107], [BB, p. 11], [MI, pp. 290–294, 301–303, 305–306, 310–311], [MO, pp. 78–80], [MPF, pp. 15–16]; [Ku, Ku & Zhang].

Symmetrical Rearrangement Inequalities
See **Spherical Rearrangement Inequalities**.

Symmetrization Inequalities [PÓLYA & SZEGÖ; STEINER] (a) Symmetrization of a domain in $\mathbb{R}^2$ with respect to a line leaves the area unchanged but decreases the perimeter, polar moment of inertia about the centre of gravity, capacity and principal frequency.
(b) Symmetrization of a domain in $\mathbb{R}^3$ with respect to a plane leaves the volume unchanged but decreases the surface area and capacity.

COMMENTS (i) Symmetrization of a domain in $\mathbb{R}^2$ with respect to a line is called *Schwarz symmetrization*, while symmetrization of a domain in $\mathbb{R}^3$ with respect to a plane is called *Steiner symmetrization*. These symmetrizations can be defined in $\mathbb{R}^n, n \geq 1$.
(ii) The importance of symmetrization lies in the fact that the processe reduces certain functionals. So by obtaining a sequence of symmetrizations that converge to a ball, on which the minimum occurs and on which the functional can be calculated, inequalities can be proved.
(iii) See also **Capacity Inequalities, Frequency Inequalities, Isodiametric Inequalities, Isoperimetric Inequalities, Moment of Inertia Inequalities**.

REFERENCES [Evans & Gariepy, pp. 67–70], [Lieb & Loss, pp. 79–80], [Pólya & Szegö, 1951, pp. 6–7, 153–154].

Szegö's Inequality If $\underline{a}$ is a decreasing non-negative n-tuple and if f convex on $[0, a_1]$ with $f(0) \leq 0$ then

$$\sum_{i=1}^{n}(-1)^{i-1}f(a_i) \geq f\left(\sum_{i=1}^{n}(-1)^{i-1}a_i\right).$$

COMMENTS (i) The condition $f(0) \leq 0$ can be omitted if n is odd, the case given originally by Szegö; the above result is due to Bellman. The case n odd holds under the weaker assumption of Wright convexity; see **Convex Function Inequalities** COMMENTS (ix).
(ii) This result is a consequence of **Order Inequalities** (b), and is related to **Steffensen's Inequalities** (1).
(iii) The case of $f(x) = x^r, r > 1$, is Weinberger's inequality.

EXTENSIONS (a) [BRUNK] If f is convex on $[0, b]$ with $b \geq a_1 \geq \cdots \geq a_n \geq 0$, and if $1 \geq h_1 \geq \cdots \geq h_n \geq 0$ then

$$\sum_{i=1}^{n}(-1)^{i-1}h_i f(a_i) \geq f\left(\sum_{i=1}^{n}(-1)^{i-1}h_i a_i\right).$$

(b) [OLKIN] If $\underline{w}$ is a decreasing n-tuple with $0 \leq w_i \leq 1, 1 \leq i \leq n$, and $\underline{a}$ is a decreasing non-negative n-tuple, and if f is a convex function on $[0, a_1]$, then

$$\left(1 - \sum_{i=1}^{n}(-1)^{i-1}w_i\right)f(0) + \sum_{i=1}^{n}(-1)^{i-1}w_i f(a_i) \geq f\left(\sum_{i=1}^{n}(-1)^{i-1}w_i a_i\right).$$

COMMENTS (iv) For a related inequality see **Petrović's Inequality**.

REFERENCES [AI, pp. 112–114], [BB, pp. 47–49], [MO, p. 98], [MPF, p. 359], [PPT, pp. 156–159].

Székely, Clark & Entringer Inequality   (a) *If $\underline{a}$ is an increasing real sequence, $p \geq 1, n \geq 2$ then*

$$\sum_{i=1}^n \tilde{\Delta} a_{i-1} a_{n+1-i}^p \leq \left(\sum_{i=1}^n \tilde{\Delta} a_{i-1} a_{n+1-i}^{1/p} \right)^p,$$

where $a_0 = 0$.
(b) [INTEGRAL ANALOGUE] *If $f \in \mathcal{L}([0,1]), f \geq 0$ and $p \geq 1$, then*

$$\int_0^1 f(x) \left(\int_0^{1-x} f \right)^p dx \leq \left(\int_0^1 f(x) \left(\int_0^{1-x} f \right)^{1/p} dx \right)^p.$$

In both inequalities the constant is best possible.

COMMENTS   (i) These are extensions, due to Alzer, of the original results.
(ii) The result (a) is implied by **Alzer's Inequalities** (c).
REFERENCES   [Alzer, 1994[(3)]].

- end of S -

T

Talenti's Inequality *If $a > 0$ and f is positive and decreasing on $[a,b]$ then*

$$\log\left(1 + \frac{1}{1+af(a)}\int_a^b f\right) \leq \int_a^b \frac{f(t)}{1+tf(t)}\,dt.$$

COMMENTS This result has been extended and provided with a converse by Lemmert & Alzer.

REFERENCES [*General Inequalities, vol. 6*, pp. 441–443].

Tchakaloff's Inequality
See **Čakalov's Inequality**.

Tchebysheff's Inequality
See **Čebišev's Inequality**.

Three Circles Theorem
See **Hadamard's Three Circles Theorem**.

Three Lines Theorem
See **Phragmén-Lindelöf Inequality**.

Thunsdorff's Inequality *If f is non-negative and concave on $[a,b]$ and if $0 < r < s$, then*

$$\left(\frac{1+s}{b-a}\int_a^b f^s\right)^{1/s} \leq \left(\frac{1+r}{b-a}\int_a^b f^r\right)^{1/r}.$$

COMMENTS (i) This result has been extended to n-convex functions, and to weighted means; see [*MPF*].

(ii) This is a converse to (r;s), and a particular case is **Favard's Inequalities** (b). The case $r=1, s=2$ is the *Frank-Pick inequality*. The result is due to Thunsdorff but his proof was not published; a proof was given by Berwald. The discrete case below was proved much later.

DISCRETE ANALOGUE   If $\underline{a}$ is both increasing and concave and if $0 < r < s$ then

$$\mathfrak{M}_n^{[s]}(\underline{a};\underline{w}) \le C \mathfrak{M}_n^{[r]}(\underline{a};\underline{w}), \tag{1}$$

where if $\underline{b} = \{0, 1, \ldots, n-1\}$ then

$$C = \frac{\mathfrak{M}_n^{[s]}(\underline{b};\underline{w})}{\mathfrak{M}_n^{[r]}(\underline{b};\underline{w})}. \tag{2}$$

If $\underline{a}$ is decreasing then (1) holds with $\underline{b}$ in C replaced by $\underline{c} = \{n-1, \ldots, 1, 0\}$.
If $\underline{a}$ is convex and $0 < s < r$ then (~ 1) holds with C given by (2).

COMMENTS   (iii) The method of proof is based on **Power Mean Inequalities** EXTENSIONS (j). The result has extensions to n-convex sequences.

REFERENCES   [AI, p. 307], [MPF, pp. 50–56].

Ting's Inequalities   If $f \in C([0, a])$, is non-negative and convex, and if $\alpha > 2$ then

$$\frac{a(\alpha-1)}{\alpha+1} \le \frac{\int_0^a x^{\alpha-1} f(x)\,dx}{\int_0^a x^{\alpha-2} f(x)\,dx} \le \frac{a\alpha}{\alpha+1},$$

with equality on the left-hand side if $f(x) = a - x$, and on the right-hand side if $f(x) = x$.

COMMENTS   Ting gave the case α an integer; the general result is due to Ross, who also gave a simpler proof and other extensions of Ting's result.

REFERENCES   [General Inequalities, vol.4, pp. 119–130].

Totally Positive Function Inequalities   If $: \mathbb{R}^2 \to \mathbb{R}$ is a totally positive of order n function then for all $m, 1 \le m \le n$, and $x_1 < \cdots < x_m, y_1 < \cdots < y_m$,

$$\det\left(f(x_i, y_j)\right)_{1 \le i, j \le m} \ge 0.$$

COMMENTS   This is just the definition of this class of functions. A related class to the case $n = 2$ can be found in **Copula Inequalities**.

REFERENCES   [EM, Supp., pp. 469–470]; [Karlin, pp. 11–12], [Tong, ed., p. 54].

Trace Inequalities   [BELLMAN] If A, B are Hermitian matrices then:

(a)
$$\operatorname{tr}(AB) \le \frac{\operatorname{tr}(A^2) + \operatorname{tr}(B^2)}{2},$$

with equality if and only if $A = B$;

(b)
$$\operatorname{tr}(AB) \le \sqrt{\operatorname{tr}(A^2)\operatorname{tr}(B^2)},$$

with equality if and only if $A = kB$, for some constant k;

(c) $$\operatorname{tr}(AB) \le \sqrt{\operatorname{tr}(A^2 B^2)},$$

with equality if and only if A, B commute.

COMMENTS   The *trace* of a square matrix is the sum of its diagonal elements.
REFERENCES   [MPF, pp. 224-225]; [General Inequalities, vol.2, pp. 89–90].

Triangle Inequality
(a) If $\underline{a}, \underline{b}$ are real n-tuples then

$$|\underline{a} + \underline{b}| \le |\underline{a}| + |\underline{b}|, \tag{1}$$

with equality if and only if $\underline{a} \sim^+ \underline{b}$.

(b) If $\underline{a}, \underline{b}$ are real n-tuples then

$$|\underline{a} - \underline{b}| \ge ||\underline{a}| + |\underline{b}||,$$

with equality if and only if $\underline{a} \sim^+ \underline{b}$.

(c) If $\underline{a}, \underline{b}, \underline{c}$ are real n-tuples then

$$|\underline{a} - \underline{c}| \le |\underline{a} - \underline{b}| + |\underline{b} - \underline{c}|, \tag{T}$$

with equality only if $\underline{b}$ is between $\underline{a}$ and $\underline{c}$.

COMMENTS (i) We say that $\underline{b}$ is between $\underline{a}$ and $\underline{c}$ if for some $\lambda, 0 \le \lambda \le 1$, we have $\underline{b} = (1 - \lambda)\underline{a} + \lambda\underline{c}$.
(ii) (a) is the $p = 2$ case of (M), and a proof is given in any book on linear algebra.
(iii) (c) is the *Triangle Inequality*, and is a simple deduction from (1), which in turn follows form (T). As a result of this equivalence (1) is also referred to as (T).
(iv) Its name follows from the geometric interpretation of $|\underline{a} - \underline{b}|$ as the distance between $\underline{a}$ and $\underline{b}$.

EUCLID BOOK 1 PROPOSITION 20 THEOREM   Any two sides of a triangle are together greater than the third side.

EXTENSIONS   (a) If $\underline{a}_j, 1 \le j \le m$ are real n-tuples then

(i) $$\left|\sum_{j=1}^{m} \underline{a}_j\right| \le \sum_{j=1}^{m} |\underline{a}_j|,$$

with equality if and only if all are on the same ray from the origin;

(ii) $$|\underline{a}_m - \underline{a}_1| \le \sum_{j=2}^{m} |\underline{a}_j - \underline{a}_{j-1}|,$$

with equality if and only if $\underline{a}_2, \ldots, \underline{a}_{m-1}$ lie in order between $\underline{a}_1$ and $\underline{a}_m$.

(b) [RYSER] If $\underline{a}, \underline{b}$ are real n-tuples and if $\underline{m} = \min\{\underline{a}, \underline{b}\}, \underline{M} = \max\{\underline{a}, \underline{b}\}$ then

$$|\underline{a} + \underline{b}| \le |\underline{M}| + |\underline{m}| \le |\underline{a}| + |\underline{b}|.$$

COMMENTS   (v) An integral analogue of (a)(i) is **Integral Inequalities** (a).

CONVERSE INEQUALITIES [JANOUS]   If $\underline{a}, \underline{b}$ are real non-zero n-tuples then

$$\underline{a}.\underline{b}\,\frac{|\underline{a}| + |\underline{b}|}{|\underline{a}|\,|\underline{b}|} \leq |\underline{a} + \underline{b}|.$$

COMMENTS   (vi) For another converse inequality see **Wilf's Inequality**. In certain geometries the converse inequality replaces the triangle inequality as the basic inequality.
(vii) The triangle inequality occurs in various guises and in many generalizations; see **Absolute Value Inequalities** (1)–(3), **Complex Number Inequalities** (1), (2) (3), **Metric Inequalities** (1), **Norm Inequalities** (1), (2).

REFERENCES   [AI, pp. 170–171], [EM, vol.7, p. 363], [MPF, pp. 473–517].

Trigonometric Function Inequalities   (a) If $0 < |x| \leq \pi/2$ then

$$\cos x < \frac{\sin x}{x} < 1. \tag{1}$$

(b) If $0 \leq x < y \leq \pi/2$ then

$$\frac{x}{y} \leq \frac{\sin x}{\sin y} \leq \left(\frac{\pi}{2}\right)\frac{x}{y}.$$

(c) If $x \geq \sqrt{3}$ then

$$(x+1)\cos\frac{\pi}{x+1} - x\cos\frac{\pi}{x} > 1.$$

(d) [KUBO] If $0 \leq x \leq \pi$ then

$$\sin x(1 + \cos x) \leq \left(\sin\frac{x+\pi}{4}\right)\left(1 + \cos\frac{x+\pi}{4}\right).$$

(e) [CALDERÓN] If $1 < p \leq 2$, $-\pi/2 \leq \theta \leq \pi/2$ then for some positive λ,

$$\left(\frac{1 + \lambda\cos p\theta}{1 + \lambda}\right)^{1/p} \leq \lambda\cos\theta.$$

(f) If $x \in \mathbb{R}, n \geq 1$ then

$$\left|\left(\frac{\sin x}{x}\right)^{(n)}\right| \leq \frac{1}{n+1}, \quad \text{and} \quad \left|\left(\frac{1-\cos x}{x}\right)^{(n)}\right| \leq \frac{1}{n+1},$$

with equality in the first inequality only if n is even and $x = 0$, and in the second only if n is odd and $x = 0$.
(g) If $0 < x < \pi/2$ then

$$\log\sec x < \frac{1}{2}\sin x \tan x.$$

(h) If $0 \leq x, y \leq 1$ or $1 \leq x, y < \pi/2$

$$\tan x \tan y \leq \tan 1 \tan xy.$$

(j) If $0 < x < \pi/2$ then

$$\frac{x}{\pi - 2x} < \frac{\pi}{4} \tan x.$$

(k) If $\leq x, y \leq 1$

$$\arcsin x \arcsin y \leq \frac{1}{2} \arcsin xy.$$

(ℓ) [SHAFER] If $x > 0$ then

$$\arctan x > \frac{8\sqrt{3}x}{3\sqrt{3} + \sqrt{75 + 80x^2}}.$$

(m) If $a > b$ then for all $t \in \mathbb{R}, t \neq 0$,

$$\frac{b-a}{\sqrt{(a^2+t)(b^2+t)}} \leq \frac{1}{t}\left(\arctan\frac{b}{t} - \arctan\frac{a}{t}\right).$$

(n) Let us write

$$S_{2n-1}(x) = \sum_{i=1}^{n} \frac{(-1)^{i-1}x^{2i-1}}{(2i-1)!}, \quad C_{2n}(x) = \sum_{i=0}^{n} \frac{(-1)^{i-1}x^{2i}}{(2i)!};$$

then for all x, n,

$$(-1)^{n+1}\left(\cos x - C_{2n}(x)\right) \geq 0; \quad (-1)^{n+1}x\left(\sin x - S_{2n+1}(x)\right) \geq 0.$$

(o) If $0 < x < y < \sqrt{6}$ then, with the above notation,

$$\frac{S_{4n-1}(y)}{S_{4n-1}(x)} < \frac{\sin y}{\sin x} < \frac{S_{4n+1}(y)}{S_{4n+1}(x)}.$$

(p) If $n \geq 1, |x| < \pi/2$ then

$$\operatorname{cosec}^2 x - \frac{1}{2n+1} < \sum_{i=-n}^{n} \frac{1}{(x-i\pi)^2} < \operatorname{cosec}^2 x,$$

while if $0 < x \leq \pi/2$

$$\operatorname{cosec} x - \frac{x}{4n+1} < \sum_{i=-2n}^{} 2n\frac{(-1)^i}{x-i\pi} < \operatorname{cosec} x + \frac{x}{4n+2}.$$

(q) If $0 < a_i < \pi, 1 \leq i \leq n$ then

$$\mathfrak{G}_n(\sin \underline{a}) \leq \sin \mathfrak{A}_n(\underline{a}).$$

(r) If $z \in \mathbb{C}$ and $0 < |z| < 1$ then

$$|\cos z| < 2 \quad \text{and} \quad |\sin z| < \frac{6|z|}{5}.$$

COMMENTS (i) A geometric proof of (1) can be found in any elementary calculus book where it is needed to show that $\lim_{x \to 0} \sin x/x = 1$; a variant is given in **Jordan's Inequality**.
(ii) Kubo's result, (d), is a particular case of **Function Inequalities** (b).
(iii) Calderón's inequality, (e), was used in his proof of M.Riesz's theorem; see **Conjugate harmonic Function Inequalities** (a).
(iv) The inequalities in (f) follow from the identities

$$\left(\frac{\sin x}{x}\right)^{(n)} = \frac{1}{x^{n+1}} \int_0^x y^n \sin\left(y + \frac{(n+1)\pi}{2}\right) dy,$$

$$\left(\frac{1-\cos x}{x}\right)^{(n)} = \frac{1}{x^{n+1}} \int_0^x y^n \sin\left(y + \frac{n\pi}{2}\right) dy.$$

(v) (h) and (k) are simple deductions from (Č).
(vi) (n) is a special case of **Gerber's Inequality** EXTENSIONS.
(vii) Inequality (o) follows from ($\sim J$) and the concavity of sin on the interval $[0, \pi]$. Similar inequalities for the cosine and tangent functions can be obtained by using their strict concavity on the intervals $[-\pi/2, \pi/2]$, and $[0, \pi/2[$ respectively.
(viii) (p) follows from the strict concavity of sin on $[0, \pi]$.
(ix) (q) is a consequence of the log-concavity of $1/\sin x$; it should be compared with **Bessel Function Inequalities** COROLLARIES (a).
(x) (r) follows from **Integral Inequalities** (a) applied to the function complex function $(x+ti)^{-2}$.
(xi) See also **Enveloping Series Inequalities** COMMENTS (ii), **Function Inequalities** (a), **Jordan's Inequality**, **Lochs' Inequality**, **Ostrowski's Inequality** COMMENTS (iii), **Trigonometric Polynomial Inequalities**.

REFERENCES [AI, pp. 235, 240, 243–249], [MI, p. 232], [MPF, pp. 476–477]; [Abramowicz & Stegun, p. 75], [General Inequalities, vol.2, pp. 161–162; vol. 5, p. 143], [Niven, 1981, pp. 92–110], [Rudin, 1966, pp. 345–346]; [Kubo, T], [Shafer].

Trigonometric Integral Inequalities [OSTROWSKI] If $f \in \mathcal{L}([a,b])$ is monotone then

$$\left|\int_a^b f(x) \cos x \, dx\right| \leq 2(|f(a) - f(b)| + |f(b)|).$$

COMMENTS (i) This follows from **Bounded Variation Function Inequalities**, EXTENSIONS.
(ii) See also **Fourier Transform Inequalities**.
REFERENCES [AI, p. 301].

Trigonometric Polynomial Inequalities

A *trigonometric polynomial of degree at most n* is anything of the form

$$\sum_{k=0}^{n} a_k \sin kx + b_i \cos kx,$$

where $n \in \mathbb{N}, a_k, b_k \in \mathbb{R}, 0 \leq i \leq n$; that is, it is a partial sum of the first $n+1$ terms of a trigonometric series.
The same name is also used for either $\sum_{k=0}^{n} c_k e^{ikx}$, $c_k \in \mathbb{C}, 0 \leq k \leq n$, or $\sum_{k=-n}^{n} c_k e^{ikx}$, $c_k \in \mathbb{C}, -n \leq k \leq n$.

(a) If $0 \leq x_i \leq \pi$, $1 \leq k \leq n$, then

$$\sum_{k=1}^{n} \sin x_k \leq n \sin\left(\frac{1}{n}\sum_{k=1}^{n} x_k\right).$$

with equality if and only if $x_1 = \cdots = x_n$.

(b) If $n \geq 1, x \in \mathbb{R}$ then

$$0 < \sum_{k=1}^{n} \frac{\sin kx}{k} < \int_0^\pi \frac{\sin x}{x} \, dx = 1.8519\ldots < 1 + \frac{\pi}{2}.$$

(c) If $n \geq 1, x \in \mathbb{R}$ then

$$\left|\sum_{k=1}^{n} \frac{(-1)^k}{k} \sin kx\right| < \sqrt{2}|x|.$$

(d) If $0 < x < 2\pi$ then

$$\left|\sum_{k=m+1}^{m+n} e^{ikx}\right| < \operatorname{cosec}\frac{x}{2}.$$

(e) If $0 < \beta < 1, 0 < x \leq \pi$ then

$$\left|\sum_{k=1}^{n} \frac{\cos kx}{k^\beta}\right| \leq C_\beta x^{\beta-1}; \quad \left|\sum_{k=1}^{n} \frac{\sin kx}{k^\beta}\right| \leq C_\beta x^{\beta-1}.$$

(f) If $0 < x \leq \pi$ then

$$\left|\sum_{k=1}^{n} \frac{\cos kx}{k}\right| \leq C - \log x.$$

(g) [BARI[1]] If T_n is a trigonometric polynomial of degree at most n then
(i):
$$\|T_n(x)\|_{\infty,[-\pi,\pi]} \leq (n+1)\|T_n(x)\sin x\|_{\infty,[-\pi,\pi]};$$
(ii): if $1 \leq p < q < \infty$ then
$$\|T_n\|_{q,[-\pi,\pi]} \leq 2^{1+1/p-1/q}\|T_n\|_{p,[-\pi,\pi]}.$$

(h)
$$\sum_{k=1}^{n}(n+1-k)\sin ix > 0; \qquad \sum_{k=1}^{n}\sin kx + \frac{1}{2}\sin(n+1)x \geq 0.$$

COMMENTS (i) (a) is an immediate consequence of (J).
(ii) (b), (c) should be compared with the **Fejér-Jackson Inequality**. (1), (2).
(iii) See also **Bernšteĭn's Inequality** (a), **Bohr-Favard Inequality**, **Dirichlet Kernel Inequalities**, **Fejér Kernel Inequalities**, **Jackon's Inequality**, **Integral Mean Value Theorems** EXTENSIONS (b), **Littlewood's Problem**, **Nikol'skiĭ's Inequality**, **Poisson Kernel Inequalities**, **Young's Inequalities** (d), and EXTENSIONS.

REFERENCES [MPF, pp. 579–580, 583–584, 613–614]; [Zygmund, vol.I, p. 191]; [Alzer, 1992[(3)]], [Gluchoff & Hartmann].

Trudinger's Inequality If $\Omega \subseteq \mathbb{R}^n, n \geq 2$ is an open domain, $|\Omega| < \infty$ and $f \in \mathcal{W}^{1,n}(\Omega)$ with $f = 0$ on $\partial\Omega$, $\int_\Omega |\nabla f|^n \leq 1$, and if $\alpha \leq \alpha_n = n a_n^{1/n-1}$, then

$$\int_R e^{\alpha f^{n'}} \leq C_n |\Omega|,$$

where n' is the conjugate index, and C_n is a constant that depends on n.

COMMENTS (i) The definition of $\mathcal{W}^{1,n}(\Omega)$ is in **Sobolev's Inequalities**. The inequality is a limiting case of **Sobolev's Inequalities** (b).
(ii) If $\alpha > \alpha_n$ the integral is still finite but can be arbitrarily large.
(iii) The result is Trudinger's, but the exact range of α, the value of α_n, is due to Moser.
(iii) The cases of $f \in \mathcal{W}^{1,p}(\Omega), p \neq n$, are easier than this limiting case; see [Moser].
REFERENCES [EM, Supp., pp. 377–378]; [Moser].

Turán's Inequalities (a) If $a_k, z_k \in \mathbb{C}, 1 \leq k \leq n$, with $\min_{1 \leq k \leq n} |z_k| = 1$ and if $p \in \mathbb{R}$ then

$$\max_{p=m+1,\ldots,m+n} \left|\sum_{k=1}^{n} a_k z_k^p\right| \geq \left(\frac{n}{2e(m+n)}\right)^n \left|\sum_{k=1}^{n} a_k\right|. \qquad (1)$$

---
[1] Н Бари.

(b) With the notation of (a), but now assuming that $1 = |z_1| \geq |z_2| \geq \cdots \geq |z_n|$,

$$\max_{p=m+1,\ldots,m+n} \left| \sum_{k=1}^{n} a_k z_k^p \right| \geq 2 \left(\frac{n}{4e(m+n)} \right)^n \min_{1 \leq j \leq n} \left| \sum_{k=1}^{j} a_j \right|. \qquad (2)$$

(c) With the notation of (a), but now assuming that $z_k \neq 0, 1 \leq k \leq n$, and $\max_{1 \leq k \leq n} |z_k| = \Lambda$, $\min_{1 \leq k, j \leq n, k \neq j} |z_k - z_j| = \lambda > 0$ then

$$\max_{p=m+1,\ldots,m+n} \frac{\left| \sum_{k=1}^{n} a_k z_k^p \right|}{\sum_{k=1}^{n} |a_k| |z_k|^p} \geq \frac{1}{n} \left(\frac{\lambda}{2\Lambda} \right)^{n-1}.$$

COMMENTS (i) The best value of the constant on the right-hand side of (1), due to Makai and de Bruijn, is

$$\left(\sum_{k=0}^{n-1} \binom{m+k}{k} 2^k \right)^{-1}.$$

(ii) The constant in (2) is best possible and is due to Kolesnik & Straus.

SPECIAL CASES (a) If $z_k \in \mathbb{C}, 1 \leq k \leq n$, with $\min_{1 \leq k \leq n} |z_k| = 1$ and if $p \in \mathbb{N}$ then

$$\max_{1 \leq p \leq n} \left| \sum_{k=1}^{n} z_k^p \right| \geq 1,$$

with equality if and only if the z_k are vertices of a regular polygon on the unit circle.

(b) [TURÁN, ATKINSON] If $z_k \in \mathbb{C}, |z_k| \leq 1, 1 \leq k \leq n, z_n = 1$ and if $p \in \mathbb{N}$ then

$$\max_{1 \leq p \leq n} \left| \sum_{k=1}^{n} z_k^p \right| > \frac{1}{3}. \qquad (3)$$

COMMENTS (iii) The special cases do not follow directly from the main results.
(iv) The best value of the constant on the right-hand side of (3) is not known.

REFERENCES [AI, pp. 122–125], [MPF, pp. 651–659].

- end of T -

U-V

Ultraspherical Polynomial Inequalities (a) [LORCH] If $P_n^{(\alpha)}$ is an ultraspherical, Gegenbauer, polynomial and if $0 < \alpha < 1$, then

$$\sin^\alpha \theta |P_n^{(\alpha)}(\cos\theta)| \leq \frac{1}{(\alpha-1)!} \left(\frac{2}{n}\right)^{1-\alpha}, \quad 0 \leq \theta \leq \pi.$$

The constant is best possible.

(b) If $x^2 \leq 1 - \left(\dfrac{1-\alpha}{n+\alpha}\right)^2$ then

$$\left(P_{n+1}^{(\alpha)}(x)\right)^2 \geq P_n^{(\alpha)}(x) P_{n+2}^{(\alpha)}(x).$$

COMMENTS (i) The inequality in (b) should be compared with **Bessel Function Inequalities** (c), and **Legendre Polynomial Inequalities** (b).

EXTENSIONS [LORCH] With the above notation,

$$\sin^\alpha \theta |P_n^{(\alpha)}(\cos\theta)| \leq \frac{1}{(\alpha-1)!} \left(\frac{2}{n+\alpha}\right)^{1-\alpha}, \quad 0 \leq \theta \leq \pi.$$

COMMENTS (ii) A particular case of this is the first inequality in **Martin's Inequalities**. Extensions of the second Martin inequality have been given by Common.

REFERENCES [General Inequalities, vol. 1, pp. 35–38], [Szegö, p. 171]; [Common].

Univalent Function Inequalities
See **Area Theorems, Bieberbach's Conjecture, Distortion Theorems, Entire Function Inequalities, Grunsky's Inequalities, Rotation Theorems**.

Upper and Lower Limit Inequalities
If $\underline{a}, \underline{b}$ are real sequences then

$$\limsup \underline{a} + \liminf \underline{b} \leq \limsup(\underline{a} + \underline{b}) \leq \limsup \underline{a} + \liminf \underline{b},$$

provided the terms in the inequalities are defined.

(b) If $\underline{a}, \underline{b}$ are non-negative sequences then

$$\limsup \underline{a} \liminf \underline{b} \leq \limsup \underline{a}\,\underline{b} \leq \limsup \underline{a} \limsup \underline{b},$$

provided the terms in the inequalities are defined.

COMMENTS (i) These are simple consequences of **Inf and Sup Inequalities**.
(ii) The operations on these inequalities are understood to be in $\overline{\mathbb{R}}$. In addition we could assume the terms of the sequences to be taking values in $\overline{\mathbb{R}}$.
(iii) There are analogous results for functions taking values in $\overline{\mathbb{R}}$; see also **L'Hôpital's Rule**.
REFERENCES [Bourbaki,1960, p. 164].

Upper Semi-continuous Function Inequalities
See **Semi-continuous Function Inequalities**.

van der Waerden's Conjecture
If S is an $n \times n$ doubly stochastic matrix then
$$\mathrm{per}S \geq \frac{n!}{n^n},$$
with equality if and only if all the entries in S are equal to $1/n$.

COMMENTS (i) This conjecture was made in 1926 and caused much interest; it was settled by Falikman, and Egoričev[1] in 1980-81. The standard reference on permanents, [Minc], was written two years before this solution was published.
(ii) For a definition of a doubly stochastic matrix see **Order Inequalities** COMMENTS (i).
(iii) See also **Permanent Inequalities**.
REFERENCES [EM, vol.7, p. 127]; [General Inequalities, vol.3, pp. 23–40], [Minc, pp. 11, 73–102].

Variation Inequalities
If $f : [a,b] \to \mathbb{R}$ and $\delta > 0$ define
$$V(f;\delta) = \sup\left\{\sum |f(y) - f(x)|\right\},$$
where the sup is taken over all collections of nonoverlapping intervals $[x,y]$ with $\sum |y - x| < \delta$, then
$$V(f^*;\delta) \leq V(f;\delta).$$

COMMENTS See also **Bounded Variation Function Inequalities**.
REFERENCES [Yanagihara].

Vietoris's Inequality
If
$$J(x,m,n) = \frac{1}{B(m,n)} \int_x^1 t^{m-1}(1-t)^{n-1}\,dt,$$
then
$$J\left(\frac{r}{r+s}, r, s+1\right) < \frac{1}{2} < J\left(\frac{r}{r+s}, r+1, s\right).$$

COMMENTS J is called the *incomplete Beta function*; B denotes the Beta function, for a definition see **Beta Function Inequalities**.
REFERENCES [Lochs].

---

[1] Д И Фаликман, Г П Егоричев.

von Neumann & Jordan Inequality *If X is a Banach space there is a unique constant $C, 1 \leq C \leq 2$, such that if $x, y \in X$, not both zero, then*

$$\frac{1}{C} \leq \frac{||x+y||^2 + ||x-y||^2}{2(||x||^2 + ||y||^2)} \leq C.$$

COMMENTS   (i) By use of the **Clarkson Inequalities** it can be shown that if $X = \ell_p$ or $\mathcal{L}^p$ then $C = 2^{(2-p)/p}$ if $1 \leq p \leq 2$, and $C = 2^{(p-2)/p}$ if $p \geq 2$.
(ii) $C = 1$ if and only if X is finite-dimensional or a Hilbert space.

REFERENCES   [MPF, p. 550]; [Mitrinović & Vasić, 1977].

- end of U-V -

W

Wagner's Inequality
See **Cauchy's Inequality**, EXTENSIONS (c).

Walker's Inequality

If $a, b, c, x, y, z > 0$ where

$$x = b + c - a, \quad y = c + a - b, \quad z = a + b - c,$$

then

$$\frac{1}{a} + \frac{1}{b} + \frac{1}{c} \leq \frac{1}{xyz}.$$

COMMENTS This is an example of **Order Inequalities** (a), (b) since $(a, b, c) = (x, y, z)S$, where S is doubly stochastic.

REFERENCES [MO, p. 72].

Wallis's Inequality If $n \geq 1$ then

$$\frac{1}{\sqrt{\pi(n+1/2)}} < \frac{(2n-1)!!}{(2n)!!} < \frac{1}{\sqrt{\pi(n+1/4)}}.$$

COMMENTS (i) For the notation see **Factorial Function Inequalities** (f).
(ii) In Wallis's original result the right-hand side was $1/\sqrt{\pi n}$; the improvement is due to Kazarinoff.
Incidentally Wallis introduced the notation ∞.
(iii) The important use of this result is to give *Wallis's Formula*:

$$\frac{\pi}{2} = \lim_{n \to \infty} \frac{(2n)!!^2}{(2n-1)!!^2 (2n+1)}.$$

REFERENCES [AI, pp. 192–193, 287], [EM, vol.9, p. 441]; [Apostol, vol.II, p. 450–453], [Borwein & Borwein, pp. 338, 343], [Kazarinoff, pp. 47–48, 65–67].

Walsh's Inequality If $\underline{a}, \underline{w}$ are positive n-tuples

$$\mathfrak{A}_n(\underline{a}; \underline{w})\mathfrak{A}_n(\underline{a}^{-1}; \underline{w}) \geq 1,$$

with equality if and only if $\underline{a}$ is constant.

COMMENTS (i) The proof is by induction, the case $n = 1$ being trivial.
(ii) This inequality is easily seen to be equivalent to (HA). It can be used to prove the apparently stronger (GA).
(iii) For a converse inequality see **Kantorović's Inequality**.

REFERENCES   [AI, pp. 206–207], [MI, pp. 66–67].

Weak Young Inequality
See **Hardy-Littlewood-Sobolev Inequality** COMMENTS (iv).

Weierstrass's Inequalities (a) If $0 < a_i < 1$, $w_i \geq 1$, $1 \leq i \leq n$, then

$$\prod_{i=1}^{n}(1 - a_i)^{w_i} > 1 - \sum_{i=1}^{n} w_i a_i,$$

$$1 + \sum_{i=1}^{n} w_i a_i < \prod_{i=1}^{n}(1 + a_i)^{w_i} < \frac{1}{\prod_{i=1}^{n}(1 - a_i)^{w_i}}.$$

(b) If in addition $\sum_{i=1}^{n} w_i a_i < 1$ then

$$\prod_{i=1}^{n}(1 + a_i)^{w_i} < \frac{1}{1 - \sum_{i=1}^{n} w_i a_i},$$

$$\prod_{i=1}^{n}(1 - a_i)^{w_i} < \frac{1}{1 + \sum_{i=1}^{n} w_i a_i}.$$

COMMENTS (i) Taking $w_i = 1, 1 \leq i \leq n$, and combining the above results we get, under the appropriate conditions, the classical *Weierstrass Inequalities*:

$$1 + \sum_{i=1}^{n} a_i < \prod_{i=1}^{n}(1 + a_i) < \frac{1}{1 - \sum_{i=1}^{n} a_i}, \qquad (1)$$

$$1 - \sum_{i=1}^{n} a_i < \prod_{i=1}^{n}(1 - a_i) < \frac{1}{1 + \sum_{i=1}^{n} a_i}.$$

(ii) A proof of (1) can be given using **Chong's Inequalities** (2).

RELATED INEQUALITIES (a) If $0 \leq a_i \leq 1/2$, $1 \leq i \leq n$, $n \geq 2$, then

$$\frac{1}{2^n} \leq \frac{\prod_{i=1}^n (1+a_i)}{\prod_{i=1}^n (1+(1-a_i))} \leq \frac{1+\sum_{i=1}^n a_i}{1+\sum_{i=1}^n (1-a_i)} \leq \frac{1+\prod_{i=1}^n a_i}{1+\prod_{i=1}^n (1-a_i)} \leq 1,$$

equalities occur if and only if $a_1 = \cdots = a_n = 1/2$ or 0.

(b) [FLANDERS] If $0 < \underline{a} < 1$ then

$$\frac{1+\mathfrak{A}_n(\underline{a};\underline{w})}{1-\mathfrak{A}_n(\underline{a};\underline{w})} \leq \mathfrak{G}_n\left(\frac{1+\underline{a}}{1-\underline{a}};\underline{w}\right), \tag{2}$$

with equality if and only if $\underline{a}$ is constant.

COMMENTS (iii) In the case of equal weights (2) reduces to **Geometric Mean Inequalities** (2).
(iv) See also **Geometric-Arithmetic Mean Inequality** EXTENSIONS (b), **Hölder's Inequality** COMMENTS (vii), **Ky Fan's Inequality**, **Myer's Inequality**.

REFERENCES [AI, pp. 35, 210], [HLP, p. 60], [MI, p. 21], [MPF, pp. 69–77]; [Bourbaki, 1960, pp. 176–177].

Weinberger's Inequality

See **Szegö's Inequality** COMMENTS (iii).

Weyl's Inequalities

(a) If A, B are $n \times n$ Hermitian matrices then

$$\lambda_{(k)}(A) + \lambda_{(1)}(B) \leq \lambda_{(k)}(A+B) \leq \lambda_{(k)}(A) + \lambda_{(n)}(B), \quad 1 \leq k \leq n.$$

(b) If A is a square matrix then

$$\big(\log|\lambda_1(A)|,\ldots,\log|\lambda_n(A)|\big) \prec \big(\log|\sigma_1(A)|,\ldots,\log|\sigma_n(A)|\big),$$

where $\sigma_s(A)$, $1 \leq s \leq n$ denotes the singular values of A.

(c) With the notation of (b), and $p > 0$,

$$\sum_{j=1}^n |\lambda_j(A)|^p \leq \sum_{j=1}^n |\sigma_j(A)|^p.$$

COMMENTS (i) The singular values of a matrix A are the eigenvalues of $(AA^*)^{1/2}$.
(ii) (c) is a corollary of (b) and **Order Inequalities** (b).
(iii) For another inequality of Weyl see the **Heisenberg-Weyl Inequality**.

REFERENCES [MO; pp. 231–233]; [General Inequalities, vol. 4, pp. 213–219], [Horn & Johnson, pp. 181–182], [König, p. 24], [Marcus & Minc, pp. 116–117].

Whiteley Mean Inequalities

A natural generalization of the symmetric and complete symmetric functions are the *Whiteley symmetric functions*, $t_n^{[k,s]}, n = 1, 2, \ldots, k \in \mathbb{N}, s \in \mathbb{R}, s \neq 0$, that are generated by:

$$\sum_{k=0}^{\infty} t_n^{[k,s]}(\underline{a}) x^k = \begin{cases} \prod_{i=1}^{n}(1 + a_i x)^s, & \text{if } s > 0, \\ \prod_{i=1}^{n}(1 - a_i x)^s, & \text{if } s < 0; \end{cases}$$

where $\underline{a}$ is a positive n-tuple.

The *Whiteley means*, of order n, of the positive n-tuple $\underline{a}$ are:

$$\mathfrak{W}_n^{[r,s]}(\underline{a}) = \begin{cases} \left(\dfrac{t_n^{[r,s]}(\underline{a})}{\binom{ns}{r}} \right)^{1/r}, & \text{if } s > 0, \\ \left(\dfrac{t_n^{[r,s]}(\underline{a})}{(-1)^r \binom{ns}{r}} \right)^{1/r}, & \text{if } s < 0. \end{cases}$$

(a) If k is an integer, $1 \le k \le n, \ne 0$, then

$$\min \underline{a} \le \mathfrak{W}_n^{[k,s]}(\underline{a}) \le \max \underline{a},$$

with equality if and only if $\underline{a}$ is constant.

(b) If $s > 0$, k an integer, $1 \le k < s$, when s is not an integer, or $1 \le k < ns$, when s is an integer,

$$\left(\mathfrak{W}_n^{[k,s]}(\underline{a}) \right)^2 \ge \mathfrak{W}_n^{[k-1,s]}(\underline{a}) \mathfrak{W}_n^{[k+1,s]}(\underline{a}). \tag{1}$$

If $s < 0$ then ($\sim$1) holds. In both cases there is equality if and only if $\underline{a}$ is constant.

(c) If $s > 0$, k, ℓ are integers, with $1 \le k < \ell < s+1$, when s is not an integer, and $1 \le k < \ell < ns$, when s is an integer, then

$$\mathfrak{W}_n^{[\ell,s]}(\underline{a}) \le \mathfrak{W}_n^{[k,s]}(\underline{a}). \tag{2}$$

If $s < 0$ then ($\sim$2) holds. In both cases there is equality if and only if $\underline{a}$ is constant.

(d) If $s > 0, k \in \mathbb{N}$, and if $k < s+1$ if s is not an integer, then

$$\mathfrak{W}_n^{[k,s]}(\underline{a} + \underline{b}) \ge \mathfrak{W}_n^{[k,s]}(\underline{a}) + \mathfrak{W}_n^{[k,s]}(\underline{b}).$$

If $s < 0$ this inequality is reversed. The inequality is strict unless either $k = 1$ or $\underline{a} \sim \underline{b}$.

COMMENTS (i) These results generalize the analogous results in **Complete Symmetric Mean Inequalities, Elementary Symmetric Function Inequalities, Symmetric Mean Inequalities**. They have been further generalized by Whiteley and Menon.

REFERENCES [BB, pp. 35–36], [MI, pp. 317–333], [MO, p. 82], [MPF, pp. 166–168].

Wilf's Inequality *If $\alpha \in \mathbb{R}, 0 < \theta < \pi/2$, and if $\alpha - \theta \leq \arg z_k \leq \alpha + \theta, 1 \leq k \leq n$ then*

$$\left| \sum_{k=1}^{n} z_k \right| \geq \cos \theta \sum_{k=1}^{n} |z_k|.$$

COMMENTS This is a converse of **Complex Number Inequalities** EXTENSIONS (a). It seems to have been first proved by Petrović.

EXTENSIONS [JANIĆ, KEČKIĆ & VASIĆ] *If $\alpha \in \mathbb{R}, 0 < \theta < \pi/2$, $\alpha \leq \arg z_k \leq \alpha + \theta$, $1 \leq k \leq n$, then*

$$\left| \sum_{k=1}^{n} z_k \right| \geq \max \left\{ \cos \theta, \frac{1}{\sqrt{2}} \right\} \sum_{k=1}^{n} |z_k|.$$

INTEGRAL ANALOGUES *If $f : [a, b] \to \mathbb{C}$ is integrable and if for some θ, $0 < \theta < \pi/2$, $-\theta \leq \arg f(x) \leq \theta$, then*

$$\left| \int_a^b f \right| \geq \cos \theta \int_a^b |f|.$$

REFERENCES [AI, pp. 310–311], [MI, p. 167], [MPF, pp. 492–497]; [PPT, pp. 128–132].

Wilson's Inequalities
See **Nanson's Inequality** (2).

Wirtinger's Inequality
If f is a function of period 2π with $f, f' \in \mathcal{L}^2([0, 2\pi])$ and $\int_0^{2\pi} f = 0$, then

$$\int_0^{2\pi} f^2 \leq \int_0^{2\pi} f'^2,$$

with equality if and only if $f(x) = A \cos x + B \sin x$.

COMMENTS (i) A discrete analogue can be found in **Ky Fan-Taussky-Todd Inequalities** and [Pachpatte]; see also **Ozeki's Inequalities** COMMENTS (i).
(ii) This inequality has given rise to considerable research. A full discussion of its history of can be found in [AI]. Higher dimensional analogues are **Friederich's Inequality** and **Poincaré's Inequality**.

EXTENSIONS (a) [TANANIKA] *If $f' \in \mathcal{L}^2([0, 1]), f \in \mathcal{L}^p([0, 1]), p \geq 1$, and $\int_0^{2\pi} f = 0$, then*

$$\|f\|_{p,[01]} \leq \left(\sqrt{\frac{p}{\pi}} \right) \left(\frac{1}{2^{(p-1)/p}(p+2)^{(p+2)/2p}} \right) \left(\frac{((p+2)/2p)!}{(1/p)!} \right) \|f'\|_{2,[0,1]}.$$

(b) [VORNICESCU] *If $f \in C^1([0, 2\pi])$ with $f(0) = f(2\pi) = 0$ and $\int_0^{2\pi} f = 0$ then*

$$\frac{1}{\pi} \sum_{n=2}^{\infty} (n^2 - 1)(a_n^2 + b_n^2) + \int_0^{2\pi} f^2 \leq \int_0^{2\pi} f'^2,$$

where a_n, b_n are the Fourier coefficients of f.
(c) [ALZER] If f is a real valued continuously differentiable function of period 2π with $\int_0^{2\pi} f = 0$ then

$$\max_{0 \leq x \leq 2\pi} f^2(x) \leq \frac{\pi}{6} \int_0^{2\pi} f'^2.$$

COMMENTS   (iv) Vornicescu has given a discrete analogue of his result; his proof depends on applying Wirtinger's inequality to a suitable function. For another inequality involving Fourier coefficients see **Boas's Inequality**.

(v) See also **Benson's Inequalities** COMMENTS, **Opial's Inequality**.

REFERENCES   [AI, pp. 141–154], [BB, pp. 177–178], [HLP, pp. 84–187]; [General Inequalities, vol. 7, pp. 153–155], [Zwillinger, p. 207]; [Alzer, 1992[(2)]], [Pachpatte, 1987[(2)]], [Vornicescu].

Wright Convex Function Inequalities
See **Convex Function Inequalities** COMMENTS (viii), **Szegö's Inequality** COMMENTS (i).

- end of W -

Y-Z

Young's[1] Convolution Inequality If $g \in \mathcal{L}^r(\mathbb{R}^n), h \in \mathcal{L}^s(\mathbb{R}^n), 1 \leq r, s, \leq \infty$, and if $\dfrac{1}{r} + \dfrac{1}{s} = 1 + \dfrac{1}{t} \geq 0$ then $g \star h \in \mathcal{L}^t(\mathbb{R}^n)$ and

$$||g \star h||_t \leq K_1 ||f||_r ||g||_s; \qquad (1)$$

equivalently: if $f \in \mathcal{L}^r(\mathbb{R}^n), g \in \mathcal{L}^s(\mathbb{R}^n), h \in \mathcal{L}^t(\mathbb{R}^n), 1 \leq r, s, t \leq \infty$, where $\dfrac{1}{r} + \dfrac{1}{s} + \dfrac{1}{t} = 2$ then

$$\left| \int_{\mathbb{R}} f(\underline{x}) g \star h(\underline{x}) \, \mathrm{d}\underline{x} \right| \leq K_2 ||f||_r ||g||_s ||h||_t. \qquad (2)$$

The constants are

$$K_1 = (C_r C_s C_{t'})^n, \quad K_2 = (C_r C_s C_t)^n \quad \text{where} \quad C_p^2 = \frac{p^{1/p}}{p'^{1/p'}},$$

the primed numbers being the conjugates of the respective unprimed numbers. The inequalities are strict when $r, s, t > 1$ unless almost everywhere

$$\begin{aligned} f(x) &= A \exp\{-r'\langle \underline{x} - \underline{a}, (\underline{x} - \underline{a})H \rangle + i\underline{d}.\underline{x}\}, \\ g(x) &= B \exp\{-s'\langle \underline{x} - \underline{b}, (\underline{x} - \underline{b})H \rangle + i\underline{d}.\underline{x}\}, \\ h(x) &= C \exp\{-t'\langle \underline{x} - \underline{c}, (\underline{x} - \underline{c})H \rangle + i\underline{d}.\underline{x}\}, \end{aligned}$$

where $A, B, C \in \mathbb{C}$, $\underline{a}, \underline{b}, \underline{c}, \underline{d} \in \mathbb{R}^n$, and H is a real, symmetric, positive-definite matrix.

COMMENTS (i) Inequalities (1), (2) with $K_1 = K_2 = 1$ were proved by Young. The exact values of the constants, and the cases of equality, are due to Bruscamp & Lieb.
(ii) There are various other forms of this result depending on the function spaces used.
(iii) The case of (1) with $t = \infty$, when r, s are conjugate indices and $f \star g$ is uniformly continuous, has been called a *backward Hölder's inequality*:

$$\sup_{a \leq t \leq b} \int_a^b f(x - t) g(t) \, \mathrm{d}x \leq ||f||_r ||g||_s.$$

---

[1] This is W H Young.

DISCRETE ANALOGUE If $\underline{a} \in \ell_r$, $\underline{b} \in \ell_s$ and if t is defined as in (1) then $\underline{a} \star \underline{b} \in \ell_t$ and
$$\|\underline{a} \star \underline{b}\|_t \le \|\underline{a}\|_r \|\underline{b}\|_s,$$
with equality if and only if $\underline{a} = \underline{0}$, $\underline{b} = \underline{0}$ or $\underline{a}$ and $\underline{b}$ differ from $\underline{0}$ in one entry only.

COMMENTS (iv) For another convolution result see the **Hardy-Littlewood-Sobolev Inequality**.

REFERENCES [EM, vol.2, pp. 427–428], [HLP, pp. 200–202], [MPF, pp. 178–181]; [Hewitt & Stromberg, pp. 396–400], [Hirschman, pp. 168–169], [Lieb & Loss, pp. 90–97], [Zwillinger, p. 207], [Zygmund, vol.I, pp. 37–38].

Young's[1] Inequalities (a) If f is a strictly increasing continuous function on $[0, c]$ with $f(0) = 0$ and $0 \le a, b \le c$ then

$$ab \le \int_0^a f + \int_0^b f^{-1}. \tag{1}$$

There is equality if and only if $b = f(a)$.

(b) Under the same conditions

$$ab \le af(a) + bf^{-1}(b).$$

(c) If $p, q, r > 0$ with $1/p + 1/q + 1/r = 1$, and if p', q' are, respectively, the conjugate indices of p, q and if $f \in \mathcal{L}^{p'}(\mathbb{R}), g \in \mathcal{L}^{q'}(\mathbb{R})$ then $fg \in \mathcal{L}(\mathbb{R})$ and

$$\int_{\mathbb{R}} |fg| \le \left(\int_{\mathbb{R}} |f|^{p'} |g|^{q'} \right)^{1/r} \left(\int_{\mathbb{R}} |f|^{p'} \right)^{1/q} \left(\int_{\mathbb{R}} |g|^{q'} \right)^{1/p}. \tag{2}$$

(d) If $0 < \theta < \pi$ and $-1 < \alpha \le 0$, then

$$\sum_{k=0}^{n} \frac{\cos k\theta}{k + \alpha} > 0.$$

(e) If R is the outer radius of a bounded domain in $\mathbb{R}^n, n \ge 2$, of diameter D then

$$R \le D \sqrt{\frac{n}{n+2}}.$$

COMMENTS (i) Inequality (1), which holds under slightly wider conditions, has a very simple geometric proof.

(ii) Applying (1) to the function $f(x) = x^p, p > 0$ gives a proof of (B) in the form given in **Geometric- Arithmetic Mean Inequality** (2).

(iii) For another application see **Logarithmic Function Inequalities** (e).

(iv) Inequality (2) is a simple deduction from the integral analogue of **Hölder's Inequality** (2), in the case $n = 3$, and $\rho_3 = 1$.

(v) A definition of outer radius is given in **Isodiametric Inequality** COMMENTS (ii).

---

[1] This is W H Young.

EXTENSIONS (a) [BROWN & KOUMANDOS] If $0 < \theta < \pi$ and $\alpha \geq 1$ then the function

$$\cos\theta/2 \left(\sum_{k=1}^{n} \frac{\cos k\theta}{k^\alpha} \right)$$

is strictly decreasing. In particular

$$\frac{5}{6} + \sum_{k=1}^{n} \frac{\cos k\theta}{k} > 0, \qquad 0 \leq \theta \leq \pi.$$

(b) [HYLTÉN-CAVALLIUS] If $0 < \theta \leq \pi$ then

$$1 - \log \circ \sin x/2 + \frac{\pi - x}{2} \geq 1 + \sum_{i=1}^{n} \frac{\cos i\theta}{i} > 0.$$

COMMENTS (vi) See also **Conjugate Convex Function Inequalities** (b), **Gale's Inequality**, **N-function Inequalities**.

REFERENCES [AI. pp. 48–50], [BB, p. 7], [EM, vol.5, p. 204], [HLP, pp. 111–113] [MPF, pp. 379–389, 615], [PPT, pp. 239–246]; [Titchmarsh, 1948, p. 97]; [Brown & Koumandos].

Zagier's Inequality Let f, g be monotone decreasing non-negative functions on $[0, \infty[$ then for any integrable $F, G : [0, \infty[\to [0, 1]$,

$$\int_0^\infty fg \geq \frac{\int_0^\infty fF \int_0^\infty Gg}{\max\{\int_0^\infty F, \int_0^\infty G\}}.$$

In particular if $f, g : [0, \infty[\to [0, 1]$ be monotone decreasing integrable functions

$$\int_0^\infty fg \geq \frac{\int_0^\infty f^2 \int_0^\infty g^2}{\max\{\int_0^\infty f, \int_0^\infty g\}}.$$

COMMENTS (i) This is a converse of the integral analogue of (C).

DISCRETE ANALOGUES [ALZER, PEČARIĆ] If $0 < a_i, b_i, c_i, d_i \leq 1$, $1 \leq i \leq n$, and if both $\underline{a}, \underline{b}$ are decreasing then

$$\frac{\sum_{i=1}^{n} a_i c_i \sum_{i=1}^{n} b_i d_i}{\max\left\{ \sum_{i=1}^{n} c_i, \sum_{i=1}^{n} d_i \right\}} \leq \sum_{i=1}^{n} a_i b_i.$$

In particular

$$\frac{\sum_{i=1}^{n} a_i^2 \sum_{i=1}^{n} b_i^2}{\max\left\{ \sum_{i=1}^{n} a_i, \sum_{i=1}^{n} b_i \right\}} \leq \sum_{i=1}^{n} a_i b_i.$$

COMMENTS (ii) For another result of Zagier see **Difference Means of Gini**.

REFERENCES [MPF, pp. 95–96, 155]; [Pečarić, 1994], [Zagier].

Zeta Function Inequalities (a) If $s > 1, n \geq 1$ then

$$\frac{1}{s-1} < n^{s-1} \sum_{i \geq n} \frac{1}{i^s} \leq \zeta(s),$$

with equality on the right if and only if $n = 1$.
(b) If $s > 1$, $\gamma \geq 0$ then

$$\sum_{i \geq n} \frac{1}{i^s} \leq \frac{n^\gamma(s+\gamma)}{(s-1)(n^{s+\gamma} - (n-1)^{s+\gamma})}.$$

(c) If $s > 1$ then

$$\frac{1}{n[(n^{s-1} - (n-1)^{s-1})]} < \sum_{i \geq n} \frac{1}{i^s} < \frac{(n+1)^{s-1}}{n^s((n+1)^{s-1} - n^{s-1})}.$$

REFERENCES   [Bennett, 1996, p.14].

- end of Y-Z -

REFERENCES

BASIC REFERENCES

[AI] MITRINOVIĆ, D S, WITH VASIĆ, P M *Analytic Inequalities*, Springer-Verlag, Berlin, 1970.

[BB] BECKENBACH, E F & BELLMAN, R *Inequalities*, Springer-Verlag, Berlin, 1961.

[HLP] HARDY, G H , LITTLEWOOD, J E & PÓLYA, G *Inequalities*, Cambridge University Press, Cambridge, 1934.

[EM] *Encyclopedia of Mathematics, Volumes 1–10, Supplement*, Kluwer Academic Press, Dordrecht, 1988–1997.

[MI] BULLEN, P S, MITRINOVIĆ, D S & VASIĆ, P M *Means and Their Inequalities*, D Reidel, Dordrecht, 1988.

[MO] MARSHALL, A W & INGRAM, O *Inequalities: Theory of Majorization and Its Applications*, Academic Press, New York-London, 1979.

[MPF] MITRINOVIĆ, D S, PEČARIĆ, J E & FINK, A M *Classical and New Inequalities in Analysis*, D Reidel, Dordrecht, 1993.

[PPT] PEČARIĆ, J E, PROSCHAN, F & TONG, Y L *Convex Functions, Partial Orderings and Statistical Applications*, Academic Press Inc., 1992.

BOOKS

ABRAMOWICZ, M & STEGUN, I A *Handbook of Mathematical Functions*, Dover Publications, Inc., New York, 1965.

ADAMS, R A *Sobolev Spaces*, Academic Press, New York, 1975.

AGARWAL, R P EDITOR *Inequalities and Applications*, World Scientific Series in Applicable Analysis, 3; World Scientific Publishing Company Inc., River Edge NJ, 1994.

AHLFORS, L V *Complex Analysis*, McGraw-Hill, New York, 1966.

AHLFORS, L V *Conformal Invariants Topics in Geometric Function Theory*, McGraw-Hill, New York, 1973.

APOSTOL, T M *Calculus, Volumes I, II*, Blaisdell Publishing Company, New York, 1961.

APOSTOL, T M ET AL, EDITORS.   *A Century of Calculus Parts I, II*,   Mathematical Association of America, 1992.

BAINOV, D & SIMEONOV, P [1]   *Integral Inequalities and Applications*, English translation: Hoksbergen, R A M & Kovachev, V, Mathematics and its Applications (East European Series), 57, Kluwer Academic Publishers Group, Dordrecht, 1992.

BALEY, G P   *Multivariable Analysis*, Springer-Verlag, Berlin, 1984.

BENNETT, G   *Factorizing the Classical Inequalities*, Memoir American Mathematical Society 120, # 576, 1996.

BILER, P & WITKOWSKI, A   *Problems in Mathematical Analysis*, Dekker, New York, 1990.

BOBKOV, S G, & HOUDRÉ, C   *Some Connections between Isoperimetric and Sobolev-type Inequalities*,   Memoir American Mathematical Society 129, # 616, 1997.

BORWEIN, J M & BORWEIN, P B   *Pi and the AGM. A Study in Analytic Number Theory and Computational Complexity*, John Wiley and Sons, New York, 1987.

BOTTEMA, O, DJORDJEVIĆ, R Ž, JANIĆ, R R & MITRINOVIĆ, D S   *Geometric Inequalities*,   Groningen, 1969

BOURBAKI, N   *Fonctions d'une Variable Réelle (Théorie Élémentaire), Chapitres I–III*, Hermann & Cie., Paris, 1949.

BOURBAKI, N   *Topologie Générale, Chapitres III–IV, 3ième édition, revue et augmentée*, Hermann & Cie., Paris, 1960.

BOURBAKI, N   *Topologie Générale, Chapitres V–VIII, 2ième édition*, Hermann & Cie., Paris, 1955.

BROMWICH, T J I'A   *An Introduction to the Theory of Infinite Series*, 2nd edition, revised, MacMillan, London, 1965.

CLOUD, M J, & DRACHMAN, B C   *Inequalities with Applications to Engineering*, Springer-Verlag, Berlin, 1998.

CONWAY, J B   *Functions of One Complex Variable I, II*, Springer-Verlag, Berlin, 1973.

COURANT, R, & HILBERT, D   *Methods of Mathematical Physics, Volume I*, Interscience Publishers Inc., New York, 1953.

DICTIONARY   *Dictionary of Mathematics*, Itô, K editor; 2nd edition, English translation, MIT Press, Cambridge Massachusetts, 1977.

DIENES, P   *The Taylor Series*, reprint, New York, 1957.

DRAGOMIR, S S   *The Gronwall Type Lemmas and Applications*,   Monografi Matematice 29, Timoşoara, 1987.

DUNFORD, N & SCHWARTZ, J T   *Linear Operators Part I; General Theory*, Interscience Publishers, New York, 1967.

EVANS, L C, & GARIEPY, R F   *Measure Theory and Fine Properties of Functions*, CRC Press, London, 1992.

---

[1] Д Баинов, П Симеонов.

EVERITT, W N EDITOR   *Inequalities. Fifty Years on from Hardy, Littlewood and Pólya*, Proceedings of an International Conference, Birmingham. U.K., 1987; Lecture Notes in Pure and Applied Mathematics, 129, Marcel Dekker, New York, 1991.

FELLER, W   *An Introduction to Probability Theory and its Applications, Volumes I, II*, John Wiley & Sons, New York, 1957, 1966.

GALAMBOS, J & SIMONELLI, I   *Bonferroni-type Inequalities with Applications*, Springer-Verlag, Berlin, 1996.

GELBAUM, B R & OLMSTED, J M H   *Theorems and Counterexamples in Mathematics*, Springer-Verlag, Berlin, 1990.

GENERAL INEQUALITIES   *Volumes 1–7, Proceedings of First-Seventh, International Conferences on General Inequalities, Oberwolfach, 1976, 1978, 1981, 1984, 1986, 1990, 1995;* Beckenbach, E F, Walter, W, and Bandle, C, Everitt, W N, Losonczi, L, editors., International Series of Numerical Mathematics, 41, 47, 64, 71, 80, 103, 123, Birkhaüser Verlag, Basel, 1978, 1980, 1983, 1986, 1987, 1992, 1997.

GEORGE, C   *Exercises in Integration*, Springer-Verlag, Berlin, 1984.

GINI, C WITH BARBENSI, G, GALVANI, L, GATTI, S & PIZZETTI, E   *Le Medie*, Milan, 1958.

GROSSE-ERDMANN, K-G   *The Blocking Technique, Weighted Mean Operators and Hardy's Inequality*, Lecture Notes in Mathematics # 1679, Springer-Verlag, Berlin, 1998.

HÁJOS, G , ET AL   *Hungarian Problem Book II*, Random House, New York, 1963.

HALMOS, P R   *Problems for Mathematicians Young and Old*, Dolciani Mathematical Expositions # 12, Mathematical Association of America, 1991.

HEINS, M   *Complex Function Theory*, Academic Press, New York, 1968.

HEWITT, E AND STROMBERG, K   *Real and Abstract Analysis*, Springer-Verlag, Berlin, 1965.

HIRSCHMAN, JR. I I EDITOR   *Studies in Real and Complex Analysis*, Mathematical Association of America, 1965.

HLAWKA, E   *Ungleichungen*, Vorlesungen über Mathematik, Wien: MANZ Verlags- und Universitätsbuchhandlung, 1990.

HORN, R A & JOHNSON, C R   *Topics in Matrix Analysis*, Cambridge University Press, 1991.

INEQUALITIES   *Inequalities, Inequalities II, Inequalities III, Proceedings of Symposia in Inequalities, 1965, 1967, 1969;* Shisha, O editor, Academic Press, New York, 1967, 1970, 1972.

JEFFREYS, H & JEFFREYS, B S   *Methods of Mathematical Physics*, Cambridge University Press, Cambridge, 1950.

KACZMARZ, S & STEINHAUS, H   *Theorie der Orthogonalreihen*, Chelsea Publishing Company, New York, 1951.

KARLIN, S   *Total Positivity*, Stanford University Press, Stanford, 1968.

KAWOHL, B *Rearrangements and Convexity of Level Sets in PDE*, Lecture Notes in Mathematics # 1150, Springer-Verlag, Berlin, 1985

KAZARINOFF, N D *Analytic Inequalities*, Holt, Rhinehart and Winston, New York, 1961.

KHINCHIN, A YA [2] *Continued Fractions*, University of Chicago Press, 1961

KNOPP, K *Theory and Application of Infinite Series*, Blackie and Son Ltd., London, 1928.

KÖNIG, H *Eigenvalue Distribution of Compact Operators*, Birkhaüser, Basel, 1986.

KOROVKIN, P P *Inequalities*, New York, 1961.

KRASNOSEL'SKIĬ, M A & RUTICKIĬ, YA B [3] *Convex Functions and Orlicz Spaces*, English translation: Boron, L F, Noordhoff Ltd., Groningen, 1961.

KWONG, M K & ZETTL, A *Norm Inequalities for Derivatives and Differences*, Lecture Notes in Mathematics, #1536, Springer-Verlag, Berlin. 1992.

LEVIN, V I & STEČKIN, S B [4] *Inequalities*, American Mathematical Society Translations, (2) 14 (1960), 1–29.

LIEB, E H, & LOSS, M *Analysis*, Graduate Studies in Mathematics, American Mathematical Society, # 14, 1997.

LOÈVE, M *Probability Theory*, Van Nostrand Company Inc., Princeton, 1963.

MAMEDOV, YU D, AŠIROV S, & ATDAEV, S [5] Теоремы о Неравенствах, "Ylm" Ashk habad, 1980.

MARCUS, M & MINC, H *A Survey of Matrix Theory and Matrix Inequalities*, Allyn & Bacon, Boston, 1964.

MARTINOK, A A, LAKSHMIKANTHAM V & LEELA, S [6] УстойчивостьДвужения: Метод интегральных Неравенцтв, "Naukova Dumka", Kiev, 1989.

MELZAK, Z A *Mathematical Ideas, Modelling and Applications*, New York, 1976.

MINC, H *Permanents*, Encyclopedia of Mathematics and Its Applications, Volume 6, Addison-Wesley, New York, 1978.

MILOVANOVIĆ, G V, MITRINOVIĆ, D S & RASSIAS, TH M *Topics in Polynomials: Extremal Problems, Inequalities, Zeros*, World Scientific Publishing Company Inc., River Edge, 1994.

MITRINOVIĆ, D S & PEČARIĆ, J E *Diferencijalne i Integralne Nejednakosti*, Matematički Problem i Ekspozicije # 13, "Naučna Kniga", Beograd 1988.

MITRINOVIĆ, D S & PEČARIĆ, J E *Monotone Funkcije i Njihove Nejednakosti*, Matematički Problem i Ekspozicije # 17, "Naučna Kniga", Beograd 1990.

---

[2] А Я Хинчин.Also transliterated as Khintchine, Hinčin.

[3] М А Краснокельский, Я Б Рутицкий.

[4] В И Левин, С Б Стечкин.This is a translation of some material from an appendix to the Russion version of [HLP].

[5] Ю Д Мамедов, С Аширов, С Атдаев.

[6] А А Мартынок, В Лакшмиктам, С Лила.

MITRINOVIĆ, D S & PEČARIĆ, J E *Hölderova i Srodne Nejednakosti*, Matematički Problem i Ekspozicije # 18, "Naučna Kniga", Beograd 1990[2].

MITRINOVIĆ, D S & PEČARIĆ, J E *Cikličene Nejednakosti i Cikličene Funkcionalne Jednačine*, Matematički Problem i Ekspozicije # 19, "Naučna Kniga", Beograd 1991.

MITRINOVIĆ, D S & PEČARIĆ, J E *Nejednakosti i Norme*, Matematički Problem i Ekspozicije # 20, "Naučna Kniga", Beograd 1991[2].

MITRINOVIĆ, D S & PEČARIĆ, J E *Inequalities involving Functions and their Integrals and Derivatives*, Mathematics and Applications (East European Series), 53, Kluwer Academic Publishers Group, Dordrecht, 1991[3].

MITRINOVIĆ, D S, PEČARIĆ, J E, & VOLENEC, V *Recent Advances in Geometric Inequalities*, Kluwer Academic Publishers Group, Dordrecht, 1989.

MITRINOVIĆ, D S & POPADIĆ, M S *Inequalities in Number Theory*, Naučni Podmladak, Niš, 1978.

MITRINOVIĆ, D S & VASIĆ, P M *Sredine*, Uvodjenje Mladih u Naučni Rad V, Belgrade, 1968.

MITROVIĆ, D & ŽUBRINIĆ, D *Fundamentals of Applied Functional Analysis*, Pitman Monographs and Surveys in Pure and Applied Mathematics, 91, Longman Scientific & Technical, Harlow, 1998.

NIVEN, I *Maxima and Minima Without Calculus*, Mathematical Expositions No.6, Mathematical Association of America, 1981.

OPIC, B & KUFNER, A *Hardy-type Inequalities*, Pitman Research Notes in Mathematics Series, 219, Longman Scientific & Technical, Harlow, 1990.

PACHPATTE, B G *Inequalities for Differential and Integral Equations*, Academic Press Inc, 1998.

PEČARIĆ, J E *Konvekse Funkcije: Nejednakosti*, Matematičiki Problemi i Ekspozicije 12, "Naučna Kniga", Beograd 1987.

PÓLYA, G & SZEGÖ, G *Isoperimetric Inequalities in Mathematical Physics*, Annals of Mathematics Studies 27, Princeton University Press, Princeton, 1951.

PÓLYA, G & SZEGÖ, G *Problems and Theorems in Analysis I*, Springer-Verlag, Berlin, 1972.

POPOVICIU, T *Les Fonctions Convexes*, Hermann & Cie., Paris, 1945.

PROTTER, M H & WEINBERGER, H F *Maximum Principles in Differential Equations*, Prentice-Hall, New Jersey, 1967.

ROBERTS, A W & VARBERG, D E *Convex Functions*, Academic Press, New York-London, 1973.

ROCKAFELLAR, R T *Convex Analysis*, Princeton University Press, Princeton, 1970.

RUDIN, W *Principles of Mathematical Analysis*, McGraw-Hill Book Company, New York-San Francisco-Toronto-London, 1964.

RUDIN, W *Real and Complex Analysis*, McGraw-Hill Book Company, New York-San Francisco-Toronto-London, 1966.

RUDIN, W *Functional Analysis*, McGraw-Hill Book Company, New York-San Francisco-Toronto-London, 1973.

SAKS, ST *Theory of the Integral, second revised edition*, Warsaw, 1937.

SHAKED, M & TONG, Y L, EDITORS *Stochastic Inequalities*, Institute Mathematical Statistics, 1992.

SZEGÖ, G *Orthogonal Polynomials*, American Mathematical Society Colloquium Publications XXIII, 1939.

TITCHMARSH, E C *The Theory of Functions*, Oxford University Press, Oxford, 1939.

TITCHMARSH, E C *Introduction to the Theory of Fourier Integrals*, Oxford University Press, Oxford, 1948.

TONG, Y L *Probability Inequalities in Multivariate Distributions*, Academic Press, New York, 1979.

TONG, Y L EDITOR *Inequalities in Statistics and Probability*, Institute of Mathematical Statistics, Lecture Notes -Monograph Series, Volume 5, 1984.

WALTER, W *Differential- und Integral- Ungleichung*, Springer-Verlag, Berlin, 1964.

WIDDER, D V *The Laplace Transform*, Princeton University Press, Princeton, 1946.

ZYGMUND, A *Trigonometric Series, Volumes I, II*, Cambridge University Press, Cambridge, 1959.

ZWILLINGER, D *Handbook of Integration*, Jones and Bartlett Publishers, London, 1992.

PAPERS

ALZER, H Über Lehmers Mittelwertfamilie, *Elemente der Mathematik. Revue de Mathématiques Élementaires*, 43 (1988), 50–54; [103].[7]

ALZER, H Sharpenings of the arithmetic mean-geometric mean inequality, *Congressus Numerantium. Utilitas Mathematica*, 75 (1990), 63–66; [68, 98].

ALZER, H An extension of an inequality of G. Pólya, *Institutlui Politehnic din Iaşi. Buletinul. Secţia I. Mecanică. Fizică*, 36(40) (1990)$^{(2)}$, 1–4, 17–18; [206].

ALZER, H On an inequality of Gauss, *Revista Matemática de la Universidad Complutense de Madrid*, 4 (1991), 179–183; [94].

ALZER, H On an integral inequality of R.Bellman, *Tamkang Journal of Mathematics*, 22(1991)$^{(2)}$, 187–191; [49].

ALZER, H A refinement of Tchebyschef's inequality, *Nieuw Archief voor Wiskunde*, (4)10 (1992), 7–9; [51].

ALZER, H A continuous and a discrete variant of Wirtinger's inequality, *Mathematica Pannonica*, 3 (1992)$^{(2)}$, 83–89; [152, 249, 264].

---

[7] This denotes the page, or pages, where the reference is used.

References

ALZER, H A short note on two inequalities for sine polynomials, *Tamkang Journal of Mathematics*, 23 (1992)$^{(3)}$, 161–163; [254].

ALZER, H On Carleman's inequality, *Portugaliæ Mathematica*, 50 (1993), 331–334; [45].

ALZER, H A converse of an inequality of G.Bennett, *Glasgow Mathematical Journal*, (35), 45 (1993)$^{(2)}$, 269–273; [27].

ALZER, H On an inequality of H. Minc and L. Sathre, *Journal of Mathematical Analysis and Applications*, 179 (1993)$^{(3)}$, 396–402; [85].

ALZER, H A note on a lemma of G. Bennett, *Quarterly Journal of Mathematics*, Oxford, (2), 45 (1994), 267–268; [27].

ALZER, H Refinement of an inequality of G. Bennett, *Discrete Mathematics*. 135 (1994)$^{(2)}$, 39–46; [27].

ALZER, H An inequality for increasing sequences and its integral analogue, *Discrete Mathematics*, 133 (1994)$^{(3)}$, 279–283; [19, 246].

ALZER, H The inequality of Ky Fan and related results, *Acta Applicandæ Mathematicæ. An International Survey Journal on Applying Mathematics and Mathematical Applications*, 38 (1995), 305–354; [151].

ALZER, H A note on an inequality involving $(n!)^{1/n}$, *Acta Mathematica Universitatis Comenianæ. New Series*, 64 (1995)$^{(2)}$, 283–285; [85].

ALZER, H On some inequalities that arise in measure theory, *unpublished*; [41].

ALZER, H & BRENNER, J L Integral inequalities for concave functions with applications to special functions, *Research Report, University of South Africa*, #94/90(4), (1990), 1–31; [123].

ALZER, H & BRENNER, J L On a double inequality of Schlömilch-Lemonnier, *Journal of Mathematical Analysis and Applications*, 168 (1992), 319–328; [228].

ALZER, H, BRENNER, J L & RUEHR, O G Inequalities for the tails of some elementary series, *Journal of Mathematical Analysis and Applications*, 179 (1993), 500–506; [83].

ANASTASSIOU, G A Ostrowski type inequalities, *Proceedings of the American Mathematical Society*, 123 (1995), 3775–3881; [200].

BEARDON, A F The Schwarz-Pick lemma for derivatives, *Proceedings of the American Mathematical Society*, 125 (1997), 3255–3256; [231].

BECKER, M. & STARK, E.L An extremal inequality for the Fourier coefficients of positive cosine polynomials, *Univerzitet u Beogradu. Publikacije Elektrotehničkog Fakulteta. Serija Matematika*, No.577–No.598, (1977), 57–58; [143].

BENNETT, G An inequality suggested by Littlewood, *Proceedings of the American Mathematical Society*, 100 (1987), 474–476; [158].

BENNETT, G Some elementary inequalities II, *Quarterly Journal of Mathematics*, Oxford, (2), 39 (1988), 385–400; [27].

BENYON, M J, BROWN, B M & EVANS, W M On an inequality of Kolmogorov type for a second-order differential expression, *Proceedings of the Royal Society. London. Series A. Mathematical and Physical Sciences*, 442 (1993), 555–569; [*113*].

BOAS, R Absolute convergence and integrability of trigonometric series, *Journal of Rational Mechanics and Analysis*, 5 (1956), 631–632; [*38*].

BOAS, R Inequalities for a collection, *Mathematics Magazine*, 52 (1979), 28–31; [*92*].

BOROGOVAC, M & ARSLANAGOĆ, Š Generalisation and improvement of two series inequalities, *Periodica Mathematica Hungarica. Journal of the János Bolyai Mathematical Society*, 25 (1992), 221–226; [*115*].

BORWEIN, D & BORWEIN, J M A note on alternating series, *American Mathematical Monthly*, 93 (1986), 531–539; [*18*].

BORWEIN, J M & BORWEIN, P B The way of all means, *American Mathematical Monthly*, 94 (1987)$^{(2)}$, 519–522; [*22*].

BORWEIN, P B Exact inequalities for the norms of factors of polynomials, *Canadian Journal of Mathematics*. 46 (1994), 687–698; [*147*].

BRASCAMP, H J & LIEB, E H On extensions of the Brunn-Minkowski and Prékopa-Leindler theorem, including inequalities for log concave functions, and with an application to the diffusion equation, *Journal of Functional Analalysis*, 22 (1976), 366–389; [*215*].

BRENNER, J L & ALZER, H Integral inequalities for concave functions with applications to special functions, *Proceedings of the Royal Society. Edinburgh. Series A. Mathematical and Physical Sciences* Q 118 (1991), 173–192; [*86*].

BROWN, G Some inequalities that arise in measure theory, *Journal of the Australian Mathematical Society. Series A*, 45 (1988), 83–94; [*41*].

BROWN, G & KOUMANDOS, S On a monotonic trigonometric sum, *Monatshefte für Mathematik*, 123 (1997), 109–119; [*267*].

BULLEN, P S An inequality for variations, *American Mathematical Monthly*, 90 (1983), 560; [*41*].

BULLEN, P S Inequalities of T.S. Nanjundiah, *to appear*, (1997); [*45, 188, 189, 213*].

BULLEN, P S The Jensen-Steffensen inequality, *Mathematical Inequalities & Applications*, 1 (1998), to appear; [*98, 142*].

CARLEN, E A & LOSS, M Sharp constant in Nash's inequality, *International Mathematics Research Notices*, #7 (1993), 213–215; [*190*].

CATER, F S Lengths of rectifiable curves in 2-space, *Real Analysis Exchange*, 12 (1986–1987), 282–293; [*20*].

CHOLLET, JOHN Some inequalities for principal submatrices, *American Mathematical Monthly*, 104 (1997), 609–617; [*63,182*].

COMMON, A K Uniform inequalities for ultraspherical polynomials and Bessel functions, *Journal of Approximation Theory*, 49 (1987), 331–339; [*168, 256*].

DAVIES, G S & PETERSEN, G M On an inequality of Hardy's II, *Quarterly Journal of Mathematics, Oxford*, (2) 15 (1964), 35–40; [*69*].

DIANANDA, P M Power mean inequality, *James Cook Mathematical Notes*, 7 (1995), 7004–7005, 7028–7029; [213].

DOSTANIĆ, M R On an inequality of Friederich's type, *Proceedings of the American Mathematical Society*, 125 (1997), 2115–2118; [91].

DRAGOMIR, S S A refinement of Cauchy-Schwarz inequality, *Gaz. Mat. Perfecţ. Metod. Metodol. Mat. Inf.*, 8 (1987). 94–94; [49].

DRAGOMIR, S S & CRSTICI, B A note on Jensen's discrete inequality, *Univerzitet u Beogradu. Publikacije Elektrotehničkog Fakulteta. Serija Matematika*,4 (1993), 28–34; [141].

DRAGOMIR, S S & MOND, B On a property of Gram's determinant, *Extracta Mathematicæ*, 11 (1996), 282–287; [105].

DUFF, G F D Integral inequalities for equimeasurable rearrangements, *Canadian Journal of Mathematics*. 22 (1970), 408–430; [74].

EMERY, M & YUKICH, J E A simple proof of the logarithmic Sobolev inequality, *Séminaire de Probabilités*, XXI, 173–175; *Lecture Notes in Mathematics*, 1247 (1987), Springer-Verlag, Berlin; [163].

FAĬZIEV, R F [8] A series of new general inequalities, *Doklady Akademiĭ Nauk SSSR*[9], 89, 577–581; [85].

FINK, A M Bounds on the deviation of a function from its average, *Čehoslovackaja Akademija Nauk. Čehoslovackiĭ Matematičeskiĭ Žurnal. Czechoslovak Mathematical Journal*, 42(117) (1992), 289–310; [200].

FURUTA, T & YANAGIDA, M Generalized means and convexity of inversion for positive operators, *American Mathematical Monthly*, 105 (1998), 258–259; [170].

GERBER, L The parallelogram inequality, *Univerzitet u Beogradu. Publikacije Elektrotehničkog Fakulteta. Serija Matematika*, No.461–No.497, (1974), 107–109; [202].

GLUCHOFF, A & HARTMANN, F Univalent polynomials and non-negative trigonometric sums, *American Mathematical Monthly*, 106 (1998), 508–522; [87, 254].

GROSS, L Logarithmic Sobolev inequalities, *American Journal of Mathematics*, 97 (1975), 1061–1083; [163].

HAJELA, D Inequalities between integral means of a function, *Bulletin of the Australian Mathematical Society*, 41 (1990), 245–248; [109].

IGARI, S On Kakeya's maximal function, *Japan Academy. Proceedings.Series A. Mathematical Sciences*, 62 (1986), 292–293; [144].

JANOUS, W An inequality for complex numbers, *Univerzitet u Beogradu. Publikacije Elektrotehničkog Fakulteta. Serija Matematika*, 4, (1993), 79–80; [59].

KALAJDŽIĆ, G On some particular inequalities, *Univerzitet u Beogradu. Publikacije Elektrotehničkog Fakulteta. Serija Matematika*, No. 498–No. 541, (1975), 141–143 ; [14].

---

[8] Р Ф Файзиев.

[9] Доклады Академий Наук СССР.

KEDLAYA, K Proof of a mixed arithmetic-mean geometric-mean inequality, *American Mathematical Monthly*, 101 (1954), 355–357; [*189*].

KEMP, A.W Certain inequalities involving fractional powers, *Journal of the Australian Mathematical Society, Series A*, 53 (1992), 131–136; [*41*].

KIVINUKK, A Some inequalities for convex functions, *Mitteilungen der Mathematischen Gesellschaft in Hamburg*, 15 (1996), 31–34; [*19*].

KLAMKIN, M S Problem E2428, *American Mathematical Monthly*, 82 (1975), 401; [*239*].

KLAMKIN, M S Extensions of some geometric inequalities, *Mathematics Magazine*, 49 (1976), 28–30; [*17*].

KLAMKIN, M S Extensions of an inequality, *Univerzitet u Beogradu. Publikacije Elektrotehničkog Fakulteta. Serija Matematika*, 7, (1996), 72–73; [*17*].

KLAMKIN, M S & NEWMAN, D J An inequality for the sum of unit vectors, *Univerzitet u Beogradu. Publikacije Elektrotehničkog Fakulteta. Serija Matematika*, No.338–No.352, (1971), 47–48; [*103*].

KOMAROFF, N Rearrangements and matrix inequalities, *Linear Algebra and its Applications*, 140 (1990), 155–161; [*76*].

KU, H-T, KU, M-C & ZHANG X-M Inequalities for symmetric means, symmetric harmonic means, and their applications, *unpublished*; [*77, 244*].

KUBO, F On Hilbert inequality, *Recent Advances in Mathematical Theory of Systems, Control Network and Signal Processing, I, Kobe, 1991*, 19–23, Mita, Tokyo, 1992; [*124*].

KUBO, T Inequalities of the form $f\big(g(x)\big) \geq f(x)$, *Mathematics Magazine*, 63 (1990), 346–348; [*92, 250*].

LEE, C M On a discrete analogue of inequalities of Opial and Yang, *Canadian Mathematical Bulletin*, 11 (1968), 73–77; [*197*].

LEINDLER, L Two theorems of Hardy-Bennett-type, *Acta Mathematica Hungarica*, 79 (1998). 341–350; [*111*].

LIN, M & TRUDINGER, N S on some inequalities for elementary symmetric functions, *Bulletin of the Australian Mathematical Society*, 50 (1994), 317–326; [*77*].

LOCHS, G Abschätzung spezieller Werte der unvollständigen Betafunktion, *Österreichische Akademie der Wissenschaften. Mathematische-Naturwissenschaftliche Klasse. Anzeiger*, (1986), 59–63; [*257*].

LONGINETTI, M An inequality for quasi-convex functions, *Applicable Analysis. An International Journal*, 113 (1982), 93–96; [*219*].

LORCH, L & MULDOON, M E An inequality for concave functions with applications to Bessel functions, *Facta Universitatis. Series: Mathematics and Informatics*, 2 (1987), 29–34; [*165*].

LOSONCZI, L & PÁLES, Z Minkowski's inequality for two variable Gini means, *Acta Universitatis Szegediensis. Acta Scientiarum Mathematicarum*, 62 (1996), 413–425; [103].

LOVE, E R Inequalities related to those of Hardy and of Cochrane and Lee, *Mathematical Proceedings of the Cambridge Philosophical Society*, 99 (1986), 395–408; [120].

LOVE, E R Inequalities for Laguerre Functions, *Journal of Inequalities and Applications*, 1 (1997), 293–299; [153].

LUPAŞ, A On an inequality, *Univerzitet u Beogradu. Publikacije Elektrotehničkog Fakulteta. Serija Matematika*, No.716–No.734, (1981), 32–34; [23].

MALEŠEVIĆ, J V [10] Варијанте доказа неких ставова класичне анализе, [Variants on the proofs of some theorems in classical analysis], Саопшено Конгресу МФА Југославије, [*Proceeding of the Congress of MFA Yugoslavia*], 1970; [37].

MALIGRANDA, L, PEČARIĆ, J E & PERSSON, L E On some inequalities of the Grüss-Barnes and Borell type, *Journal of Mathematical Analysis and Applications*, 187 (1994), 306–323; [86, 107].

MALIGRANDA, L, PEČARIĆ, J E & PERSSON, L E Stolarsky's inequality with general weights, *Proceedings of the American Mathematical Society*, 123 (1995), 2113–2118; [240].

MALLOWS, C L & RICHTER, D Inequalities of Chebyshev type, *Annals of Mathematical Statistics.*, 40 (1969), 1922–1932; [239].

MÁTÉ, A Inequalities for derivatives of polynomials with restricted zeros, *Proceedings of the American Mathematical Society*, 82 (1981), 221–225; [80].

MATSUDA, T An inductive proof of a mixed arithmetic-geometric mean inequality, *American Mathematical Monthly*, 102 (1995), 634–637; [189].

MERCER. A MCD A note on a paper "An elementary inequality, *Internat. J. Math. Math. Sci.*, 2 (1979). 531–535; [MR 81b:26013]" by S. Haber, *International Journal of Mathematics and Mathematical Sciences* 6 (1983). 609–611; [108].

MERKLE, M Conditions for convexity of a derivative and some applications to the Gamma function, *Æquationes Mathematicæ*, 55 (1998), 273–280; [85].

MESHKOV, V Z [11] An inequality of Carleman's type and some applications, *Doklady Akademiĭ Nauk SSSR*, 288 (1986), 46–49. English translation; *Soviet Mathematics. Doklady*[9], 33 (1986), 608–611; [45].

MILISAVLJEVIĆ, B M Remark on an elementary inequality, *Univerzitet u Beogradu. Publikacije Elektrotehničkog Fakulteta. Serija Matematika*, No.498–No.541, (1975), 179–180; [243].

MILOVANOVIĆ, G V & MILOVANOVIĆ, I Ž Discrete inequalities of Wirtinger's type for higher differences, *Journal of Inequalities and Applications*, 1 (1997), 301–310; [152].

---

[10] Ј В Малешевић.

[11] В З Мешков.

[9] Доклады Акадademий Наук СССР.

MINGARELLI A note on some differential inequalities, *Bulletin of the Institute of Mathematics. Academia Sinica*, 14 (1986) 287–288; [175].

MITRINOVIĆ, D S A cyclic inequality, *Univerzitet u Beogradu. Publikacije Elektrotehničkog Fakulteta. Serija Matematika*, No.247–No.273, (1969), 15–20; [232].

MITRINOVIĆ, D S & PEČARIĆ, J E A general integral inequality for the derivative of an equimeasurable rearrangement, *La Société Royale du Canada. L'Academie des Sciences. Comptes Rendus Mathematiques (Mathematical Reports)*, 11 (1989), 201–205; [74].

MITRINOVIĆ, D S & PEČARIĆ, J E Note on a class of functions of Godunova and Levin, *La Société Royale du Canada. L'Academie des Sciences. Comptes Rendus Mathematiques (Mathematical Reports)*, 12 $(1990)^{(3)}$, 33–36; [216].

MITRINOVIĆ, D S & PEČARIĆ, J E Bernoulli's inequality, *Rendiconti del Circolo Matematico di Palermo*, 42 (1993), 317–337; [29].

MITRINOVIĆ, D S & VASIĆ, P M Addenda to the monograph "Analytic Inequalities" I, *Univerzitet u Beogradu. Publikacije Elektrotehničkog Fakulteta. Serija Matematika*, No.577–No.598, (1977), 3–10; [258].

MOHUPATRA, R N & VAJRAVELU, K Integral inequalities resembling Copson's inequality, *Journal of the Australian Mathematical Society, Ser.A*, 48 (1990), 124–132; [112].

MOND, B & PEČARIĆ, J E A mixed mean inequality, *Australian Mathematical Society Gazette*, 23 (1996), 67–70; [189].

MOSER, J A sharp form of an inequality by N Trudinger, *Indiana University Mathematics Journal*, 20 (1971), 1077–1092; [254].

MULHOLLAND, H P Inequalities between the geometric mean differences and the polar moments of a plane distribution, *Journal of the London Mathematical Society*, 33 (1958), 261–270; [185].

MÜLLER, D A note on Kakeya's maximal function, *Archiv der Mathematik. Archives of Mathematics. Archives Mathématiques*, 49 (1987), 66–71; [144].

NEUGEBAUER, C J Weighted norm inequalities for averaging operators of monotone functions, *Publicacions Matemàtiques*, 35 (1991), 429–447; [112].

NEUMAN, E On Hadamard's inequality for convex functions, *International Journal of Mathematical Education in Science and Technology*, 19 (1988), 753–755; [123].

NEUMAN, E The weighted logarithmic mean, *Journal of Mathematical Analysis and Applications*, 188 (1994), 885–900; [162].

OPPENHEIM, A Some inequalities for triangles, *Univerzitet u Beogradu. Publikacije Elektrotehničkog Fakulteta. Serija Matematika*, No.357–No.380, (1971), 21–28; [100].

PACHPATTE, B G On multidimensional Opial-type inequalities, *Journal of Mathematical Analysis and Applications*, 126 (1987), 85–89; [197].

PACHPATTE, B G A note on Opial and Wirtinger type discrete inequalities, *Journal of Mathematical Analysis and Applications*, 127 $(1987)^{(2)}$, 470–474; [197, 264].

PACHPATTE, B G On an integral inequality involving functions and their derivatives, *Soochow Journal of Mathematics*, 13 ($1987)^{(3)}$, 805–808; [197].

PACHPATTE, B G on some integral inequalities similar to Hardy's inequality, *Journal of Mathematical Analysis and Applications*, 129 (1988), 596–606; [112].

PACHPATTE, B G A note on Askey and Karlin type Inequalities, *Tamkang Journal of Mathematics*, 19, #3 (1988)$^{(2)}$, 29–33; [23].

PACHPATTE, B G A note on Poincaré's inequality, *Analleli Ştiinţifice ale Universităţii "Al. I. Cuza" din Iaşi. Seria Nouă Secţiunea I a Matematică*, 32 (1988)$^{(3)}$, 35–36; [205].

PACHPATTE, B G On some extension of Rellich's inequality, *Tamkang Journal of Mathematics*, 22 (1991), 259–265; [225].

PACHPATTE, B G A note on Rellich type inequality, *Libertas Mathematica*, 11 (1991)$^{(2)}$, 105–115; [225].

PACHPATTE, B G A note on some series inequalities, *Tamkang Journal of Mathematics*, 27 (1996), 77–79; [201].

PACHPATTE, B G & LOVE, E R On some new inequalities related to Hardy's inequality, *Journal of Mathematical Analysis and Aplications*, 149 (1990), 17–25; [112].

PÁLES, Z Inequalities for sums of powers, *Journal of Mathematical Analysis and Applications*, 131 (1988), 265–270; [103].

PÁLES, Z Inequalities for differences of powers, *Journal of Mathematical Analysis and Applications*, 131 (1988)$^{(2)}$, 271–281; [132, 162].

PEARCE, C E M & PEČARIĆ, J E A remark on the Lo-Keng Hua inequality, *Journal of Mathematical Analysis and Aplications*, 188 (1994), 700–702; [131].

PEARCE, C E M & PEČARIĆ, J E A continuous analogue and extension of Rado's formula, *Bulletin of the Australian Mathematical Society*, 53 (1996), 229–233; [221].

PEARSON, J M A logarithmic Sobolev inequality on the real line, *Proceedings of the American Mathematical Society*, 125 (1997), 3339–3345; [163].

PEČARIĆ, J E A weighted version of Zagier's inequality, *Nieuw Archief voor Wiskunde*, 12 (1994), 125–127; [267].

PEČARIĆ, J E & PERSSON, L E On Bergh's inequality for quasi-monotone functions, *Journal of Mathematical Analysis and Applications*, 195 (1995), 393–400; [28].

PEČARIĆ, J E & VAROŠANEC, S Remarks on Gauss-Winckler's and Stolarsky's inequality, *Utilitas Mathematica*, 48 (1995), 233–241; [94, 240].

POMMERENKE, CH. The Bieberbach conjecture, *The Mathematical Intelligencer*, 7 (1985), 23–25; [34].

POPA, A An inequality for triple integrals, *Buletinul Institutului Polytehnic Bucureşti Seria Electrotehnică*, 54 (1992), 13–16; [209].

POPOVICIU, E Contributions à l'analyse de certains allures, *"Babeş-Bolyai" University, Faculty of Mathematics, Research Seminar*, 7 (1986), 227–230; [preprint]; [219].

REDHEFFER, R Recurrent inequalities, *Proceedings of the London Mathematical Society*, (3) 17 (1967), 683–699; [14, 23, 145, 224, 225].

REDHEFFER, R & VOIGT, A An elementary inequality for which equality holds in an infinite-dimensional set, *SIAM Journal on Mathematical Analysis*, 18 (1987), 486–489; [*101, 221*].

RENÉ, M Une inégalité intégrale concernant certaines fonctions log-concaves, *Séminaire de Théorie Spectrale et Géométrie, #6, Année 1987–1988*, Université de Grenoble I, Saint-Martin-d'Hères, 1988; [*80*].

Ross, D Copson type inequalities with weighted means, *Real Analysis Exchange*, 18 (1992–1993), 63–69; [*112*].

ROSSET, S Normalized symmetric functions, Newton's inequalities, and a new set of stronger inequalities, *American Mathematical Monthly*, 96 (1989), 85–89; [*77*].

RUSSELL, A M A commutative algebra of functions of generalized variation, *Pacific Journal of Mathematics*, 84 (1979), 455–463; [*41*].

RUSSELL, D C A note on Mathieu's inequality, *Æquationes Mathematicæ*, 36 (1988), 294–302; [*168*].

SÁNDOR, J On the identric and logarithmic means, *Æqationes Mathematicæ*, 40 (1990), 261–270; [*162*].

SAVOV, T Sur une inégalité considerée par D.S.Mitrinović, *Univerzitet u Beogradu. Publikacije Elektrotehničkog Fakulteta. Serija Matematika*, No.381–No.409, (1972), 47–50; [*37*].

SHAFER, ROBERT E Analytic inequalities obtained by quadratic approximation, *Univerzitet u Beogradu. Publikacije Elektrotehničkog Fakulteta. Serija Matematika*, No.577–No.598, (1977), 96–97; [*250*].

SINNADURAI, J ST-C A proof of Cauchy's inequality and a new generalization, *Mathematical Gazette*, 47 (1963), 36–37; [*105*].

SKALSKY, M A note on non-negative polynomials, *Univerzitet u Beogradu. Publikacije Elektrotehničkog Fakulteta. Serija Matematika*, No. 302–No. 319, (1970), 99–100; [*233*].

SLAVIĆ, D V On inequalities for $\Gamma(x+1)/\Gamma(x+1/2)$, *Univerzitet u Beogradu. Publikacije Elektrotehničkog Fakulteta. Serija Matematika*, No.498–No.591, (1975), 17–20; [*85*].

STOLARSKY, K B From Wythoff's Nim to Chebyshev's inequality, *American Mathematical Monthly*, 88 (1981), 889–900; [*33*].

TARIQ, Q M Concerning polynomials on the unit interval, *Proceedings of the American Mathematical Society*, 99 (1987), 293–296; [*153*].

TAYLOR, A E L'Hospital's Rule, *American Mathematical Monthly*, 59 (1952), 20–24; [*157*].

TRIMBLE, S Y, WELLS, J & WRIGHT, F T Superadditive functions and a statistical application, *SIAM Journal of Mathematical Analysis*, 20 (1989), 1255–1259; [*72, 232, 238*].

TSIBULIS, A B [12] A simple proof of an inequality, *Latviĭskiĭ Gosudarstvennyĭ Universitet imeni P. Stuchki. Latviĭskiĭ Matematicheskiĭ Ezhegodnik*, 33 (1989). 204–206; [83].

TUDOR, GH Compléments au traité de D.S.Mitrinović (VI). Une généralisation de l'inégalité de Fejér-Jackson, *Univerzitet u Beogradu. Publikacije Elektrotehničkog Fakulteta. Serija Matematika*, No.461–No.497, (1974), 111–114; [87].

VORNICESCU, N A note on Wirtinger's integral and discrete inequalities, *Seminar on Math. Anal., "Babeş-Bolyai" Univ., Cluj-Napoca*, (1991), 31–36; [264].

WANG C L Simple inequalities and old limits, *American Mathematical Monthly*, 90 (1983), 354–355; [83].

WARD, A J B Polynomial inequalities, *Mathematical Gazette*, 72 (1988), 16–19; [199].

YANAGIHARA, H An integral inequality for derivatives of equimeasurable rearrangements, *Journal of Mathematical Analysis and Applications*, 175 (1993), 448–457; [181, 257].

YOUNG, L C An inequality of the Hölder type connected with Stieltjes integration, *Acta Mathematica*, 67 (1936), 252–281; [164].

ZADEREĬ, P V [13] О многомерном аналоге одного результата Р. Боаса, Академия Наук Украинской ССР. Институт Математики. Украинский Математический Журнал, 39 (1987), 380–383; English translation: *Ukrainian Mathematical Journal*, , On a multidimensional analogue of a result of Boas, 39 (1987), 299–302; [38].

ZAGIER, D A converse to Cauchy's inequality, *American Mathematical Monthly*, 102 (1995), 919–920; [267].

- end of References -

---

[12] Also spelt Cibulis.

[13] П В Задерей.